AF358733

Metabolites from Phototrophic Prokaryotes and Algae Volume 2

Metabolites from Phototrophic Prokaryotes and Algae Volume 2

Editors

Carole A. Llewellyn
Rahul Vijay Kapoore

MDPI • Basel • Beijing • Wuhan • Barcelona • Belgrade • Manchester • Tokyo • Cluj • Tianjin

Editors
Carole A. Llewellyn
Swansea University
UK

Rahul Vijay Kapoore
Swansea University
UK

Editorial Office
MDPI
St. Alban-Anlage 66
4052 Basel, Switzerland

This is a reprint of articles from the Special Issue published online in the open access journal *Metabolites* (ISSN 2218-1989) (available at: https://www.mdpi.com/journal/metabolites/special_issues/metab_algae).

For citation purposes, cite each article independently as indicated on the article page online and as indicated below:

LastName, A.A.; LastName, B.B.; LastName, C.C. Article Title. *Journal Name* **Year**, *Article Number*, Page Range.

ISBN 978-3-03943-182-3 (Hbk)
ISBN 978-3-03943-183-0 (PDF)

Contents

About the Editors vii

Preface to "Metabolites from Phototrophic Prokaryotes and Algae Volume 2" ix

Thomas Sydney, Jo-Ann Marshall-Thompson, Rahul Vijay Kapoore,
Seetharaman Vaidyanathan, Jagroop Pandhal and J. Patrick A. Fairclough
The Effect of High-Intensity Ultraviolet Light to Elicit Microalgal Cell Lysis and Enhance
Lipid Extraction
Reprinted from: *Metabolites* **2018**, *8*, 65, doi:10.3390/metabo8040065 1

Rahul Vijay Kapoore and Seetharaman Vaidyanathan
Quenching for Microalgal Metabolomics: A Case Study on the Unicellular Eukaryotic Green
Alga *Chlamydomonas reinhardtii*
Reprinted from: *Metabolites* **2018**, *8*, 72, doi:10.3390/metabo8040072 15

Amornpan Klanchui, Sudarat Dulsawat, Kullapat Chaloemngam,
Supapon Cheevadhanarak, Peerada Prommeenate and Asawin Meechai
An Improved Genome-Scale Metabolic Model of *Arthrospira platensis* C1 (*i*AK888) and Its
Application in Glycogen Overproduction
Reprinted from: *Metabolites* **2018**, *8*, 84, doi:10.3390/metabo9050084 31

Daniel A. White, Paul A. Rooks, Susan Kimmance, Karen Tait, Mark Jones, Glen A. Tarran,
Charlotte Cook and Carole A. Llewellyn
Modulation of Polar Lipid Profiles in *Chlorella* sp. in Response to Nutrient Limitation
Reprinted from: *Metabolites* **2019**, *9*, 39, doi:10.3390/metabo9030039 49

Ana Molina-Márquez, Marta Vila, Javier Vigara, Ana Borrero and Rosa León
The Bacterial Phytoene Desaturase-Encoding Gene (*CRTI*) is an Efficient Selectable Marker for
the Genetic Transformation of Eukaryotic Microalgae
Reprinted from: *Metabolites* **2019**, *9*, 49, doi:10.3390/metabo9030049 69

Bethan Kultschar, Ed Dudley, Steve Wilson and Carole A. Llewellyn
Intracellular and Extracellular Metabolites from the Cyanobacterium *Chlorogloeopsis fritschii*,
PCC 6912, During 48 Hours of UV-B Exposure
Reprinted from: *Metabolites* **2019**, *9*, 74, doi:10.3390/metabo9040074 81

Nattaphorn Buayam, Matthew P. Davey, Alison G. Smith and Chayakorn Pumas
Effects of Copper and pH on the Growth and Physiology of *Desmodesmus* sp. AARLG074
Reprinted from: *Metabolites* **2019**, *9*, 84, doi:10.3390/metabo8040084 97

Stefan Schade, Emma Butler, Steve Gutsell, Geoff Hodges, John K. Colbourne
and Mark R. Viant
Improved Algal Toxicity Test System for Robust *Omics*-Driven Mode-of-Action Discovery in
Chlamydomonas reinhardtii
Reprinted from: *Metabolites* **2019**, *9*, 94, doi:10.3390/metabo9050094 115

Sahutchai Inwongwan, Nicholas J. Kruger, R. George Ratcliffe and Ellis C. O'Neill
Euglena Central Metabolic Pathways and Their Subcellular Locations
Reprinted from: *Metabolites* **2019**, *9*, 115, doi:10.3390/metabo9060115 137

Alla Silkina, Bethan Kultschar and Carole A. Llewellyn
Far-Red Light Acclimation for Improved Mass Cultivation of Cyanobacteria
Reprinted from: *Metabolites* **2019**, *9*, 170, doi:10.3390/metabo9080170 . . . . **161**

About the Editors

Carole A. Llewellyn's interests are in microalgae and cyanobacteria, and how they function in their natural environment, especially in relation to photophysiology. She is also interested in how algae can be used to help tackle society's big challenges. These big challenges relate to climate change, human health, bioenergy, food-security, aquaculture, waste-water and pollution bioremediation, industrial biotechnology and the circular economy. Her early research focused on the study of chlorophyll and carotenoid pigments to understand phytoplankton community composition and function, in relation to the carbon cycle and climate change. From this, she developed an interest in algal biotechnology using her knowledge on microalgal carotenoids and UV sunscreen compounds, working with industry to develop personal care products for anti-aging and cosmetics. This has led to her wider interest in understanding metabolism in microalgae and the large scale cultivation of microalgae for industrially useful products, including for food and for sustainable chemicals, to replace existing petroleum-based chemicals. She has led a number of projects funded by the Research Councils, Innovate-UK and Europe, often working with industry to develop sustainable solutions using microalgae. She currently leads the EU Interreg North West Europe ALG-AD project (ALG-AD), developing the circular economy by linking using waste from the anaerobic digestion of food and farm waste to the cultivation of algae for animal feed and high value products.

Rahul Vijay Kapoore's main interests are in microalgal biotechnology, -omics sciences (metabolomics), bioanalytical techniques, microbial consortia, circular economy, biorefineries, bioremediation and high value products from microalgae. Originally from Nashik (India), Rahul obtained a first class degree in Bachelor of Pharmacy in 2006 from M.G.V's College of Pharmacy, Nashik (Pune University). Subsequent to the completion of his degree, Rahul came to the UK in 2006 to pursue his MSc in Pharmacology and Biotechnology at Sheffield Hallam University, where he graduated in November 2007. Rahul holds a PhD in Metabolomics (2010–2014) from The University of Sheffield (TUOS), and his doctoral research was based on the development and optimisation of mass spectrometry based hyphenated techniques for microalgal and mammalian metabolomics. Rahul joined the Department of Chemical and Biological Engineering (CBE) at TUOS (2014–2017) as a Postdoctoral Research Associate (PDRA) on a BBSRC-DBT funded (1.22 million GBP) project (BB/K020633/1): "Sustainable Bioenergy and Biofuels from Microalgae: A Systems Perspective". The research involved GC-MS and LC-MS based metabolome level characterisations of microalgal strains, and thereby developing a systems level understanding in combination with other systems biology approaches (proteomics and transcriptomics), that will lead to sustainable processes for bio-energy generation from microalgae. Later, he joined Swansea University (2018–present) as a research officer involved in the ALG-AD project (circular economy): a strategic initiative of the INTERREG North West Europe Programme led by Swansea University. We propose to use unwanted nutrients from anaerobic digestion facilities to produce algal biomass for sustainable animal feeds and other high-value products.

Preface to "Metabolites from Phototrophic Prokaryotes and Algae Volume 2"

Algae (here including phototrophic prokaryotes) are a polyphyletic collection of aquatic organisms, with an enormous diversity in terms of form and function. Ubiquitous in fresh and marine environments, their contribution to global primary production approximates that of terrestrial organisms, and their role in regulating carbon and nitrogen cycles is essential to maintaining life on our planet.

In addition to the important ecological role that algae play in global carbon and nitrogen cycles, these organisms are increasingly emerging as being important in biotechnology. Their ability to fix carbon through photosynthesis, their high productivities compared to plants and the production of some unique groups of metabolites make them an attractive proposition to fulfilling the drive towards a sustainable and low carbon bio-based circular economy.

However, our understanding of metabolism and metabolite production in algae currently lags behind that in plants. From an ecological perspective, the better understanding of metabolic pathways and metabolite production (both intracellular and extracellular) will lead to an improved understanding of the role of algae in cycling elements within aquatic environments, and to improved assessments of aquatic primary production. Additionally, from a biotechnological perspective, a better understanding will lead to improved yield and the ability to manipulate algal growth for bio-production purposes. There is also potential to apply our understanding in the manipulation of metabolite pathways, and production to plant and other non-plant systems.

From both an ecological and biotechnological perspective, we need an improved understanding of acclimation and adaptation strategies to both abiotic (e.g., nutrients, light, temperature and salinity) and biotic factors (e.g., microbial consortia interactions, predator-prey interactions and bacterial/viral infection). In particular, we need an improved understanding of shifts in the allocation between primary and secondary metabolism and metabolites, and on carbon and nitrogen allocation.

This book represents research papers based on metabolomics, to improve the knowledge of metabolome and metabolism in algae, with a focus on carbon and nitrogen resource allocation. It also describes many bioanalytical techniques and emphasizes their usefulness in microalgal biotechnology. Other aspects from an ecological, biotechnological and waste-water remediation perspective are also covered. Numerous references are enlisted for those who wish to go further. We hope that this book will be useful for students, researchers, and lecturers.

Carole A. Llewellyn, Rahul Vijay Kapoore
Editors

 *metabolites*

Article

The Effect of High-Intensity Ultraviolet Light to Elicit Microalgal Cell Lysis and Enhance Lipid Extraction

Thomas Sydney [1], Jo-Ann Marshall-Thompson [2], Rahul Vijay Kapoore [2,3], Seetharaman Vaidyanathan [2], Jagroop Pandhal [2] and J. Patrick A. Fairclough [4,*]

[1] Department of Chemistry, The University of Sheffield, Sheffield S3 7HF, UK; thomas_sydney@hotmail.co.uk
[2] Department of Chemical and Biological Engineering, ChELSI Institute, Advanced Biomanufacturing Centre, The University of Sheffield, Sheffield S1 3JD, UK; jo-ann.marshall@svgcc.vc (J.-A.M.-T.); R.V.Kapoore@swansea.ac.uk (R.V.K.); s.vaidyanathan@sheffield.ac.uk (S.V.); j.pandhal@sheffield.ac.uk (J.P.)
[3] Department of Biosciences, College of Science, Swansea University, Swansea SA2 8PP, UK
[4] Department of Mechanical Engineering, The University of Sheffield, Sheffield S3 7HQ, UK
*   Correspondence: p.fairclough@sheffield.ac.uk; Tel.: +44-(0)-114-222-7798

Received: 27 August 2018; Accepted: 11 October 2018; Published: 15 October 2018

Abstract: Currently, the energy required to produce biofuel from algae is 1.38 times the energy available from the fuel. Current methods do not deliver scalable, commercially viable cell wall disruption, which creates a bottleneck on downstream processing. This is primarily due to the methods depositing energy within the water as opposed to within the algae. This study investigates ultraviolet B (UVB) as a disruption method for the green algae *Chlamydomonas reinhardtii, Dunaliella salina* and *Micractinium inermum* to enhance solvent lipid extraction. After 232 seconds of UVB exposure at 1.5 W/cm^2, cultures of *C. reinhardtii* (culture density 0.7 mg/mL) showed 90% disruption, measured using cell counting, correlating to an energy consumption of 5.6 MJ/L algae. Small-scale laboratory tests on *C. reinhardtii* showed bead beating achieving 45.3 mg/L fatty acid methyl esters (FAME) and UV irradiation achieving 79.9 mg/L (lipids solvent extracted and converted to FAME for measurement). The alga *M. inermum* required a larger dosage of UVB due to its thicker cell wall, achieving a FAME yield of 226 mg/L, compared with 208 mg/L for bead beating. This indicates that UV disruption had a higher efficiency when used for solvent lipid extraction. This study serves as a proof of concept for UV irradiation as a method for algal cell disruption.

Keywords: microalgae; cell disruption; ultraviolet light; biodiesel; *Chlamydomonas reinhardtii; Dunaliella salina; Micractinium inermum*

---

1. Introduction

Algal biofuels are the subject of large investment and a great deal of interest due to the promise of renewable, affordable, sustainable energy. Thus far, no company has achieved the full commercial potential that microalgae promise as a fuel source. Current processes to produce conventionally usable fuel from algae require numerous conversion steps. In particular, production of biodiesel from algae is severely limited due to the energy associated with cell disruption and lipid extraction. These processes can account for up to 26.2% [1] and 52% [2] of the energy input, respectively. The consequence is that the energy derived from the fuel is less than the external energy input required to process the algae [3]. Optimisation of algal cell disruption is thus an important step in the processing of algae for biochemical extraction, especially in biofuel applications [4]. Whilst various methods achieve cell disruption, they scale poorly on an industrial level. A low cost, scalable method for disruption is still sought. Such a technique would need to be low-energy, low-cost, continuously operable and, importantly, maintain the quality of the desired compounds extracted.

Conventional disruption techniques used to lyse algal cells ready for processing, such as homogenisation, microwave, sonication and bead milling, have a high associated energy cost and remain relatively expensive [5,6]. Techniques that disrupt algal cells must rupture the rigid cell wall in order to extract the commercially interesting compounds such as lipids and proteins. Additionally, these traditional techniques do not have the same disruption efficiencies on all species. Algae such as *Chlorella vulgaris* possess a thick cell wall which is highly resistive to mechanical stress, and as such, bead beating is less effective [7].

Lee et al. [8] compared five different methods, (including autoclaving, bead-beating, microwaves, sonication and a 10% NaCl solution) for cell disruption and concluded microwaves were the most effective method as this led to the largest lipid content extraction (44 mg/L). Unfortunately, the energy consumed amounted to 420 MJ per kilogram of dry algal mass which is a factor of over 4 times that of homogenisation [5]. This technique also proves difficult to optimise as the microwave energy is wasted on heating the extracellular water in the culture medium. Lee *et al.* [5] also compared bead milling, a simple technique for cell disruption. Beads are added to a culture, which is then vigorously shaken causing collisions between cells and beads; the beads erode the cell surface and cause lysis. This is slightly less effective than microwaves at cell disruption and higher in energy consumption at 504 MJ per kilogram of algal dry mass [5].

Many cell disruption methods deposit energy in the water rather than within the algae and are thus difficult to scale commercially. Bead beating is limited by losses due to viscous heating; microwave, sonication and other physico–thermal methods are limited by heating the water. Thus, this research explores the development of a low energy method to disrupt algal cells using ultraviolet (UV) light.

The short wavelength, high energy photons of ultraviolet B (UVB) and ultraviolet C (UVC) can lead to significant cell damage and are the most damaging wavelengths of UV light [9,10]. Ultraviolet A (UVA) is less effective, causing indirect damage to cells through the production of reactive oxygen species that may damage DNA, proteins and lipids [10]. UVB and UVC cause direct DNA damage through absorption of photons by DNA bases, resulting in chemical quenching and the formation of pyrimidine dimers in the sequence [11].

Few if any studies, to our knowledge, have looked at the effect of high intensity UV radiation for cell lysis, especially for the extraction of chemicals. Moharikar et al. [12] did study the effectiveness of UVC to induce apoptotic or necrotic pathways, but not from a biotechnological viewpoint. They conclude that UVC causes apoptotic and necrotic pathways in *C. reinhardtii* following sufficient exposure to UVC. Furthermore, although high doses of UV light lead to cell lysis, its use as a method of cell disruption in algae for lipid extraction has not received significant interest.

1.1. Cellular Signalling after Ultraviolet Irradiation

UV light induced cell lysis is due to the absorption of UV radiation by DNA, RNA, protein and lipids which can lead to structural damage and signalling/metabolic disorder [9]. DNA replication may be defective following UV exposure if the correct repair does not take place and may lead to mistakes in transcription and translation which result in protein synthesis with incorrect sequencing and misfolds [11]. DNA damage can be repaired after initial exposure; photorepair will occur if the algae are placed back into natural light, through the enzyme cyclobutane pyrimidine dimer photolyase [11]. In the work presented here, algae were stored in the dark following irradiation in order to reduce photorepair and maximise disruption. At low UV doses this repair mechanism can prevent lysis [11]. Additionally, other DNA repair methods such as excision repair and recombination repair are possible but cannot be prevented by dark storage [9]. At industrial scale, a sufficiently high dosage would avoid any repair that cells could undergo as they would be irreparably damaged. For instance, at 300 s at 9 (UVA) and 1.5 (UVB) W/cm^2 irradiation, cell counts before dark storage (data not presented) indicated approximately the same number of non-viable cells counted 24 h later.

The cellular signalling cascade that leads to cell death following UVB exposure is complex and has not been fully investigated in algae. UVB radiation causes the formation of pyrimidine dimers in DNA which often lead to mutation of a cell's genome. In mammalian cells, and shown here in algal cells, if sufficient mutations occur the cell may lyse through necrosis or apoptosis (see Figure 1); two pathways that operate differently but ultimately lead to cell death [13]. It has been previously shown that algae have a similar process [14]. Under intense trauma, the cell undergoes necrosis, the uncontrolled release of intracellular components. Cells which undergo this premature death rupture, and this is usually caused by mutations in genes which regulate key cellular processes. In contrast, apoptosis involves the regulated release of intracellular components and DNA fragmentation. It is often called programmed cell death due to its highly regulated nature and is activated if sufficient DNA mutations occur in particular genes. During apoptosis cells "pack" intracellular components into apoptotic bodies which are then released into the culture medium; some of which have high lipid content and can be referred to as lipid bodies.

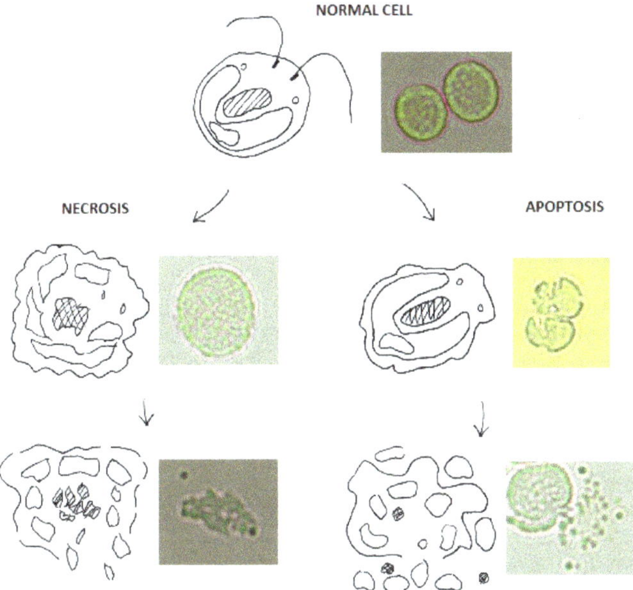

Figure 1. UV induced necrosis and apoptosis of *C. reinhardtii* cells visualised using Olympus BX50 microscope, 50× objective. Control image of no irradiation of a normal cell. UVB intensity at 1.5 W/cm^2 for 300 s (equivalent to 450 J/cm^2 UVB) for both necrosis and apoptosis. A 3 mL sample of *C. reinhardtii* in a quartz cuvette was exposed to 300 s of irradiation at 9 W/cm^2 UVA (320–395 nm) and 1.5 W/cm^2 UVB (280–320 nm) at a path length of 3 cm using a BlueWave 75 UV Curing Spot Lamp.

In this work, the structural markers of apoptosis, necrosis and lipid bodies can be visualised using light microscopy and were present during cell counting experiments (see Figure 1). Mutations arising from UV radiation can elicit both necrotic and apoptotic pathways due to the random nature of base mutation. Importantly, the mechanism of cell death initiated by UV radiation should work consistently with any species of algae, as it attacks an organism's DNA. There will be varying degrees of effectiveness due to different cellular characteristics and some species may have developed more complex defences to UV. However, given a sufficiently high UV dose these defences can be overcome.

1.2. Ultraviolet Light Compared to Conventional Disruption

Ultraviolet light as a method for algal cell disruption is a non-mechanical technique which differs from most conventional methods. There are other disruption techniques that utilise non-mechanical means such as chemicals, enzymes or microwaves but ultraviolet light is particularly effective because it targets DNA specifically, effectively shutting off an organism's ability to function [13]. Additionally, water is highly transparent to UVB and hence no energy is wasted in treating the water; only the algae are affected [15].

This technology is readily scalable as UV sterilisation for water treatment is commonly employed on a commercial scale, with 8000 municipal systems currently operational; the largest of which is situated in the USA, sterilising 2.24 billion gallons per day [16].

Irradiating a non-dewatered culture of algae for disruption could be as simple as running a culture past a large UV source. This could be optimised by controlling the flow rate and maximising surface area through the use of mirrored surfaces and large surface area plates. Another example of industrial UV treatment of water is the bottled water industry which requires as little as 10 kWh per million litres [17].

The use of ultraviolet light as a method of cell disruption on algal cells is explored here. Three species with contrasting cell wall characteristics (*C. reinhardtii*, *D. salina* and *M. inermum*) were irradiated with UV light at various durations to determine cell disruption efficiency via light microscopy. *D. salina* lacks a cell wall, *C. reinhardtii* has a reasonably durable multi-layered glycoprotein-based cell wall and *M. inermum* has a thick cell wall [18,19]. Lipid extraction using solvents and transesterification on irradiated samples was also undertaken as a suitable measurement method to translate product yields to cell disruption.

2. Results and Discussion

2.1. C. reinhardtii Irradiation

Following irradiation, morphological changes to *C. reinhardtii* cells were apparent and showed that the majority of cells were no longer viable at the longer exposure durations. In particular, loss of a well-defined cell wall at the boundary between the cell and extracellular media indicated cells were not viable and damaged beyond repair. Often cells appeared as clusters of beads which is most likely an indication of the formation of apoptotic bodies (Figure 1). Trypan blue was used as a stain, thereby identifying cells with compromised walls, though sometimes cells were not stained but were no longer viable due to fragmentation.

Viable and non-viable cell counts were not possible with the trypan blue stain as it did not always bind to cells. Thus, an effort was made to only count cells as non-viable if it was unmistakable as not to overestimate the efficacy of UV disruption. Additionally, most intact cells were swollen, indicating a loss of osmotic control leading to an influx of surrounding media, perhaps due to cell wall damage or even a critical mutation in the cells homeostatic regulatory genes.

Highly bleached cells indicative of a loss of chlorophyll were also present. Both swollen and bleached cells were likely non-viable due to imminent cell lysis through accumulation of damage beyond repair. However, it should be noted that in general, morphological changes were distinct enough to rule out any concern over bias reporting; Figure 1 shows unstained cells with distinct changes.

At higher exposure times of 150 s and more so at 300 s, a large amount of cell debris was present which indicated many cells had broken apart completely. This is likely due to cells undergoing necrosis or apoptosis, as can be seen in Figure 1.

C. reinhardtii cells were compared at different growth phases to determine if there was a difference in UV cell disruption efficacy. Experiments with *C. reinhardtii* in stationary phase showed an exponential decrease in cell viability as exposure time increases, as shown in Figure 2. Lysis of 50% of the cells occurred at 71 s and 90% occurred after 232 s (equivalent to 348 J/cm^2 UVB).

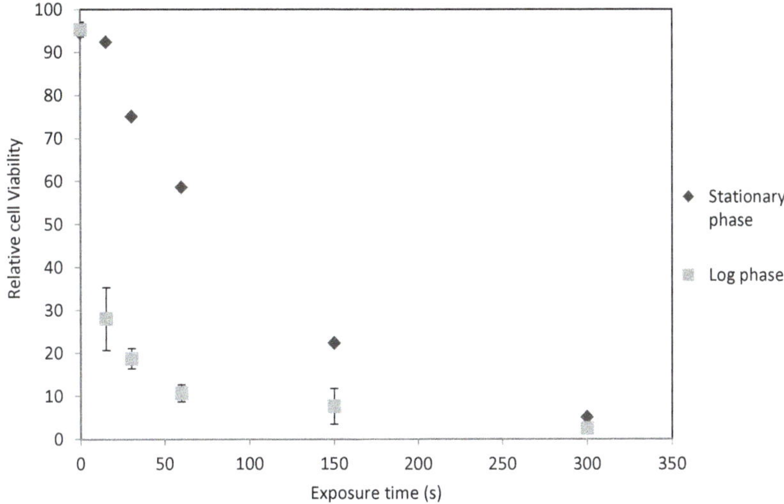

Figure 2. Relative cell viability of *C. reinhardtii* at various exposure times to ultraviolet light in the stationary and active growth phases. Stationary phase $n = 1$, log phase $n = 3$. Error bars represent the range for log phase. Control at 0 s of irradiation. UVB intensity at 1.5 W/cm^2. Carried out as described in Materials and Methods. Culture densities approximately 0.7 mg/mL dry weight.

A log phase *C. reinhardtii* culture under UV irradiation had a severe decline in cell viability with increasing exposure time. In this case, there was a prompt initial decline in cell viability as seen in Figure 2. Here 50% of algae cells were non-viable after 34 s and 90% after 127 s (equivalent to 190.5 J/cm^2 UVB).

The more severe decline of log phase cells compared to stationary phase may be due to increased vulnerability of algal DNA during cellular fission, although this has not been investigated by the scientific community thus far.

2.2. D. salina Irradiation

Viable and non-viable cells of *D. salina* were more distinguishable between one another than *C. reinhardtii* cells. Once again clusters of apoptotic bodies, swollen cells and a loss of a well-defined cell wall were visible, as seen in UV exposed *C. reinhardtii*. However, it was not possible to use trypan blue as a stain as it became apparent it had low miscibility with the saline media and unfortunately caused non-viable cells to aggregate.

The lack of a suitable stain proved insignificant as the cellular markers for viability were evident. *D. salina* cells are typically bright green and resemble a tear drop. They also are highly motile, and possess large flagella. These marked characteristics are easy to recognize and any changes due to UV radiation were distinct. After exposure, cells were more spherical and at high exposures there was a discernible reduction in motility, often rendering them non-motile. Increased cell debris was present at high exposure times as in the case of *C. reinhardtii*, indicating many cells have been completely destroyed and thus do not show up in cell counts.

D. salina was irradiated in its stationary phase and Figure 3 shows the relationship between UV exposure and cell viability. Viability was similar to that of *C. reinhardtii*, with *D. salina* appearing marginally more susceptible to UV radiation.

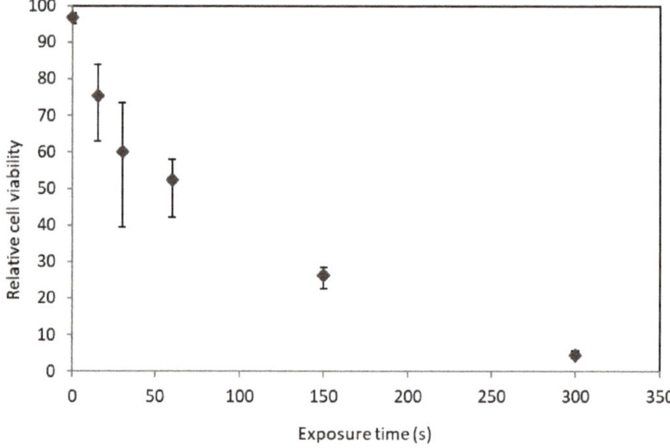

Figure 3. Relative cell viability of *D. salina* at various exposure times to ultraviolet light in the stationary phase. n = 3. Error bars represent the range. Control at 0 s of irradiation. UVB intensity at 1.5 W/cm^2. Carried out as in Materials and Methods. Culture densities approximately 0.7 mg/mL dry weight.

It was anticipated that *D. salina* would be far more susceptible to UV radiation than *C. reinhardtii* due to the difference in cell wall chemistry. However, due to the mechanism of UV disruption, cell wall characteristics may not affect its disruption efficacy. UV light does not need to interfere with the cell wall to cause DNA damage (although it can damage cell wall lipids and proteins as well) which can cause death through necrosis or apoptosis. This indicates that even thick cell walled genera such as *Chlorella* could be candidates for algal biotechnology or biodiesel production using UV radiation as a disruption method [18].

2.3. UV Radiation as a Disruption Method for Biodiesel Production

While the UV light source used in this experiment contains UVA and UVB, UVA wavelengths do not cause direct DNA damage and hence have a much-reduced effect on cell disruption. As discussed in the Introduction, UVB and UVC can lead to significant cell damage and are the most damaging wavelengths of UV light. Therefore, this study highlights UVB light as the source for eliciting microalgal cell lysis.

To this end, samples of *C. reinhardtii* were irradiated with UVB radiation (1.5 W/cm^2), alongside bead beaten samples as a control, before lipid extraction and transesterification to determine efficacy of UV radiation as a disruption method for biodiesel production. A detailed explanation of the experiment is within the methods section. Samples of *C. reinhardtii* were irradiated for 300 s or bead beaten. The samples were then solvent extracted with methanol:chloroform mixture (1:2) before heating at 80 °C for 90 min with 10% BF$_3$/methanol. Following this the samples in hexane were submitted for GC-FID based FAME analysis using a TR-FAME capillary column.

FAMEs produced from *C. reinhardtii* samples that underwent lipid extraction and transesterification are shown in Figure 4, demonstrating that the new UVB radiation method was more effective than bead beating, bead beating having 45.3 mg/L culture and UV irradiation having 79.9 mg/L culture (approximately 0.7 mg/mL algal culture density). Meaning an approximate FAME yield of 7.7% and 11.3% for bead beating and UV irradiation, respectively. This indicates that ultraviolet light is an effective method to disrupt algal cells for biodiesel production. The simplicity of using light for disruption has great potential for scaling the technology to an industrial level. In fact, flowing a culture past an ultraviolet bulb could be a simple continuous way to disrupt a culture.

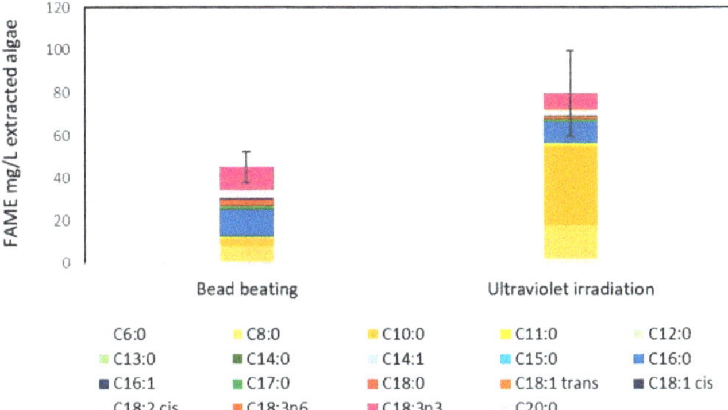

Figure 4. Comparison of biodiesel yield from disruption methods used on nitrate stressed *C. reinhardtii*. Fatty acid methyl ester yields from both bead beating and ultraviolet light irradiation disruption is shown. Yields represent transesterified lipids from 1 L of culture. UVB intensity at 1.5 W/cm^2 for 300 s (equivalent to 450 J/cm^2 UVB). Irradiation and bead beating carried out as in Materials and Methods. *n* = 2. Error bars represent the range. The legend indicates different FAME chain lengths.

2.4. Effect of Nitrate Stress on FAME Yield

Comparing UV irradiated samples and bead beaten samples from a culture containing nitrate (Figure 5) with a nitrate stressed culture (a common method used to increase lipid yield) (Figure 4), FAME yield was considerably higher in the nitrate stressed system (Figure 4), UV irradiated samples having higher yields. The major difference between the FAME yield was an increase in C8 and particularly C10 chain lengths. Interestingly this difference is not the same in the nitrate rich system (Figure 5), though as the FAME levels detected in were low, it is difficult to draw any firm conclusions other than that both UV irradiation and bead beating have a similar lipid extraction efficiency when intracellular lipid content is low. Typically FAME produced from algae is high in the C16 and C18 chains, which are the most suitable chain lengths for biodiesel [20]. The FAME profiles of UV irradiated algal cells generally follow this trend where the majority FAME peaks are C16 and C18 (as shown in Figure 6 and unreported data).

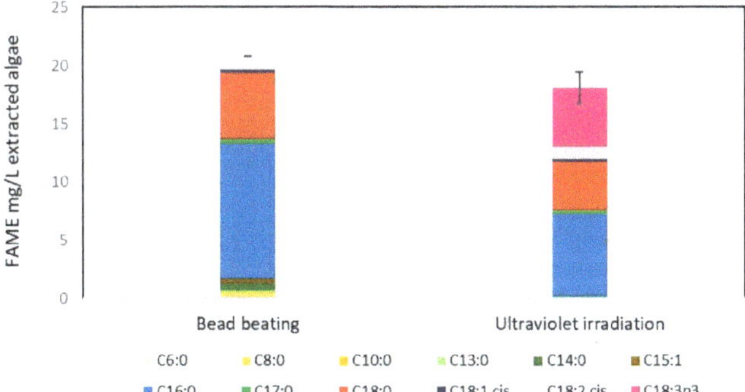

Figure 5. Comparison of biodiesel yield from disruption methods of nitrate rich *C. reinhardtii* culture. Fatty acid methyl ester yields from both bead beating and ultraviolet light irradiation disruption is shown. Yields represent transesterified lipids from 1 L of culture. UVB intensity at 1.5 W/cm^2 for 300 s (equivalent to 450 J/cm^2 UVB). Irradiation and bead beating carried out as in Materials and Methods. *n* = 1 for bead beading, *n* = 2 for UV irradiated samples. Error bars represent the range. The legend indicates different FAME chain lengths.

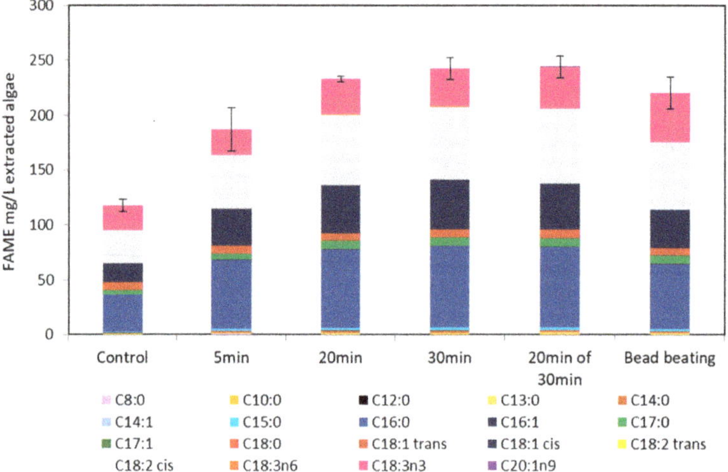

Figure 6. Comparison of biodiesel yield from UV light exposed *M. inermum*. Fatty acid methyl ester yields from both bead beating and ultraviolet light irradiation disruption is shown above. Yields are scaled to represent transesterified lipids from 1 L of culture. UVB intensity at 1.5 W/cm^2. 20 min over 30 min represents 40 s of irradiation per min for 30 min. Irradiation, bead beating, and control carried out as in Materials and Methods. Error bars represent the range. n = 2. The legend indicates different FAME chain lengths.

2.5. Ultraviolet Light Irradiation of M. inermum

Following the irradiation of stationary phase *C. reinhardtii*, it became apparent that whilst it is a useful species for study due to its growth characteristics and lipid yields, its electrostatically linked glycoprotein cell wall is not as robust as other species of algae [18]. In particular, species such as *Chlorella* possess a thick covalently linked cell wall that often has to be freeze-thawed several times to lyse [18]. *M. inermum* is another species which is similar to *Chlorella* and has a thick cell wall [19]. Previous figures (1, 2 and 3) demonstrate that ultraviolet light is capable of disrupting both *C. reinhardtii* and *D. salina*, however, both lack thick cell walls that are difficult to disrupt. Therefore *M. inermum* was exposed to ultraviolet light at various dosages to determine its efficacy for cell disruption and for FAME yield following lipid extraction and transesterification.

Figure 6 shows that with increasing UV irradiance, FAME yield from *M. inermum* increases. The duration of exposure required to maximise lipid extraction is much longer than seen with *C. reinhardtii*. This is likely due to the increased cell wall thickness and covalent nature. Bead beating was performed for 2 min cycles fifteen times as opposed to 1 min, to ensure the cells were sufficiently disrupted.

Furthermore, Figure 6 indicates the irradiation duration required was much higher but ultimately successful due to the mechanism by which UV light causes cell disruption, primarily through DNA damage. The smaller, thick cell walled *M. inermum* cells were successfully disrupted using UV light, given a high enough duration. The data presented in the cases of both *C. reinhardtii* and *M. inermum* suggests UV radiation to be a more effective cell disruption method than bead beating for FAME yield. Specifically, the FAME yields from *M. inermum* indicated UV radiation (30 min) achieved a FAME yield following transesterification of extracted lipid of 226 mg/L extracted algae compared to 208 mg/L extracted algae for bead beating.

2.6. Ultraviolet Light Cell Disruption Energy Efficiency

In order to determine the efficacy of a cellular disruption technique, the energy cost must be weighed against the disruption efficiency, especially for a biofuels application. The Bluewave DYMAX curer light source draws 75 W of power to function. Approximately 225 s of irradiation results in 90% cell lysis of a 3 mL aliquot of *C. reinhardtii*; this translates to an energy consumption of 5.6 MJ/L algae or 8 KJ/mg algae. It should be noted this is for a UV irradiation path length of only 3 cm with an approximate culture density of 0.7 mg/mL. Should a culture be concentrated through centrifugation, it is probable that higher efficiencies could be achieved using appropriate intensity, duration and path length. Eventually, a limit will be received where path length or culture density becomes too great for UV to pass through the culture, and this needs to be determined in future to allow scale up of the technique.

This is considerably lower energy (approximately 13 times) than other methods as shown by McMillan et al. [21], such as microwave treatment, which achieved $94.92 \pm 1.38\%$ lysis with an energy consumption of 74.6 MJ/L algae (1.8×10^8 cells/mL). It should be noted that the above ~95% disruption efficiency was achieved with *Nannochloropsis oculata*, a smaller alga with a resistive cell wall. However, the fact that UV irradiation does not need to mediate the cell wall to cause disruption suggests disruption efficiencies will be similar across many species. Disruption efficiencies for *D. salina* and *C. reinhardtii* are comparable in their stationary phase and support that theory. This energy efficiency could further be improved upon through the use of reflective surfaces, more efficient bulbs or balancing irradiance intensity against duration.

2.7. Limitations of the Study

This study aims to present the use of UV light as a proof of concept method for cell disruption of algae to enhance lipid extraction. This novel method will need to scale in order to be industrially viable, however, this study was not aimed at exploring scalability. To this end, additional data is required in order to determine industrial relevance. Said data must detail the effect of UV light on cell disruption as a function of increasing cell density with costs and efficacy compared. A methodology designed with that in mind will further elucidate UV light's industrial potential as an algal cell disruption method for biodiesel production.

Additionally, as this study is focused on biodiesel production, chain length and degree of saturation of fatty acids are an important measure of FAME quality. Due to the proof of concept nature of the study these aspects have not been directly addressed, however, they are important factors when determining scalability and industrial relevance of the novel method. While this study does not directly focus on FAME quality, is it of note that the FAMEs obtained using UV disruption of *M. inermum* do not have any notable difference in FAME quality (Figure 6). This is in contrast to *C. reinhardtii* FAMEs (Figures 4 and 5), where there is a distinct increase in C8 and C10 esters, and in the degree of unsauration. Therefore, at this stage it is difficult to conclude whether UV disruption has an effect on FAME quality, but it is clear that additional studies need to explore the quality of biodiesel produced.

2.8. Future Work

Following this work various other studies will further investigate the use of UV light as an algal disruption method to enhance lipid extraction for biodiesel production. The initial experiments using UV light disruption were performed on *C. reinhardtii* as it is a model species; in general, it does not produce high lipid yields, however, for the purposes of determining if the technique was viable, it was a suitable candidate.

This proof of concept study was focussed on the viability of UV light as a disruption method rather than on maximising lipid yield using high cell density cultures (and therefore high lipid/FAME content). In order to investigate UV disruption for higher FAME yields, higher cell density cultures would be required which present different scalability challenges that are outside the scope of a proof of concept approach.

Future research should be aimed at determining the effectiveness of UV disruption for lipid extraction using high lipid producing algae with high cell densities. Such research would be more suitable in a study that is aimed at determining the scalability of the technique as well as optimising the irradiation duration, concentration and efficacy over a wider range of algae.

Additional work will also include the use of concentrated solar radiation as the ultraviolet light source for disruption of microalgal cells. Another study will explore the use of novel recyclable magnetic nanoparticles that can extract the disrupted algae and lipids following UV light disruption.

3. Materials and Methods

3.1. Algae Strains and Cultivation

Chlamydomonas reinhardtii (wild type CCAP 11/32A), *Dunaliella salina* (wild type CCAP 19/30) and *Micractinium inermum* were grown in 250 mL conical flasks with TAP medium, Vonshak's medium [22] and Bold's Basal Medium, respectively, at 25 °C under magnetic stirring and positioned approximately 10 cm from 36 W fluorescent tube light sources [22,23]. *Micractinium inermum* was isolated from an environmental sample at the University of Sheffield [24].

Cells were grown up to approximately 0.7 mg/mL dry weight. These species were selected as suitable organisms for UV irradiation studies as their broad background literature and contrasting cell wall characteristics would give insight into the effectiveness of UV radiation to lyse cells regardless of cell wall strength.

Cells were nitrate stressed, when required, by re-inoculating a late log phase culture in nitrogen free TAP medium, which was made by omitting ammonium chloride from the TAP medium recipe [25].

3.2. Ultraviolet Light Irradiation of Algae

Various 3 mL samples of *C. reinhardtii* (24 samples), *D. salina* (15 samples) and *M. inermum* (8 samples) in a quartz cuvette were exposed to a series of irradiation periods ranging from 15 s to 300 s (5–30 min for *M. inermum*) with 9 W/cm^2 UVA (320–395 nm) and 1.5 W/cm^2 UVB (280–320 nm) at a path length of 3 cm using a BlueWave 75 UV Curing Spot Lamp. The area of irradiation was approximately 1 cm^2. Following this, the samples were left for 24 h in the dark. A cell count distinguishing between viable and non-viable cells was then undertaken using trypan blue as detailed below by means of an Olympus BX50 microscope (50× objective) and a Neubauer haemocytometer as described by Wu et al. [26]. Briefly, 5 µL trypan blue stained samples (1:1 ratio) of control (unirradiated cells) and UV irradiated samples were loaded into the haemocytometer. Cells were observed for morphological changes by counting live and dead cells in 16 small squares within each of the 4 large squares and averaging the live and dead cells before comparing the total live to dead ratio to calculate the percentage viability, with each experiment repeated in triplicate. Disruption efficiency was then calculated for each irradiation duration (number of viable cells as a percentage of total cell count).

Trypan blue vitality stain was used to stain *C. reinhardtii* cells but could not be used with *D. salina* as the stain had poor miscibility in the saline media. However distinct morphological changes in *D. salina* following irradiation meant determining cell viability was possible without a stain. Additionally, Lugol's solution was applied to *D. salina* to inhibit their motility during counting [27].

3.3. Lipid Extraction

As described above, samples of *C. reinhardtii* (4 samples) and *M. inermum* (8 samples) were irradiated with UV. An additional control (2 samples of *M. inermum* where only a solvent extraction was performed with no method of cellular disruption) and bead beaten samples (3 samples of *C. reinhardtii*, 2 samples of *M. inermum*) were prepared. Samples were then centrifuged (8500 rpm) and pellets retained in Eppendorfs with a 1.2 mL methanol:chloroform mixture (1:2). Supernatants were also collected to determine free lipid content.

UV exposure and control samples were kept on ice and bead beating samples were bead beaten using a cell disruptor (Genie, VWR, U.K.) for 15 cycles (1 min disruption followed by 1 min in ice, or 2 min cycles in the case of *M. inermum*) [28]. 400 µL chloroform and 400 µL ultrapure water were added to each sample before centrifuging (8500 rpm). Bottom organic phases were retained in Eppendorfs. Samples were then evaporated to dryness with nitrogen gas and reconstituted with 250 µL of methanol:chloroform mixture (1:1).

3.4. Transesterification

Following this, samples were transferred to glass vials and 100 µL 10% BF_3/methanol added before incubation at 80 °C for 90 min (14 samples of *C. reinhardtii* and 24 samples of *M. inermum*, inclusive of collected supernatant samples). Samples were cooled for 10 min before addition of 300 µL ultrapure water and 600 µL hexane, then vortexed for one min. The aqueous top layers were discarded before evaporating the remaining organic phase to dryness under inert nitrogen gas using six port mini-vap evaporator (Sigma-Aldrich, Dorset, UK) [29,30]. Samples were reconstituted in 100 µL hexane and submitted for gas chromatography flame ionisation detection (GC-FID) (Thermo Finnigan TRACE 1300 GC-FID System, Thermo Scientific, Hertfordshire, UK) coupled with a TRACE™ TR-FAME GC column (25 m × 0.32 mm ID × 0.25 µm film, Thermo Scientific, Hertfordshire, UK). The experiment was repeated in duplicate.

4. Conclusions

Compared to a traditional cell disruption method such as bead beating, UV disruption was more efficient. Approximately 225 s of irradiation (equivalent to 335.7 J/cm^2 UVB) resulted in 90% cell lysis of a 3 mL aliquot of *C. reinhardtii*, translating to an energy consumption of 5.6 MJ/L algae (UVB intensity at 1.5 W/cm^2). Furthermore, GC-FID data indicated UVB radiation was more effective than bead beating for FAME yield, bead beating yielding 45.3 mg/L culture and UV irradiation yielding 79.9 mg/L culture. UV irradiation also achieved a higher FAME yield than bead beating for the thick covalently bonded cell walled species *M. inermum*, achieving a FAME yield following transesterification of extracted lipid of 226 mg/L culture compared to 208 mg/L culture for bead beating. This indicates that UV irradiation is a viable method for cell disruption to enhance lipid extraction and should be further pursued. This technology is especially promising due to its scalability because UV sterilisation for water treatment is well-established commercially.

Author Contributions: T.S. and J.-A.M.-T. contributed to the data acquisition. T.S., P.F., R.V.K., S.V. and J.P. contributed to the design and drafting of the article. All authors give their final approval of the submitted manuscript.

Funding: This work was supported through the Project Sunshine Centre for Doctoral Training which is jointly funded by a gift from the Inge Sugden Will Trust and investment from the University of Sheffield.

Conflicts of Interest: The authors declare no conflict of interest.

References

1. de Boer, K.; Moheimani, N.R.; Borowitzka, M.A.; Bahri, P.A. Extraction and conversion pathways for microalgae to biodiesel: A review focused on energy consumption. *J. Appl. Phycol.* **2012**, *24*, 1681–1698. [CrossRef]
2. Shimako, A.H.; Tiruta-Barna, L.; Pigné, Y.; Benetto, E.; Navarrete-Gutiérrez, T.; Guiraud, P.; Ahmadi, A. Environmental assessment of bioenergy production from microalgae based systems. *J. Clean. Prod.* **2016**, *139*, 51–60. [CrossRef]
3. Slade, R.; Bauen, A. Micro-algae cultivation for biofuels: Cost, energy balance, environmental impacts and future prospects. *Biomass Bioenergy* **2013**, *53*, 29–38. [CrossRef]
4. Show, K.Y.; Lee, D.J.; Tay, J.H.; Lee, T.M.; Chang, J.S. Microalgal drying and cell disruption-Recent advances. *Bioresour. Technol.* **2015**, *184*, 258–266. [CrossRef] [PubMed]
5. Lee, A.K.; Lewis, D.M.; Ashman, P.J. Disruption of microalgal cells for the extraction of lipids for biofuels: Processes and specific energy requirements. *Biomass Bioenergy* **2012**, *46*, 89–101. [CrossRef]
6. Kapoore, R.V.; Butler, T.; Pandhal, J.; Vaidyanathan, S. Microwave-Assisted Extraction for Microalgae: From Biofuels to Biorefinery. *Biology* **2018**, *7*, 18. [CrossRef] [PubMed]

7. Huang, Y.; Qin, S.; Zhang, D.; Li, L.; Mu, Y. Evaluation of Cell Disruption of Chlorella Vulgaris by Pressure-Assisted Ozonation and Ultrasonication. *Energies* **2016**, *9*, 173. [CrossRef]

8. Lee, J.Y.; Yoo, C.; Jun, S.Y.; Ahn, C.Y.; Oh, H.M. Comparison of several methods for effective lipid extraction from microalgae. *Bioresour. Technol.* **2010**, *101*, S75–S77. [CrossRef] [PubMed]

9. Karentz, D. Ecological considerations of Antarctic ozone depletion. *Antarct. Sci.* **1991**, *3*, 3–11. [CrossRef]

10. Kuluncsics, Z.; Perdiz, D.; Brulay, E.; Muel, B.; Sage, E. Wavelength dependence of ultraviolet-induced DNA damage distribution: Involvement of direct or indirect mechanisms and possible artefacts. *J. Photochem. Photobiol. B Biol.* **1999**, *49*, 71–80. [CrossRef]

11. Rastogi, R.P.; Kumar, A.; Tyagi, M.B.; Sinha, R.P. Molecular Mechanisms of Ultraviolet Radiation-Induced DNA Damage and Repair. *J. Nucleic Acids* **2010**, *2010*, 592980. [CrossRef] [PubMed]

12. Moharikar, S.; D'Souza, J.S.; Kulkarni, A.B.; Rao, B.J. Apoptotic-like cell death pathway is induced in unicellular chlorophyte Chlamydomonas reinhardtii (chlorophyceae) cells following uv irradiation: Detection and functional analyses. *J. Phycol.* **2006**, *42*, 423–433. [CrossRef]

13. Fink, S.L.; Cookson, B.T. Apoptosis, pyroptosis, and necrosis: Mechanistic description of dead and dying eukaryotic cells. *Infect. Immun.* **2005**, *73*, 1907–1916. [CrossRef] [PubMed]

14. Segovia, M.; Haramaty, L.; Berges, J.A.; Falkowski, P.G. Cell death in the unicellular chlorophyte Dunaliella tertiolecta: A hypothesis on the evolution of apoptosis in higher plants and metazoans. *Plant Physiol.* **2003**, *132*, 99–105. [CrossRef] [PubMed]

15. Wozniak, B.; Dera, J. *Light Absorption in Sea Water*; Springer: New York, NY, USA, 2007.

16. Water Technology. Catskill–Delaware Ultraviolet Water Treatment Facility, New York, 2015. Available online: http://www.water-technology.net/projects/-catskill-delaware-ultraviolet-water-treatment-facility/ (accessed on 29 May 2017).

17. Gleick, P.H. *The World's Water. Volume 7: The Biennial Report on Freshwater Resources*; Island Press: Washington, DC, USA, 2011.

18. Loewus, F.A.; Runeckles, V.C. *The Structure, Biosynthesis, and Degradation of Wood, Boston, MA*; Springer: New York, NY, USA, 1977.

19. Hoshina, R.; Fujiwara, Y. Molecular characterization of Chlorella cultures of the National Institute for Environmental Studies culture collection with description of Micractinium inermum sp. nov., Didymogenes sphaerica sp. nov., and Didymogenes soliella sp. nov. (Chlorellaceae, Tr.). *Phycol. Res.* **2013**, *61*, 124–132.

20. Selvarajan, R.; Felföldi, T.; Tauber, T.; Sanniyasi, E.; Sibanda, T.; Tekere, M. Screening and Evaluation of Some Green Algal Strains (Chlorophyceae) Isolated from Freshwater and Soda Lakes for Biofuel Production. *Energies* **2015**, *8*, 7502–7521. [CrossRef]

21. McMillan, J.R.; Watson, I.A.; Ali, M.; Jaafar, W. Evaluation and comparison of algal cell disruption methods: Microwave, waterbath, blender, ultrasonic and laser treatment. *Appl. Energy* **2013**, *103*, 128–134. [CrossRef]

22. Vonshak, A. Laboratory Techniques for the cultivation of microalgae. In *Handbook of Microalgal Mass Culture*; Richmond, A., Ed.; Taylor & Francis: London, UK, 1986; p. 30.

23. Gorman, D.S.; Levine, R.P. Cytochrome f and plastocyanin: Their sequence in the photosynthetic electron transport chain of Chlamydomonas reinhardi. *Proc. Natl. Acad. Sci. USA* **1965**, *54*, 1665–1669. [CrossRef] [PubMed]

24. Smith, R.T.; Bangert, K.; Wilkinson, S.J.; Gilmour, D.J. Synergistic carbon metabolism in a fast growing mixotrophic freshwater microalgal species Micractinium inermum. *Biomass Bioenergy* **2015**, *82*, 73–86. [CrossRef]

25. Boyle, N.R.; Page, M.D.; Liu, B.; Blaby, I.K.; Casero, D.; Kropat, J.; Cokus, S.J.; Hong-hemersdorf, A.; Shaw, J.; Karpowicz, S.J.; et al. Three acyltransferases and nitrogen-responsive regulator are implicated in nitrogen starvation-induced triacylglycerol accumulation in Chlamydomonas. *J. Biol. Chem.* **2012**, *287*, 15811–15825. [CrossRef] [PubMed]

26. Wu, Z.; Shen, H.; Ondruschka, B.; Zhang, Y.; Wang, W.; Bremner, D.H. Removal of blue-green algae using the hybrid method of hydrodynamic cavitation and ozonation. *J. Hazard. Mater.* **2012**, *235*, 152–158. [CrossRef] [PubMed]

27. Preetha, K.; John, L.; Subin, C.S.; Vijayan, K.K. Phenotypic and genetic characterization of Dunaliella (Chlorophyta) from Indian salinas and their diversity. *Aquat. Biosyst.* **2012**, *8*, 27. [CrossRef] [PubMed]

28. Hounslow, E.; Kapoore, R.V.; Vaidyanathan, S.; Gilmour, D.J.; Wright, P.C. The Search for a Lipid Trigger: The Effect of Salt Stress on the Lipid Profile of the Model Microalgal Species Chlamydomonas reinhardtii for Biofuels Production. *Curr. Biotechnol.* **2016**, *5*, 305–313. [CrossRef] [PubMed]

29. Kapoore, R.V. Mass Spectrometry Based Hyphenated Techniques for Microalgal and Mammalian Metabolomics. Ph.D. Thesis, University of Sheffield, Sheffield, UK, August 2014.

30. Pandhal, J.; Choon, W.; Kapoore, R.; Russo, D.; Hanotu, J.; Wilson, I.; Desai, P.; Bailey, M.; Zimmerman, W.J.; Ferguson, A.S. Harvesting Environmental Microalgal Blooms for Remediation and Resource Recovery: A Laboratory Scale Investigation with Economic and Microbial Community Impact Assessment. *Biology* **2017**, *7*, 4. [CrossRef] [PubMed]

 *metabolites*

Article

Quenching for Microalgal Metabolomics: A Case Study on the Unicellular Eukaryotic Green Alga *Chlamydomonas reinhardtii*

Rahul Vijay Kapoore [1,2,*] and **Seetharaman Vaidyanathan** [1]

[1]   Department of Chemical and Biological Engineering, ChELSI Institute, Advanced Biomanufacturing Centre,
      The University of Sheffield, Sheffield S1 3JD, UK; S.Vaidyanathan@Sheffield.ac.uk
[2]   Department of Biosciences, College of Science, Swansea University, Swansea SA2 8PP, UK
*   Correspondence: R.V.Kapoore@Swansea.ac.uk; Tel.: +44-(0)1792-51-3671

Received: 14 October 2018; Accepted: 29 October 2018; Published: 31 October 2018

Abstract: Capturing a valid snapshot of the metabolome requires rapid quenching of enzyme activities. This is a crucial step in order to halt the constant flux of metabolism and high turnover rate of metabolites. Quenching with cold aqueous methanol is treated as a gold standard so far, however, reliability of metabolomics data obtained is in question due to potential problems connected to leakage of intracellular metabolites. Therefore, we investigated the influence of various parameters such as quenching solvents, methanol concentration, inclusion of buffer additives, quenching time and solvent to sample ratio on intracellular metabolite leakage from *Chlamydomonas reinhardtii*. We measured the recovery of twelve metabolite classes using gas chromatography mass spectrometry (GC-MS) in all possible fractions and established mass balance to trace the fate of metabolites during quenching treatments. Our data demonstrate significant loss of intracellular metabolites with the use of the conventional 60% methanol, and that an increase in methanol concentration or quenching time also resulted in higher leakage. Inclusion of various buffer additives showed 70 mM HEPES (4-(2-hydroxyethyl)-1-piperazineethanesulfonic acid) to be suitable. In summary, we recommend quenching with 60% aqueous methanol supplemented with 70 mM HEPES ($-40\,°C$) at 1:1 sample to quenching solvent ratio, as it resulted in higher recoveries for intracellular metabolites with subsequent reduction in the metabolite leakage for all metabolite classes.

Keywords: metabolomics; microalgae; quenching; *Chlamydomonas reinhardtii*; gas chromatography mass spectrometry (GC-MS)

1. Introduction

The major bottlenecks associated with sample preparation include efficient sampling, quenching, and extraction of metabolites, achieved with minimal alteration of the internal metabolome signature. In order to retain a valid snapshot of the metabolome, rapid sampling and quenching of enzyme activities is a crucial step in any metabolomics workflow. Ideally, quenching solvent should halt the constant flux of metabolism and high turnover rate of metabolites without causing any damage to the cell membrane/wall thereby avoiding any leakage of intracellular metabolites [1,2]. Quenching with 60% v/v cold methanol at $-40\,°C$ or $-50\,°C$ has been used widely in the past for microbial, fungi, yeast and plant metabolomics. However, potential problems connected to the leakage of intracellular metabolites with cold methanol quenching was reported later for yeast [3] and bacterial cells [4–7]. Various alternatives to cold methanol quenching, such as filter culture methodology [8], fast filtration [4], mass balance approach [9] and use of alternative quenching solvents (such as glycerol-saline, methanol/glycerol and methanol/NaCl) have been evaluated for bacterial metabolomics [5,10,11]. However, all suggested alternatives have advantages and disadvantages, and more importantly cannot be directly applied to a given organism, without prior evaluation [12]. In addition, these alternatives have also been shown to add difficulties in the overall metabolomics workflow. For example, the higher viscosity and hygroscopicity of glycerol has been shown to result in prolonged processing time (during separation of glycerol from cells) [13] and interference with the commonly employed silylation derivatization reaction (required for gas chromatography mass spectrometry (GC-MS) analysis) [14,15].

To date, the most widely used quenching method is that of 60% aqueous methanol, which usually results in leakage of the intracellular metabolites, and requires an accurate balancing of the quenching supernatant and sample [16]. Sampling methods optimised for prokaryotes cannot be simply adopted for eukaryotic cells due to basic differences in the cell structure. The inclusion of additives to methanol will act as a buffer or will restrict osmotic shock, leading to a decrease in leakage, as has been reported with bacterial cells [17], yeast [9,14] and mammalian cells [18,19]. The inclusion of tricine buffer in 60% aqueous methanol has been shown to result in lower recoveries for the intracellular metabolites with GC-MS because of the interference of tricine with derivatization reactions [20]. However, with other additives within the same biological system, contradictory conclusions have been reported (Table S1). A few reports have covered the influence of methanol concentration on metabolite leakage [9,21], however, even in these the approach has been kept limited to the quantification of specific metabolites. Another important factor that needs to be evaluated includes the influence of initial (before quenching) and the final temperature of methanol (sample-quenching solvent mixture). The lower the temperature, the slower the turnover rate of all the enzymes within the cell will be and hence more efficient is the quenching process [19]. However, only Canelas and co-workers [9] appear to have evaluated this factor for yeast, where the authors concluded that "leakage free quenching" can be achieved with the use of pure methanol rather than 60% methanol with quenching solvent to sample ratio of 5:1, and at −40 °C or lower. However, with a similar "leakage free quenching" method, Kim and co-workers [22] have demonstrated severe leakage in *Saccharomyces cerevisiae*, suggesting reduced membrane integrity caused by the use of extreme cold quenching conditions. The influence of contact time of sample with the quenching solvent has not been investigated yet, where higher leakage of specific classes of metabolites is likely to occur upon increasing the exposure time of sample to that of quenching solvent [4]. Finally, the influence of sample to quenching solvent ratio should also be carefully considered, as it directly alters the temperature of the quenching process and might help in minimising intracellular metabolite leakage. Schädel and co-workers [13] evaluated this factor for *Escherichia coli* metabolomics and observed no change in the impact of methanol with the temperature variation.

As discussed above, despite contradicting reports, quenching protocols have been rigorously studied and optimised for yeast, mammalian and bacterial models. Lee and Fiehn suggested yeast can be regarded as a good proxy for *Chlamydomonas reinhardtii* as both are eukaryotes and have sturdy cell walls compared to that of bacterial models which are easily prone to metabolite leakage caused by harsh quenching treatments [23]. However, it is important to note that minor differences in cellular characteristics including membrane, wall structure, size and sampling techniques employed, can influence the efficiency of quenching, recovery of different metabolite classes and the rate of metabolite leakage. Therefore, optimised quenching methods for yeast, mammalian and bacterial models cannot be simply adopted for microalga without critically evaluating them. In case of microalgal samples, we are not aware of many reports to date, on this issue. Lee and Fiehn [23], reported quenching cultures of *C. reinhardtii* with 70% aqueous methanol (−70 °C) with 1:1 ratio to sample, which reportedly resulted in minimum leakage of intracellular metabolites. In this case, the resultant final concentration of methanol was 35% and final temperature recorded was −20 °C. This finding was in contrast to their previously published report, where no leakage was reported and results concluded that *C. reinhardtii* cultures are resistant to quenching with cold methanol [24].

The objective of this investigation is to examine quantitatively, the influence of the above mentioned factors on the extent of metabolite leakage in *C. reinhardtii* cultures. To achieve our objective, we have designed and categorised the experiments into three different approaches. Approach 1 involves evaluation of selected quenching solvents (nine in total) with varying methanol concentration and with various buffer additives, on the extent of metabolite leakage. Approach 2 investigates the effect of prolonged quenching duration on metabolite leakage and approach 3 investigates the effect of the ratio of quenching solvent to culture on metabolite leakage. Furthermore, we investigated the recovery of twelve different metabolite classes using the GC-MS technique across different quenching treatments. The evaluation was based on recovery of a median number of metabolites within each class in all possible sample fractions and recovery of average metabolite peak intensity per class (derived from normalised median peak intensities of metabolites within each class).

2. Results and Discussion

To achieve our objectives in a broader sense, we have designed the experiments and categorised them into three different approaches as illustrated in Figure 1. In all the applied quenching treatments, the temperature during the quenching process was maintained below $-20\ ^\circ$C. Although metabolites with high turnover rates such as adenosine triphosphate (ATP) and adenosine diphosphate (ADP) might still remain active at $-20\ ^\circ$C, they are not usually detected with GC-MS based analysis. In most of the studies that evaluated the quenching methods for bacterial models [13,17], conclusions were drawn primarily based on ATP assay and adenylate energy charge. However, comment on quenching efficiency protocols cannot be made only based on these assays as it does not take into consideration the possible alterations in the rest of the metabolome. In the present study, we focussed primarily on GC-MS based analysis and the optimisation of quenching protocols were done by taking into consideration quantification of detectable compounds with GC-MS analysis. In all the three approaches, evaluation and comparison within different treatments were based on two response variables, where only features that were present in at least three biological replicates out of five were considered for further analysis:

- Response variable 1: Median number of metabolites recovered in;

  - (a) Cell extracts
  - (b) In quenching supernatant
  - (c) Only in cell extract and not in supernatant (unique to cells)
  - (d) Only in supernatant and not in cell extract (unique to supernatant) and
  - (e) In both the cell extracts and supernatants (common to both)

- Response variable 2: Recoveries of metabolites within twelve different classes of metabolites with respect to their normalised median peak intensities, and represented by the average metabolite peak intensity for each class.

We analysed the unquenched samples in order to get an estimate of extracellular metabolites accurately, as an unquenched sample serves as a reliable standard for comparison of the extracellular metabolome data from the quenched sample. In addition, we have also analysed the culture media as a control for appropriate quantification of an intracellular pool, to account for the leftover media components, if any.

2.1. Leakage Based on Recovery of Median Metabolite Numbers

2.1.1. Effect of Methanol Concentration and Inclusion of Buffer Additives on Metabolite Leakage (Approach 1)

In order to compare the extent of metabolite leakage and recoveries of intracellular metabolites under different methanol concentrations and additives, cells of *C. reinhardtii* were harvested and quenched as described in the experimental section with nine different quenching solvents and additives. Variation in quenching treatments was tested to investigate the different effects and the results are summarised in Table 1. For an overall comparison, the harvested cell extracts, a cell-free supernatant of both quenched and non-quenched cells along with the blank sample (culture medium) were analysed to determine the extent of leakage of intracellular metabolites during quenching. After monitoring cell-free supernatant of quenched, non-quenched cells and the blank medium, the necessary correction was done for appropriate calculation of intracellular metabolites as shown in Figure 2.

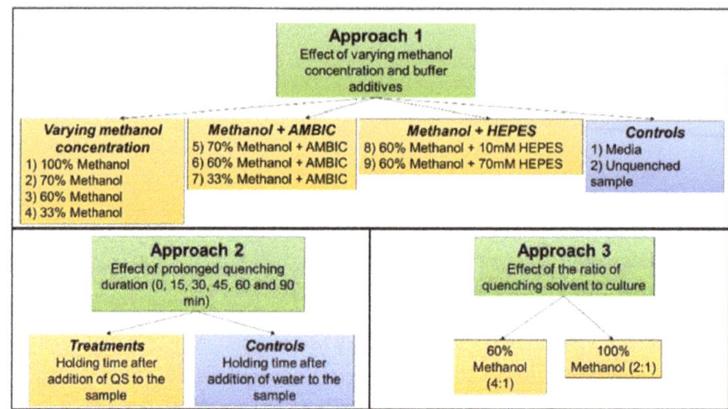

Figure 1. Experimental design and general workflow adopted for the evaluation and optimisation of quenching protocols for *Chlamydomonas reinhardtii*.

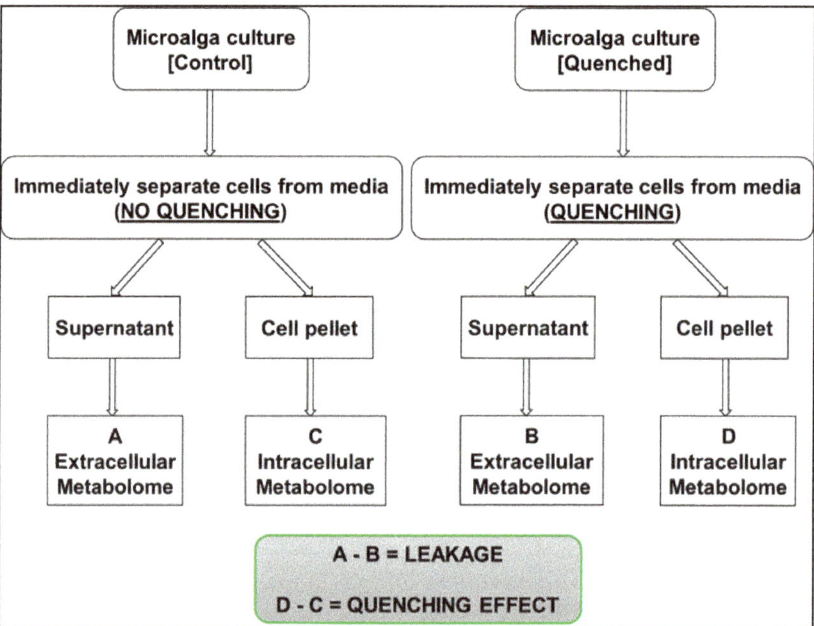

Figure 2. Schematic displaying the necessary correction required for appropriate calculation of intracellular metabolites after monitoring cell-free supernatant of quenched, non-quenched cells and the blank medium.

Table 1. Variation in quenching treatment tested to investigate the effects of varying methanol concentration, effect of inclusion of various buffer additives to quenching solution, effect of temperature and effect of sample to quenching solvent ratio on metabolite leakage.

	Method Code	Addition of Buffer	Initial Quenching Solvent Temperature (°C)	Initial Concentration of Methanol (*v/v*) (%)	Sample/Quenching Solvent Ratio	Final Concentration of Methanol after Quenching (*v/v*) (%)	Final Temperature of Resultant Mixture after Quenching (°C)
Approach 1	33M	x	−50	33	1:1	17	−20
	33A	0.85% AMBIC	−50	33	1:1	17	−20
	60M	x	−50	60	1:1	30	−30
	60A	0.85% AMBIC	−50	60	1:1	30	−30
	60H10	10 mM HEPES	−50	60	1:1	30	−30
	60H70	70 mM HEPES	−50	60	1:1	30	−30
	70M	x	−50	70	1:1	35	−35
	100M	x	−50	100	1:1	50	−40
Approach 3	60–41	x	−50	60	1:4	48	−40
	100–21	x	−50	100	1:2	77	−45
Controls	Control (Unquenched)	x	x	x	x	x	x
	Media	x	x	x	x	x	x

A summary of the unique recovery efficiency of all the nine quenching treatments is shown in Figure 3, where the metabolite class and numbers detected for various treatments are plotted. Figure 3a shows the number of metabolites detected in the cell pellets, Figure 3b those detected in the cell-free supernatant, Figure 3c indicates the number of metabolites present in the cells but not in the supernatants, Figure 3d indicates the number of metabolites present in the supernatants and not in the cells, and Figure 3e indicates the number of metabolites present in both the cell pellet and the supernatant. Higher metabolite numbers in the supernatant (Figure 3b), relatively high numbers detected in both the cells and the supernatants (Figure 3e), and corresponding low numbers unique to the cell pellets (Figure 3c) indicate high metabolite leakage. A higher proportion of metabolites detected in Figure 3e, compared to that detected in Figure 3c or Figure 3d indicates that there is an increased chance of metabolite leakage.

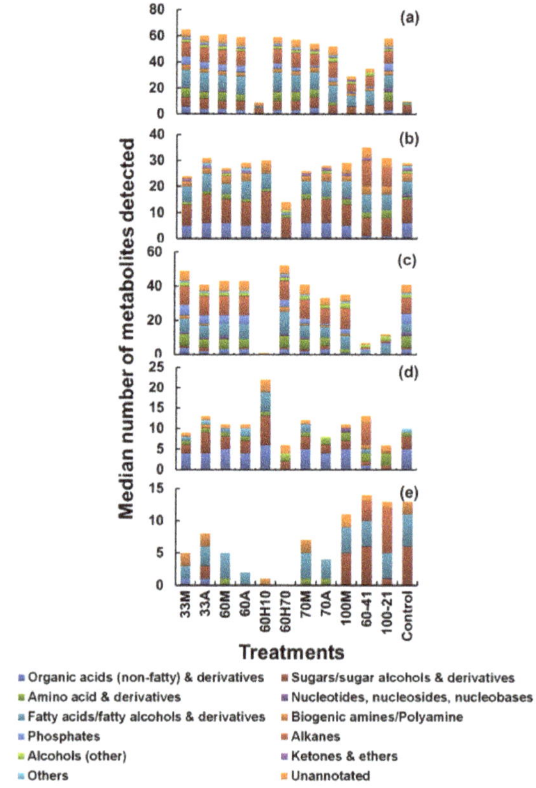

Figure 3. A summary of the unique recovery efficiency of all the nine quenching treatments involved in Approach 1 and 3. The X-axis represents different sampling treatments: 33M = 33% methanol; 33A = 33% methanol + ammonium bicarbonate (AMBIC); 60M = 60% methanol; 60A = 60% methanol + AMBIC; 60H10 = 60% methanol + 10 mM 4-(2-hydroxyethyl)-1-piperazineethanesulfonic acid (HEPES); 60H70 = 60% methanol + 70 mM HEPES; 70M = 70% methanol; 70A = 70% methanol + AMBIC; 100M = 100% methanol; 60–41 = 60% methanol with solvent to sample ratio 4:1 and 100–21 = 100% methanol with solvent to sample ratio 2:1. After all the treatments, the extracted metabolites from cell extracts, cell-free supernatant post quenching and blank samples were analysed by gas chromatography mass spectrometry (GC-MS). (**a**) Metabolites identified in cell extracts only, (**b**) metabolites identified in supernatant only, (**c**) metabolites present only in cell extract (and not in the supernatant)—unique to cells, (**d**) metabolites present only in supernatants (and not in cell extract)—unique to supernatants, (**e**) metabolites present in both the cell extract and supernatant—(common to both).

As can be seen from Figure 3c, 60% aqueous methanol supplemented with 70 mM HEPES (60H70) yielded higher recoveries of metabolites unique to cells as compared to other treatments. Correspondingly, metabolites common to both the cells and supernatant (Figure 3e) are the lowest. The number of metabolites detected in the supernatant is also the least compared to other treatments (Figure 3d). Therefore, among all the treatments, 60H70 seems to preserve the integrity of *C. reinhardtii* cells, resulting in higher recoveries of intracellular metabolites with minimal leakage into the extracellular environment.

Among other treatments, we did not observe any variations between 60M and 60A treatments. On the other hand, recovery of intracellular metabolites decreases with corresponding increase in extracellular levels as the methanol concentration increases. As can be seen from Figure 3c, higher recovery of intracellular metabolites was observed with 33M compared to 60M, 70M and 100M whereas metabolites common to both the cells and supernatants (Figure 3e) are lower with 33M compared to other treatments. Among treatments where methanol was supplemented with 0.85% AMBIC, 60A yielded higher recoveries of metabolites unique to cells (Figure 3c) compared to 33A and 70A. Correspondingly, metabolites common to both the cells and supernatants (Figure 3e) are the lowest. However, as mentioned above, we did not observe any major change in recoveries between 60M alone compared to that of 60A, therefore the use of 60M alone would be advantageous over the use of methanol supplemented with AMBIC. Moreover, comparison within treatments where non-buffered methanol was used for quenching, quenching with 33M seems to be a better option among non-buffered methanol and buffered with AMBIC. Therefore, in studies involving use of MS based hyphenated techniques (especially liquid chromatography mass spectrometry (LC-MS)), where use of salts/buffers as quenching additives introduces an additional source of variance to the experimental procedure, increases complexity of data and causes ion suppression in LC-MS, we strongly recommend use of 33M as a second alternative to 60H70. However as demonstrated earlier [25], an addition of HEPES has no apparent interference with the derivatization reactions and GC-MS analysis.

In summary, recovery of intracellular metabolites was noticed to decrease with corresponding increase in extracellular levels, as the methanol concentration in the quenching solvent increased. Our results were in agreement with that of previous reports [7,21] and in contrast to reports [9,13] where authors reported higher recovery of intracellular metabolites from *E. coli* and *S. cerevisiae* respectively with the corresponding increase in methanol concentration. In contrast, Tredwell and co-workers [20] reported that no major difference in leakage from *Pichia pastoris* was observed with the varying concentration of methanol or the inclusion of various buffer additives in quenching solvent. Furthermore, Canelas and co-workers [9] tested the influence of buffer additives (HEPES, AMBIC and/or tricine) in methanol on metabolite leakage from yeast, and reported no significant benefit in buffering or increasing the ionic strength of the quenching solvent. The authors reported slightly lower intracellular recoveries, which is completely contradictory to our finding with *C. reinhardtii*. The possible reasons behind this contradiction might be differences in the organism/sample type. Moreover, it is important to note that conclusions were drawn from an estimation of only two specific metabolites which were being used as representative examples for the respective class. Similarly, in contrast to our findings where 60H70 yielded highest intracellular recoveries with minimal leakage, Schädel and co-workers [13] reported higher leakage in *E. coli* with the inclusion of HEPES to methanol compared to conventional 60% methanol. These contradictory observations further support our previous conclusions [1], that sampling and quenching techniques are highly sample/cell dependent and needs critical evaluation and validation for every organism under investigation before being adopted for a quantitative metabolomics study.

2.1.2. Effect of Prolonged Exposure to Quenching Solvent on Metabolite Leakage (Approach 2)

Ideally, quenched cells should be processed as quickly as possible in order to avoid the leakage of intracellular metabolites into the extracellular environment, as prolonged contact time of cells with the quenching solvents might increase the chances of metabolite losses via diffusion of small metabolites through the cell membrane. In 1992, De Koning and Van Dam [26] demonstrated no leakage of intracellular metabolites from yeast sample after exposure of quenching solvent to sample for 30 min. This has been evaluated in the past for *S. cerevisiae* [9] and for the fungus *Penicillium chrysogenum* [21], however, there are no reports of such studies carried out on microalgal samples. Therefore, to test this theory for microalga, cells of *C. reinhardtii* were exposed to the quenching solvent (60% *v/v* aqueous methanol) for a prolonged period of various time intervals to evaluate the extent of metabolite leakage. Briefly, 1 mL of cell suspension was rapidly plunged into 1 mL of quenching solvent as described in the experimental section. To evaluate the extent of metabolite leakage in response to prolonged exposure to the quenching solvent, a broader range of time intervals were selected including 0, 15, 30, 45, 60 and 90 min. In the case of 0 min treatment, samples were processed immediately by centrifugation, whereas for other treatments prolonged exposure was achieved by leaving the quenched samples at $-40\,^{\circ}$C for the above specified time intervals prior to centrifugation. A parallel set of samples were processed as a control for each time interval. For example, in the case of the control sample for the 15 min time interval, 1 mL of cell suspension was harvested and allowed to stand for 15 min followed by addition of 1mL of water and subsequently centrifuged. The addition of water was done in order to account for the variations caused by dilution. Water was purposely selected in this case instead of any other buffer solution such as phosphate buffer saline (PBS), as the addition of PBS might result in an extremely higher concentration of phosphates and will interfere in subsequent GC-MS analysis or will result in overestimation of phosphates in intracellular pools. For an overall comparison, the harvested cell extracts, a cell-free supernatant of both quenched (samples) and non-quenched cells (controls) for all the above specified time intervals along with the blank sample (culture medium) were analysed to determine the extent of metabolite leakage in response to prolonged exposure to the quenching solvent. After monitoring cell-free supernatant of both quenched, non-quenched cells and the blank sample, the necessary correction was done for appropriate calculation of intracellular metabolites as shown in Figure 2. A summary of the unique recovery efficiency of all the applied treatments within this approach is shown in Figure 4, where the metabolite class and median numbers detected for various treatments are plotted.

As can be seen from Figure 4c, samples processed at the 0 time interval yielded higher recoveries of metabolites compared to all other treatments except for the 30 min interval where surprisingly higher numbers of metabolites were detected than that of the 0 min time interval. A similar trend was observed with the control samples where higher recoveries were observed with C0 treatment compared to other treatments except for C15 where again surprisingly higher recoveries were observed than that of C0. The possible reason behind higher recoveries might be degradation or inter-conversion of metabolites. Correspondingly higher number of metabolites unique to the supernatant (Figure 4d) were detected as the contact time of cell suspension was prolonged for more than 30 min with that of quenching solvent. Approximately four-fold increase in recoveries of extracellular metabolite numbers was observed with 45, 60 and 90 min treatments compared to that of the 0 min treatment (Figure 4d). No variations in recoveries of metabolites classes and numbers were observed between 0, 15 and 30 min time intervals whereas a small increase in the recoveries of metabolites was observed with control treatments as contact time was increased from 0 to 15 and then to 30 min indicating increased metabolite leakage.

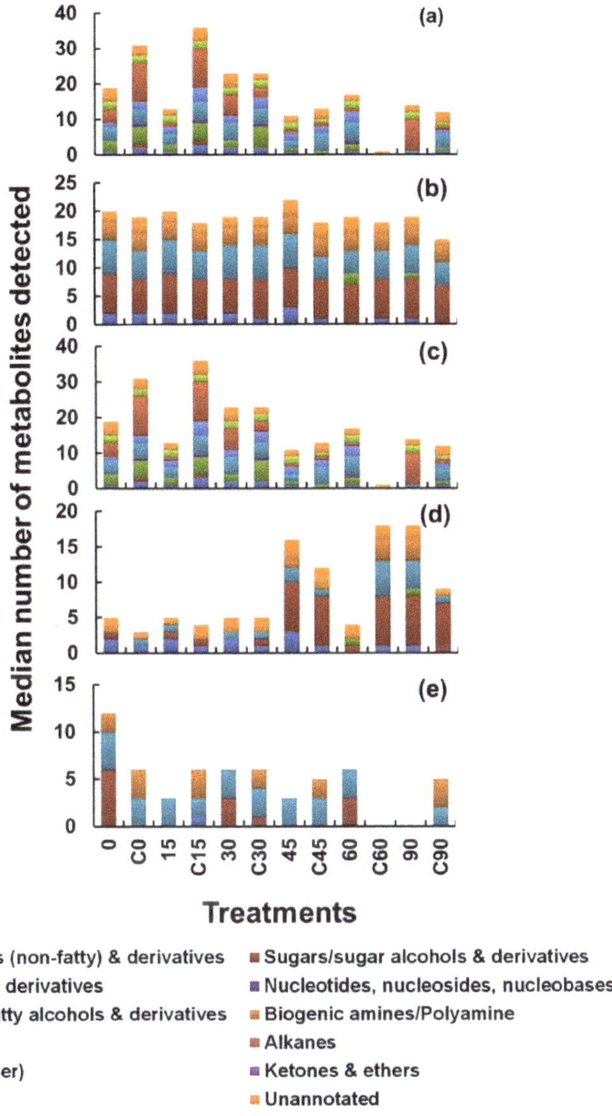

Figure 4. A summary of the unique recovery efficiency of all the six quenching treatments involved in Approach 2. The X-axis represents different time intervals along with their control samples (C), for each time interval. After all treatments the extracted metabolites from cell extracts, cell-free supernatant post quenching and blank samples were analysed by GC-MS. (**a**) Metabolites identified in cell extracts only, (**b**) metabolites identified in supernatant only, (**c**) metabolites present only in cell extract (and not in the supernatant)—unique to cells, (**d**) metabolites present only in supernatants (and not in cell extract)—unique to supernatants, (**e**) metabolites present in both the cell extract and supernatant—(common to both).

2.1.3. Effect of Quenching Solvent to Culture Ratio (Temperature Influence) (Approach 3)

Influence of sample to quenching solvent ratio was studied in the past for *S. cerevisiae* and *E. coli*, where quenching with pure methanol at sample to solvent ratio of 1:5 at −40 °C has been shown to have reduced leakage of intracellular metabolites compared to use of 60M. In order to test this parameter with microalgal samples and to study the influence of quenching solvent and sample mixture, temperature and final concentration of methanol after quenching, 1 mL culture of *C. reinhardtii* was harvested and rapidly quenched with either 100% methanol with quenching solvent to sample ratio of 2:1 (100–21) or with 60% aqueous methanol with quenching solvent to sample ratio of 4:1 (60–41). Both the quenching solvents were pre-chilled to −50 °C prior to quenching. Addition of the cells to the quenching solution increased the temperature by no more than 15 °C, thereby keeping the resulting mixture temperature below −20 °C sufficient to stop the metabolism as demonstrated in past [13,16,27]. For an overall comparison, the harvested cell extracts, a cell-free supernatant of both quenched and non-quenched cells along with the blank sample (culture medium) were analysed and the necessary correction was done for appropriate calculation of intracellular metabolites and to determine the extent of leakage of intracellular metabolites during quenching. The results of the investigation are summarised in Figure 3.

With both the treatments 60–41 and 100–21 higher metabolite numbers in the supernatant (Figure 3b), relatively high numbers detected in both the cells and the supernatants (Figure 3e), and corresponding low numbers unique to the cell pellets (Figure 3c) were observed, suggesting high metabolite leakage. A higher proportion of metabolites detected in Figure 3e, compared to that detected in Figure 3c or Figure 3d confirms that there is an increased chance of metabolite leakage with both the treatments. With this approach, we did not observe any significant improvements in preserving the cell integrity by altering the final methanol concentration in the resultant mixture after quenching. Moreover, an increment in quenching solvent to sample ratio in order to keep the temperatures of the resulting mixture below −20 °C for the effective halting of metabolism seems to have no influence on preserving the cell integrity and in minimising the metabolite leakage. In contrast, higher leakage was observed with this approach compared to approach 1.

2.2. Leakage Based on Metabolite Peak Intensities

With respect to analysis of metabolite classes recovered and the median intensities of individual metabolites identified within each class, in total, 151 unique putative metabolite identifications were made in approach 1 and 3, and 123 made in approach 2, across all treatments. In addition, there were unique features that did not match with metabolites on the database and were hence classed as "unannotated". These were 46 in approach 1 and 3, and 42 in approach 2 (Tables S2 and S3).

The results are compiled in Figure 5, where the median peak intensities of metabolites were averaged within each class and plotted for the different treatments in the three approaches, 1–3. Values above zero indicate higher detection in cell extracts than supernatants and values lower than zero indicate higher detection in supernatant than in cell extracts, suggesting leakage. For approach 1, superior recoveries of organic acids were observed with the 33M treatment, with minimal leakage. With sugars/sugar alcohols and derivatives, extracellular levels were higher for all treatments, except 100M. Since the observation is made with almost all conditions, it is possible that this is a result of leakage metabolism, rather than leakage due to quenching. Higher recoveries of amino acids were observed in the cell extracts with 60M and 33M treatments, suggesting minimal leakage of this class in these treatments. A higher leakage of amino acids was observed with 100M and 60H10 treatments. This is in agreement with the observation of Britten and McClure [28], where the authors demonstrated leakage of all free amino acids in *E. coli* upon variation in the osmolality of the surrounding medium. Nucleotides/nucleosides/nucleobases showed poor recoveries. The possible reason for lower recoveries in both the cell extracts and the supernatant might be that GC-MS is not an ideal technique for analysing this class. This has been previously suggested in a few studies. In such cases, LC-MS or capillary electrophoresis mass spectrometry (CE-MS) [29,30] might be a suitable alternative to GC-MS in order to improve the metabolome coverage with respect to nucleotides/nucleosides/nucleobases. Our results were in agreement with the results reported elsewhere [3,6] where no detectable leakage of nucleotides was observed with the cold methanol solution in the case of the yeast sample (which in our case corresponds to the 60M treatment). Fatty acids/fatty alcohols and derivatives showed minimal leakage in most treatments, and in particular in 100M and 60H70 treatments. Somewhat similar conclusions were reported in a previously published report [15], where higher peak intensities were reported for fatty acids with glycerol-saline quenching solution and extraction with 100% methanol using freeze-thaw cycles. Biogenic amines, in particular, ethanolamine, putrescine and tryptamine, showed minimal leakage in the 60H70 treatment but were also retained better in 33M, 33A and 70M treatments (Figure 5). Phosphates and alkanes were recovered in the cell extract with minimal leakage into supernatant with all the treatments except with 60H10 treatment, where negligible amounts of phosphates and alkanes were recovered in both the cell extract and the supernatant. Among all the treatments, higher recoveries were observed with 60H70. A gradual increase in the methanol concentration of quenching solvent resulted in a gradual decrease in the corresponding intracellular levels of phosphates which was in further agreement with a previously published report [9]. Alcohols were retained better in all treatments except 60H10. In the case of ketones and ethers, all the treatments (except with 60H70) showed leakage (Figure 5a).

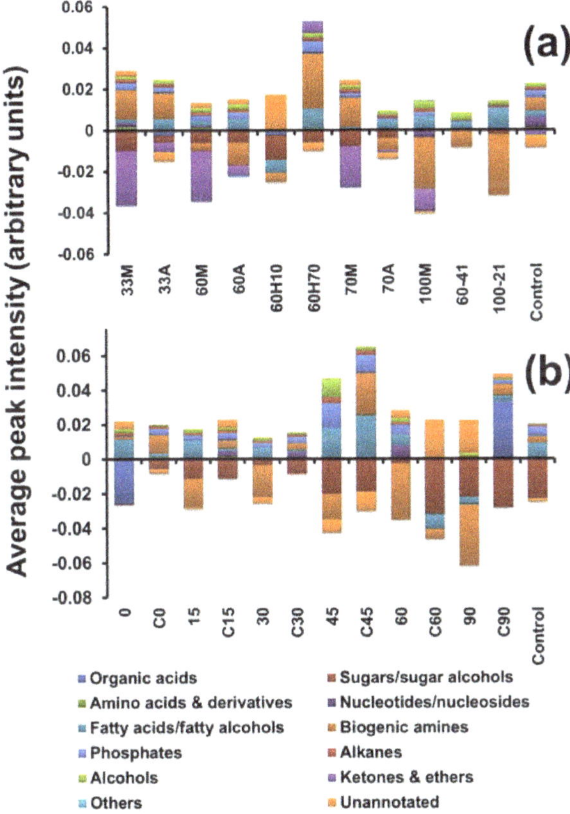

Figure 5. Difference in the average peak intensities of metabolites between cell extracts and supernatants, within each class is plotted for each treatment in approach 1 and 3 (**a**) and approach 2 (**b**).

For approach 2, total median intensities of all organic acids showed gradual decrease in cell extract recoveries from 15 min onwards up to 45 min hold time with corresponding increased recoveries in supernatant indicating leakage, whereas minor recoveries were observed in cell extracts with 45,

60 and 90 min intervals suggesting complete degradation or inter-conversion of all organic acids. In the case of sugars/sugar alcohols and derivatives lower recoveries for intracellular levels were observed as the exposure of cell suspension to the quenching solvent was increased from 0 to 30 minutes, indicating time dependant slow release of sugars into the extracellular medium, with complete leakage into the medium at 45 and 90 min exposures (Figure 5b). In the case of amino acids and derivatives, no leakage was observed up to 30 min exposure. Our findings were in agreement with a previously published report [6] where leakage of amino acids was increased with the increase in a contact time of sample with the quenching solvent. A gradual increase in extracellular levels of fatty acids/fatty alcohols and derivatives were observed as the hold time increased from 0 to 45 min (0-SN, 15-SN, 30-SN and 45-SN) indicating slower leakage of these less polar or semi-polar metabolites via diffusion. Putrescine and other biogenic amines were detected with all the applied quenching treatments. The results suggest slower time dependant leakage of all biogenic amines/polyamines, as extracellular levels were gradually increased with the corresponding decrease in the intracellular levels as the exposure to quenching solvent was increased from 0 min to 45 min. In the case of phosphates, absolutely no recoveries were observed in the quenching supernatant among all the applied treatments. A possible reason might be the larger and more polar nature of these metabolites making their diffusion through the cell wall difficult as suggested elsewhere in the case of yeast and fungi samples [9,21], leading to less leakage compared to other classes such as amino acid and organic acids which are smaller compounds and might be released more easily than the larger ones. However, despite their non-polar nature, similar trends as that of phosphates were observed with alkanes where no recoveries were observed in the quenching supernatant. The possible reason might be their larger size making their diffusion through the cell wall difficult as most of the alkanes that were recovered have molecular weight (MW) ranging from 212 to 352. A similar trend was observed with alcohols, ketones and ethers. In summary, nearly all the metabolite classes showed a gradual decrease in the intracellular levels with a corresponding increase in the extracellular levels, as the contact time of sample to quenching solvent was increased. Our findings were in agreement with previously published reports, where similar conclusions were drawn with yeast and bacterial models [6,9,21].

For approach 3, in the case of amino acids, organic acids, biogenic amine/polyamines, sugars/sugar alcohols and ketones and ethers very low recoveries were obtained in both the cell extracts and in the supernatant with both the quenching treatments compared to those in approach 1. With both the treatments, no recoveries were observed for nucleotides/nucleosides/nucleobases and other classes of metabolites. Superior recoveries of fatty acids/fatty alcohols and derivatives were obtained in the cell extract with the 100–21 treatment with the corresponding decrease in extracellular levels indicating minimal leakage (Figure 5a). Moreover, the average peak intensities recovered for this class was superior compared to all the applied treatments in approach 1. A similar trend was observed in recoveries of alcohols with 100–21 treatment.

3. Materials and Methods

3.1. Chemicals and Analytical Reagents

All chemicals and reagents were obtained from Sigma-Aldrich (Dorset, UK) unless stated otherwise.

3.2. Microalgal Cultivation

The *C. reinhardtii* strain (CC4323) was grown until the end of exponential phase (72 h) under constant illumination in a Sanyo incubator (MLR-351H, Sanyo versatile environmental test chamber, Osaka, Japan) at 25 °C in 1 L shake flasks, containing 1 L of TAP medium, under constant illumination at 85 μmol/m^2/s. The medium (per L) is composed of 2.42 g Tris; 25mL of TAP salts (15 g NH_4Cl, 4 g·$MgSO_4$·$7H_2O$, 2 g $CaCl_2$·$2H_2O$ in 1 L dH_2O); 0.375 mL of phosphate solution (28.8 g K_2HPO_4, 14.4 g KH_2PO_4 in 100 mL dH_2O); 1mL solution Hunter's trace elements purchase from the Chlamydomonas Resource Centre (St. Paul, MN, USA) and 1 mL of glacial acetic acid. Cells were harvested at an OD of 1.2 at 680 nm wavelength.

3.3. Sampling and Quenching

To evaluate and minimise the leakage of metabolites during quenching treatments, the overall design of quenching experiments has been categorised into three different approaches as summarised in Figure 1. At the incubation site, 1 mL of cell suspension was rapidly plunged into a 2 mL pre-chilled Eppendorf containing 1 mL of pre-chilled quenching solvent (-50 °C) unless stated otherwise. Addition of the cells to the quenching solvent increased the temperature by no more than 15 °C. The centrifuge was set at -9 °C and the rotor was pre-chilled at -24 °C. The quenched biomass was then centrifuged for 2 min at 2500× *g* at -9 °C. The supernatant was removed rapidly, and transferred to a 2 mL pre-chilled Eppendorf to assess the leakage of internal metabolites. The pellets and supernatant were snap frozen in liquid nitrogen and stored at -80 °C for further analysis.

3.4. Metabolite Extraction

The pellets were lyophilized at -50 °C overnight prior to extraction. Briefly, 500 µL of extraction solvent (methanol:chloroform:water, (M:C:W) (5:2:2)) was added to lyophilized cells, as suggested elsewhere [23,31,32], pre-chilled at -48 °C, along with an equal volume of glass beads (425–600 µm i.d., acid washed, from Sigma). Cells were then disrupted using a Cell disruptor, (Genie, VWR, U.K.), designed to simultaneously agitate and vortex at high speed, thereby resulting in a rapid disruption of cells in a small aliquot of extraction solvent (0.5 to 2 mL). A total of 11 cycles of disruption was performed with 2 min disruption, with an interval of 1 min on ice. The sample was then centrifuged at 13,000× rpm, at -9 °C for 15 min to remove any cell debris. The supernatant was transferred to a new pre-chilled Eppendorf tube (-20 °C) and the remaining cell debris was subjected to re-extraction (5 cycles) with 500 µL of extraction. The combined supernatant was then evaporated to dryness using a vacuum concentrator (Eppendorf 5301 vacuum concentrator, Sigma-Aldrich (Dorset, UK)). The dried extract was then stored overnight at -80 °C for further analysis.

3.5. Metabolite Derivatization and GC-MS Analysis

Metabolite derivatization was performed the next day on a stored dried extract, as suggested elsewhere [23]. Briefly, 30 µL of 20 mg/mL methoxyamine hydrochloride in pyridine was added to the dried extract and samples were shaken (200 rpm) for 80 min at 37 °C to protect the aldehyde and ketone groups. The samples were then derivatized by trimethylsilylation of acidic protons by addition of 45 µL MSTFA (N-methyl-N-trimethylsilyltrifluoroacetamide) and incubating them further in shaking condition (200 rpm) at 40 °C for 80 min. GC-MS analysis, metabolite identification and data analysis were conducted as described elsewhere [1,25,33].

4. Conclusions

In summary, we have performed a comprehensive investigation of appropriate sampling and quenching methods for *C. reinhardtii* using GC-MS, which involved untargeted quantitative analysis of several intracellular and extracellular metabolites. Our optimised miniaturised quenching method requires only 1 mL of microalgal culture, thereby subsequently reducing the volume of quenching solvent required, suitable for processing larger number of samples within a shorter period of time, ideal for metabolomics investigation where a large number of samples need processing. To date, there is no effective alternative to the cold methanol quenching method, where quenching can only be achieved by sharp temperature drift which usually leads to leakage caused by cold shock phenomenon. Hence, for reliable metabolome analysis, it is essential to measure the metabolite levels in all possible fractions and establish mass balance to trace the fate of metabolites during quenching treatment. Our results clearly showed higher losses of intracellular metabolites with the use of conventional 60% methanol for quenching. Analysis with respect to the influence of varying methanol concentration in quenching solvent on the extent of metabolite leakage clearly showed higher leakage with the increase in methanol concentration. On the other hand, analysis with respect to the inclusion of various buffer additives to quenching solvent showed 60% aqueous methanol supplemented with 70 mM HEPES to recover higher intracellular levels for nearly all metabolite classes with minimal leakage, supporting our previous findings with adherent mammalian cells [25]. An increment in quenching solvent to sample ratio in order to keep the temperatures of the resulting mixture below −20 °C for the effective halting of metabolism seems to have no influence on preserving the cell integrity and in minimising the metabolite leakage. In contrast, higher leakage was observed with this approach. The study of prolonged exposure of a sample to quenching solvent has shown a higher amount of leakage as the contact time of sample to that of quenching solvent increases. Hence for quantitative metabolomics studies in *C. reinhardtii*, we strongly recommend quenching with 60% aqueous methanol supplemented with 70 mM HEPES (−40 °C) at 1:1 sample to quenching solvent ratio, as it resulted in higher recoveries for intracellular metabolites with subsequent reduction in the metabolite leakage for all classes of metabolites. We believe the outcome of this research and the optimised quenching method can be directly applied and extended to other similar freshwater microalgal cultures or indeed cells with a similar cell envelope architecture and provides a standardised approach to metabolomics validations in other organisms.

Supplementary Materials: The following are available online at http://www.mdpi.com/2218-1989/8/4/72/s1, Table S1: Summary of the main quenching and washing solvents applied for the metabolome analysis of different biological systems. (Sing x stands for no information available or provided in the research papers surveyed). Table S2: List of putatively identified metabolites in *C. reinhardtii* extracts across different applied quenching protocols (approach 1 and 3). Class 1 = Organic acids (non-fatty) and derivatives; 2 = Sugars/sugar alcohols and derivatives; 3 = Amino acid and derivatives; 4 = Nucleotides, nucleosides, nucleobases; 5 = Fatty acids/fatty alcohols and derivatives; 6 = Biogenic amines/Polyamine; 7 = Phosphates; 8 = Alkanes; 9 = Alcohols (other); 10 = Ketones and ethers; 11 = Others and 12 = Unknowns. Table S3: List of putatively identified metabolites in *C. reinhardtii* extracts across different applied quenching protocols (approach 2). Class 1 = Organic acids (non-fatty) and derivatives; 2 = Sugars/sugar alcohols and derivatives; 3 = Amino acid and derivatives; 4 = Nucleotides, nucleosides, nucleobases; 5 = Fatty acids/fatty alcohols and derivatives; 6 = Biogenic amines/Polyamine; 7 = Phosphates; 8 = Alkanes; 9 = Alcohols (other); 10 = Ketones and ethers; 11 = Others and 12 = Unknowns.

Author Contributions: R.V.K. conceived the ideas and designed methodology in consultation and supervision of S.V. R.V.K. carried out the lab work, analysed the data, and wrote the manuscript with supervisory advice of S.V. All authors contributed critically to the drafts and gave final approval for publication.

Funding: The authors acknowledge financial support from EPSRC (EP/E036252/1) and BBSRC (BB/K020633/1).

Conflicts of Interest: The authors declare no conflict of interest.

References

1. Kapoore, R.V.; Coyle, R.; Staton, C.A.; Brown, N.J.; Vaidyanathan, S. Cell line dependence of metabolite leakage in metabolome analyses of adherent normal and cancer cell lines. *Metabolomics* **2015**, *11*, 1743–1755. [CrossRef]

2. Kapoore, R.V.; Vaidyanathan, S. Towards quantitative mass spectrometry-based metabolomics in microbial and mammalian systems. *Philos. Trans. R. Soc. A Math. Phys. Eng. Sci.* **2016**, *374*. [CrossRef] [PubMed]

3. Gonzalez, B.; François, J.; Renaud, M. A rapid and reliable method for metabolite extraction in yeast using boiling buffered ethanol. *Yeast* **1997**, *13*, 1347–1355. [CrossRef]

4. Bolten, C.J.; Kiefer, P.; Letisse, F.; Portais, J.-C.; Wittmann, C. Sampling for metabolome analysis of microorganisms. *Anal. Chem.* **2007**, *79*, 3843–3849. [CrossRef] [PubMed]

5. Villas-Bôas, S.G.; Bruheim, P. Cold glycerol–saline: The promising quenching solution for accurate intracellular metabolite analysis of microbial cells. *Anal. Biochem.* **2007**, *370*, 87–97. [CrossRef] [PubMed]

6. Villas-Bôas, S.G.; Højer-Pedersen, J.; Åkesson, M.; Smedsgaard, J.; Nielsen, J. Global metabolite analysis of yeast: Evaluation of sample preparation methods. *Yeast* **2005**, *22*, 1155–1169. [CrossRef] [PubMed]

7. Wittmann, C.; Krömer, J.O.; Kiefer, P.; Binz, T.; Heinzle, E. Impact of the cold shock phenomenon on quantification of intracellular metabolites in bacteria. *Anal. Biochem.* **2004**, *327*, 135–139. [CrossRef] [PubMed]

8. Rabinowitz, J.D. Cellular metabolomics of *Escherchia coli*. *Expert Rev. Proteom.* **2007**, *4*, 187–198. [CrossRef] [PubMed]

9. Canelas, A.B.; Ras, C.; ten Pierick, A.; van Dam, J.C.; Heijnen, J.J.; Van Gulik, W.M. Leakage-free rapid quenching technique for yeast metabolomics. *Metabolomics* **2008**, *4*, 226–239. [CrossRef]

10. Buchholz, A.; Takors, R.; Wandrey, C. Quantification of Intracellular Metabolites in *Escherichia coli* K12 Using Liquid Chromatographic-Electrospray Ionization Tandem Mass Spectrometric Techniques. *Anal. Biochem.* **2001**, *295*, 129–137. [CrossRef] [PubMed]

11. Shen, Y.; Fatemeh, T.; Tang, L.; Cai, Z. Quantitative metabolic network profiling of *Escherichia coli*: An overview of analytical methods for measurement of intracellular metabolites. *TrAC Trends Anal. Chem.* **2016**, *75*, 141–150. [CrossRef]

12. Pinu, F.R.; Villas-Boas, S.G.; Aggio, R.J.M. Analysis of Intracellular Metabolites from Microorganisms: Quenching and Extraction Protocols. *Metabolites* **2017**, *7*, 53. [CrossRef] [PubMed]

13. Schädel, F.; David, F.; Franco-Lara, E. Evaluation of cell damage caused by cold sampling and quenching for metabolome analysis. *Appl. Microbiol. Biotechnol.* **2011**, *92*, 1261–1274. [CrossRef] [PubMed]

14. Spura, J.; Christian Reimer, L.; Wieloch, P.; Schreiber, K.; Buchinger, S.; Schomburg, D. A method for enzyme quenching in microbial metabolome analysis successfully applied to gram-positive and gram-negative bacteria and yeast. *Anal. Biochem.* **2009**, *394*, 192–201. [CrossRef] [PubMed]

15. Zhao, C.; Nambou, K.; Wei, L.; Chen, J.; Imanaka, T.; Hua, Q. Evaluation of metabolome sample preparation methods regarding leakage reduction for the oleaginous yeast Yarrowia lipolytica. *Biochem. Eng. J.* **2014**, *82*, 63–70. [CrossRef]

16. Wellerdiek, M.; Winterhoff, D.; Reule, W.; Brandner, J.; Oldiges, M. Metabolic quenching of *Corynebacterium glutamicum*: Efficiency of methods and impact of cold shock. *Bioprocess Biosyst. Eng.* **2009**, *32*, 581–592. [CrossRef] [PubMed]

17. Faijes, M.; Mars, A.E.; Smid, E.J. Comparison of quenching and extraction methodologies for metabolome analysis of *Lactobacillus plantarum*. *Microb. Cell Fact.* **2007**, *6*, 27. [CrossRef] [PubMed]

18. Kronthaler, J.; Gstraunthaler, G.; Heel, C. Optimizing high-throughput metabolomic biomarker screening: A study of quenching solutions to freeze intracellular metabolism in CHO cells. *Omics J. Integr. Boil.* **2012**, *16*, 90–97. [CrossRef] [PubMed]

19. Sellick, C.A.; Hansen, R.; Maqsood, A.R.; Dunn, W.B.; Stephens, G.M.; Goodacre, R.; Dickson, A.J. Effective quenching processes for physiologically valid metabolite profiling of suspension cultured mammalian cells. *Anal. Chem.* **2008**, *81*, 174–183. [CrossRef] [PubMed]

20. Tredwell, G.D.; Edwards-Jones, B.; Leak, D.J.; Bundy, J.G. The development of metabolomic sampling procedures for *Pichia pastoris*, and baseline metabolome data. *PLoS ONE* **2011**, *6*, e16286. [CrossRef] [PubMed]

21. De Jonge, L.P.; Douma, R.D.; Heijnen, J.J.; van Gulik, W.M. Optimization of cold methanol quenching for quantitative metabolomics of *Penicillium chrysogenum*. *Metabolomics* **2012**, *8*, 727–735. [CrossRef] [PubMed]

22. Kim, S.; Lee, D.Y.; Wohlgemuth, G.; Park, H.S.; Fiehn, O.; Kim, K.H. Evaluation and optimization of metabolome sample preparation methods for *Saccharomyces cerevisiae*. *Anal. Chem.* **2013**, *85*, 2169–2176. [CrossRef] [PubMed]

23. Lee, D.Y.; Fiehn, O. High quality metabolomic data for *Chlamydomonas reinhardtii*. *Plant Methods* **2008**, *4*, 7. [CrossRef] [PubMed]

24. Bölling, C.; Fiehn, O. Metabolite profiling of *Chlamydomonas reinhardtii* under nutrient deprivation. *Plant Physiol.* **2005**, *139*, 1995–2005. [CrossRef] [PubMed]

25. Kapoore, R.V.; Coyle, R.; Staton, C.A.; Brown, N.J.; Vaidyanathan, S. Influence of washing and quenching in profiling the metabolome of adherent mammalian cells: A case study with the metastatic breast cancer cell line MDA-MB-231. *Analyst* **2017**, *142*, 2038–2049. [CrossRef] [PubMed]

26. Koning, W.d.; Dam, K.v. A method for the determination of changes of glycolytic metabolites in yeast on a subsecond time scale using extraction at neutral pH. *Anal. Biochem.* **1992**, *204*, 118–123. [CrossRef]

27. Weuster-Botz, D. *Die Rolle der Reaktionstechnik in der Mikrobiellen Verfahrensentwicklung*; Forschungszentrum Jülich, Zentralbibliothek: Jülich, Germany, 1999.

28. Britten, R.J.; McClure, F.T. The amino acid pool in *Escherichia coli*. *Bacteriol. Rev.* **1962**, *26*, 292. [PubMed]

29. Ramautar, R.; Somsen, G.W.; de Jong, G.J. CE-MS in metabolomics. *Electrophoresis* **2009**, *30*, 276–291. [CrossRef] [PubMed]

30. Soga, T.; Ohashi, Y.; Ueno, Y.; Naraoka, H.; Tomita, M.; Nishioka, T. Quantitative metabolome analysis using capillary electrophoresis mass spectrometry. *J. Proteome Res.* **2003**, *2*, 488–494. [CrossRef] [PubMed]
31. Gullberg, J.; Jonsson, P.; Nordström, A.; Sjöström, M.; Moritz, T. Design of experiments: An efficient strategy to identify factors influencing extraction and derivatization of *Arabidopsis thaliana* samples in metabolomic studies with gas chromatography/mass spectrometry. *Anal. Biochem.* **2004**, *331*, 283–295. [CrossRef] [PubMed]
32. Weckwerth, W.; Wenzel, K.; Fiehn, O. Process for the integrated extraction, identification and quantification of metabolites, proteins and RNA to reveal their co-regulation in biochemical networks. *Proteomics* **2004**, *4*, 78–83. [CrossRef] [PubMed]
33. Kapoore, R.V. Mass Spectrometry Based Hyphenated Techniques for Microalgal and Mammalian Metabolomics. Ph.D. Thesis, University of Sheffield, Sheffield, UK, 2014.

 *metabolites* 

Article

An Improved Genome-Scale Metabolic Model of *Arthrospira platensis* C1 (*i*AK888) and Its Application in Glycogen Overproduction

Amornpan Klanchui [1], Sudarat Dulsawat [2], Kullapat Chaloemngam [3],
Supapon Cheevadhanarak [4], Peerada Prommeenate [5],* and Asawin Meechai [3],*

[1] Biological Engineering Program, Faculty of Engineering, King Mongkut's University of Technology Thonburi, Bangkok 10140, Thailand; oofarto3@gmail.com
[2] Pilot Plant Development and Training Institute, King Mongkut's University of Technology Thonburi (Bang Khun Thian), Bangkok 10150, Thailand; sudarat.dul@kmutt.ac.th
[3] Department of Chemical Engineering, Faculty of Engineering, King Mongkut's University of Technology Thonburi, Bangkok 10140, Thailand; kunrapat60@gmail.com
[4] Division of Biotechnology, School of Bioresources and Technology, King Mongkut's University of Technology Thonburi, Bangkok 10150, Thailand; supapon.che@kmutt.ac.th
[5] Biochemical Engineering and Pilot Plant Research and Development (BEC) Unit, National Center for Genetic Engineering and Biotechnology, National Science and Technology Development Agency, King Mongkut's University of Technology Thonburi, Bangkok 10150, Thailand
* Correspondence: peerada.pro@biotec.or.th (P.P.); asawin.mee@kmutt.ac.th (A.M.); Tel.: +66-2-470-7508 (P.P.); +66-2-470-9616 (ext. 405) (A.M.)

Received: 21 October 2018; Accepted: 20 November 2018; Published: 26 November 2018

Abstract: Glycogen-enriched biomass of *Arthrospira platensis* has increasingly gained attention as a source for bioethanol production. To study the metabolic capabilities of glycogen production in *A. platensis* C1, a genome-scale metabolic model (GEM) could be a useful tool for predicting cellular behavior and suggesting strategies for glycogen overproduction. New experimentally validated GEM of *A. platensis* C1 namely *i*AK888, which has improved metabolic coverage and functionality was employed in this research. The *i*AK888 is a fully functional compartmentalized GEM consisting of 888 genes, 1,096 reactions, and 994 metabolites. This model was demonstrated to reasonably predict growth and glycogen fluxes under different growth conditions. In addition, *i*AK888 was further employed to predict the effect of deficiencies of NO_3^-, PO_4^{3-}, or SO_4^{2-} on the growth and glycogen production in *A. platensis* C1. The simulation results showed that these nutrient limitations led to a decrease in growth flux and an increase in glycogen flux. The experiment of *A. platensis* C1 confirmed the enhancement of glycogen fluxes after the cells being transferred from normal Zarrouk's medium to either NO_3^-, PO_4^{3-}, or SO_4^{2-}-free Zarrouk's media. Therefore, *i*AK888 could be served as a predictive model for glycogen overproduction and a valuable multidisciplinary tool for further studies of this important academic and industrial organism.

Keywords: *Arthrospira platensis* C1; bioethanol; cyanobacteria; genome-scale metabolic model; glycogen

---

1. Introduction

Due to the environmental concerns of production and utilization of fossil fuels, researches towards renewable energy are currently of great interest. Conversion of carbohydrate-enriched biomass of microalgae has become one of the promising approaches for sustainable clean energy generation [1]. Since their carbohydrates are in the form of lignin-free cellulose, starch, or glycogen, microalgae are much easier to convert to monosaccharides compared to lignocellulosic feedstocks [2]. In particular, prokaryotic microalgae, cyanobacteria, have certain advantages over eukaryotic microalgae for many reasons. They possess peptidoglycan cell wall which is easily degraded by fermentation processes [3,4]. Moreover, cyanobacteria accumulate glycogen as storage carbohydrate which is an excellent feedstock for fermentation over starch [4]. Furthermore, transformation systems in cyanobacteria have been much better developed [5–7].

Arthrospira (Spirulina) platensis, a filamentous non-nitrogen-fixing cyanobacterium, is an attractive candidate which has certain properties to be used as promising feedstocks for bioethanol production. For example, it has the capacity to accumulate large amounts of glycogen during cultivation under nutrient or environmental stresses [1]. The highest glycogen productivity of 0.29 g L^{-1} d^{-1} was reported under NO_3^- depletion and high light intensity of 700 μmol photons/m^2/s [8]. It also has the unique impressive characteristics for industrial applications, including a contaminant-free culture under the outdoor cultivation [9] and a fast bio-flocculation capability under nutrient starvation condition [10]. Moreover, Aikawa et al. proposed a low cost technology that generates the highest yield of bioethanol from carbohydrate-rich *Arthrospira* biomass by yeast fermentation [3]. Although research works have showed the potential use of *Arthrospira* to generate bioethanol, however, the glycogen productivity is still not high enough to profitably produce bioethanol at the commercial scale. To overcome this challenge, strategies for enhancement of glycogen content in *A. platensis* is needed. Among *Arthrospira*, *A. platensis* strain C1 is a good candidate for bioethanol production because it has distinct advantages over the others in term of strain improvement, non-gliding property [9], genome sequence [11], transformation systems [12], and the most comprehensive data at the molecular level, transcriptomics [13,14] proteomics [15–17] and protein–protein interactions [18]. However, the glycogen production in *A. platensis* C1 has not been investigated.

Driven by advancements in high-throughput biological and computational technologies, genome-scale metabolic model (GEM), a mathematical form of the cellular metabolic network, is currently being the indispensable tool for understanding cell phenotypes and providing rational strategies to maximize production of a desired metabolic product [19]. GEMs of various organisms across three domains of life, i.e., archaea, bacteria, and eukarya have been constructed and applied in various research areas, ranging from industrial to medical biotechnology [20,21]. Flux balance analysis (FBA) [22] is the most commonly used approach to simulate the GEMs. To investigate the metabolic activity and integrate omics data for allowing comprehensive studies at the systematic level of *A. platensis* C1, the first GEM of *A. platensis* C1, *i*AK692, was developed in 2012 [23]. This model was applied to predict optimal growth behavior, metabolic phenotypes, and essential genes under autotrophic, heterotrophic, and mixotrophic growth conditions.

Since bioinformatics tools, pathway/genome databases, and literatures have now been updated, newly curated genes and biochemical knowledge have become available. In this research, an updated GEM of *A. platensis* C1 named *i*AK888 was employed. *i*AK888 was built based on information of *i*AK692 as well as new annotated and curated genomic and biochemical knowledge. The *i*AK888, a fully compartmentalized GEM, is the most up-to-date comprehensive model for *A. platensis*. The model was demonstrated herein to be a helpful tool for proposing the rational strategies for improvement of glycogen production in *A. platensis* C1.

2. Materials and Methods

2.1. Genome-Scale Metabolic Network Reconstruction

The *i*AK888 was reconstructed by refining and updating the previously reconstructed genome-scale model for *A. platensis* C1, *i*AK692 [23]. The first step in the metabolic network reconstruction process was genome reannotation, which provided the initial set of gene–protein reaction (GPR) associations. The draft genome sequence of *A. platensis* C1 (6.089 Mb; GenBank NZ_CM001632) [11] was retrieved. A functional annotation was then performed using two independent systems: (i) The rapid annotation of microbial genomes using Subsystems Technology (RAST) annotation servers [24] supported by the SEED [25] and (ii) Kyoto Encyclopedia of Genes and Genomes (KEGG) Automatic Annotation Server (KAAS) [26]. To increase confidence in the annotation, the predicted ORFs obtained from *i*AK692, RAST [18], and KAAS [26] were compared. Only the shared ORFs were considered and any conflicting annotations were discarded. All transport genes and reactions were identified using a BLAST search [27] against the Transporter Classification Database (TCDB) [28].

Subsequently, a list of GPR associations was assembled and manually curated to reflect actual physiology of *A. platensis* C1 based on various information sources including biochemical databases [25,29,30], literatures [31–34], and curated genome-scale models, *Synechocystis* sp. PCC6803 [35] and *Escherichia coli* [36]. Manual curation was performed according to the standard reconstruction protocol [37]. A confidence scoring system [37] was employed to assign a confidence score to every network reaction. The score reflects the amount of available evidence type, biochemical data, genetic data, physiological data, sequence data, and modeling data, associated with individual reaction, ranging from 1 to 4, where 1 is the lowest and 4 is the highest evidence score. The gene expression data of *A. platensis* C1 (accession E-MTAB-2714) [14] were acquired from ArrayExpress in order to verify existence of predicted ORFs involved in the reconstructed network. Briefly, gene expression data of the control sample, in which *A. platensis* C1 was cultured in Zarrouk's medium at 35 °C under a light intensity of 100 μmol photons/m^2/s, were preprocessed and normalized. Then, the average of the log intensities between a housekeeping gene, 16S rRNA, and all genes were compared. Genes whose expression were equal to or greater than the housekeeping gene indicated the evidence for the existence of these genes and the associated reactions. Furthermore, all reactions were charge and mass balanced. The reaction directions were also assigned based on the standard Gibbs free energy of reaction ($\Delta_r G^o$) provided by the SEED [25]. Formulas and charges for all metabolites were checked against KEGG [29] and PubChem [38]. Subsequently, the cofactor specificity of the enzyme in *A. platensis* C1 was verified using organism-related species literature. Cellular compartments of candidate genes and reactions were determined based on the literature and the GEM of *Synechocystis* sp. PCC6803 [35]. Moreover, BLAST search [27] and functional domain analysis using Pfam [39] were conducted to refine missing and low confidence gene annotations. The ambiguous genes existing in the *i*AK692, but absent in the annotation results of RAST [24] and KAAS [26] were removed.

Additionally, energy-generating cycles (type-II pathways) or internal cycles (type-III pathways) were checked during the reconstruction process to ensure that ATP and NAD(P) could not be produced without nutrient consumption. For ATP, the ATP maintenance flux was optimized when CO_2 and photon uptake fluxes were set to zero. For NAD(P), an artificial reaction NAD(P)H → NAD(P) + H was added in the reconstructed network and optimized when CO_2 and photon uptakes were not available. If either ATP or NAD(P) could be produced without nutrient uptake, the reactions related to the production of these energy metabolite were checked manually.

An organism-specific biomass equation, representing cell growth in silico, was formulated and used as an objective function for simulating growth phenotypes through FBA [22]. The major macromolecular constituents of biomass synthesis consisted of protein, carbohydrate, lipid, DNA, RNA, pigments, vitamins, and minerals. The content of each constituent of *i*AK888, compared to the previously published model, *i*AK692, is shown in Table 1. The protein, carbohydrate, and lipid were estimated in the exponential growth phase of *A. platensis* C1 [40]. The stoichiometric coefficients of DNA, RNA, and protein synthesis reactions were estimated from the nucleotide and amino acid contents of the published genome of *A. platensis* C1 [11]. Moreover, the biomass equation for *i*AK888 also included pigments, vitamins and minerals, of which the coefficients were calculated based on the published information of the closest species [41–45]. In this study, all stoichiometric coefficients in the biomass equation were assumed to be constant under different environmental conditions. The detailed calculation of the biomass equation is provided in Supplementary File S1.

Table 1. Comparison of biomass constituents between *i*AK888 and *i*AK692.

Component	Content (%*w/w*)	
	*i*AK888	*i*AK692 [23]
Proteins	51.44	68
Carbohydrates	31.62	16
Lipids	4.98	11
DNA	0.88	0.88
RNA	3.12	3.12
Colorants	2.84	1
Vitamins	0.11	-
Minerals	2.79	-
Ash	2.27	-
Sum	100	100

2.2. Flux Balance Analysis

Flux balance analysis (FBA) is a widely used constraint-based optimization approach for predicting specific reaction rates (fluxes) of a large-scale network based on stoichiometry and steady-state assumption [22]. Briefly, all reactions within a reconstructed genome-scale metabolic network were converted into a stoichiometric matrix, S, in which each column and row represented one unique reaction and metabolite, respectively. The entries of S are the stoichiometric coefficients of metabolites participating in a reaction. By applying the mass balance constraints and steady state assumption, the particular metabolite's concentration change per unit time is equal to zero ($Sv = 0$), where v represents the vector of reaction fluxes. Then, other constraints such as thermodynamic and enzyme capacity were accounted, thereby, determining reaction directions (reaction directionality and reaction reversibility) [46]. Reaction directionality is typically assigned based on a negative $\Delta_r G^{\circ}$ while, reaction reversibility is a kinetic property of enzymes which are able to catalyze the reactions in the forward and backward directions. Hence, thermodynamic constraints help to decide the reaction directions in FBA. Finally, a linear programming was applied to maximize or minimize an objective function, usually the biomass reaction flux. This optimization results in optimal objective flux and optimal flux distribution of the metabolic network. In this work, the COBRA toolbox version 2 [47] with MATLAB (The MathWorks, version R2015b) was employed to model and predict cell behaviors i.e., specific growth rate and glycogen production flux.

2.3. Estimation of Glycogen Production Flux

To represent the glycogen metabolism in *A. platensis*, *i*AK888 incorporates 4 glycogen associated reactions, i.e., (i) glycogen synthesis reaction, (ii) glycogen utilization reaction for biomass growth, (iii) glycogen transport reaction for carbon storage and (iv) glycogen degradation reaction. In this work, glycogen was treated as a monomer ($C_6H_{10}O_5$, $M_W = 162.141$) in the glycogen synthesis reaction where one mole of glycogen was synthesized from one mole of glucose-1-phosphate by glycogen synthase (glgA, EC 2.4.1.21). In addition, glycogen phosphorylase (glgP, EC 2.4.1.1) that is responsible for the glycogen degradation was assumed to be inactive under autotrophic growth. Thus, the glycogen production flux was simply determined from the flux of the glycogen synthesis reaction.

2.4. Model Validation

To evaluate the accuracy of the reconstructed metabolic model, comparisons between predicted phenotypes and experimental data, including a maximal growth rate under different growth conditions and maximal carbohydrate production flux under nitrogen depletion condition were performed.

For prediction of the maximum specific growth rate, the experimental data sets in which *A. platensis* C1 was grown under autotroph, heterotroph, and mixotroph were calculated and used as the input parameters. Cells grown under autotrophic conditions were simulated and compared, including two independent sets of *A. platensis* C1 experiments, (i) setting the photon uptake rate to 100 μmol photons/m^2/s and HCO$_3^-$ uptake rate to 0.2 mmol/gDCW/h [23] and (ii) setting the photon uptake rate to 200 μmol photons/m^2/s and HCO$_3^-$ uptake rate to 0.25 mmol/gDCW/h [48]. Cells grown under a heterotrophic condition were simulated by setting the photon uptake rate to zero and glucose uptake rate to 0.017 mmol/gDCW/h [23]. For mixotrophic condition, cell growth was simulated by setting the photon uptake rate to 100 μmol photons/m^2/s, HCO$_3^-$ uptake rate to 0.2 mmol/gDCW/h, and glucose uptake rate to 0.017 mmol/gDCW/h [23].

To further assess the reliability of the model, the maximum total carbohydrate production flux under nitrogen depletion condition was predicted. The photon and HCO$_3^-$ uptake rate were set at 100 μmol photons/m^2/s and 0.2 mmol/gDCW/h, respectively. NO$_3^-$ uptake rate and specific growth rate were set to zero as observed in the experiment [49]. The objective function was accounted for maximizing the carbohydrate production flux.

For all the simulations, the uptake rates of all nutrients present in the Zarrouk's medium [50] were constrained as 0 to -1000, while the transport fluxes of CO$_2$, O$_2$, and H$_2$O were left unconstrained. In addition, all photons were assumed to be absorbed and used for driving photosynthesis without the influence of photoinhibition.

2.5. Simulation of Glycogen Production Under Nutrient-Limited Conditions

The effect of NO$_3^-$, PO$_4^{3-}$, and SO$_4^{2-}$ on growth and glycogen production were simulated by FBA [22] under the autotrophic condition. The specific uptake rate of HCO$_3^-$ was fixed at 1.6 mmol/gDCW/h for all simulations to guarantee an excess carbon condition. The maximum photon uptake flux was set to 100 μmol photons/m^2/s. The effect of each nutrient on specific growth rate and total glycogen production flux was analyzed by varying the uptake flux values of each nutrient, NO$_3^-$, PO$_4^{3-}$, and SO$_4^{2-}$. The objective function was set to maximize the flux of the glycogen synthesis reaction. Additionally, flux variability analysis (FVA) [51] was performed to determine the minimum and maximum possible fluxes under the simulation conditions. Geometric FBA [52] was also used to determine a unique optimal solution which is central to the range of possible flux distributions.

2.6. Experimental Validation

To evaluate the validity of *iAK888*, *A. platensis* C1 (PCC9438) was cultured in 1 L Erlenmeyer flasks containing 500 mL of Zarrouk's medium [50] at 35 °C under white fluorescent illumination at 100 μmol photons/m^2/s until mid-logarithmic phase. Then, the cells were transferred to three different Zarrouk's media, lacking either nitrate (NaNO$_3$ and Co(NO$_3$)·6H$_2$O), or phosphate (K$_2$HPO$_4$), or sulfur (K$_2$SO$_4$, FeSO$_4$·7H$_2$O, MgSO$_4$·7H$_2$O, ZnSO$_4$·7H$_2$O, CuSO$_4$·5H$_2$O, NiSO$_4$·7H$_2$O, Ti(SO$_4$), and K$_2$Cr$_2$(SO$_4$)$_4$·24H$_2$O), and then re-incubated under the same incubation conditions as mentioned above. The control experiments were carried out in normal Zarrouk's medium formulation [50]. Samples were collected at each time point 0, 3, 6, 12, 18, 24, 48, 72 and 96 hours after the cells were transferred to different media, and were kept frozen at -80 °C for further analysis. All experiments were repeated in triplicate.

To analyze growth of *A. platensis* C1, the cell concentration was quantified by turbidity based on the optical density at 560 nm (OD$_{560}$) using a Genesys 20 spectrophotometer (Thermo scientific, Waltham, MA) and measured for dry cell weight (DCW). The correlation between OD$_{560}$ and DCW (DCW in g/L = 1.0888xOD$_{560}$) was calculated based on triplicated experiments. Glycogen content was measured using iodine-glycogen assay [53]. Briefly, 100 μL of wet cell was mixed with 50 mg of glass beads and 500 μL of phosphate-buffered saline. The samples were vortexed at maximum speed for 5 min and incubated at 65 °C for 10 min. The samples were then centrifuged at 12,000 rpm for 10 min at 4 °C. Subsequently, 100 μL of supernatants were mixed with 5 μL of iodine solution in Greiner 96-well plate and incubated at 25 °C for 1 min. Absorbance was determined using Microplate reader (Tecan Infinite M200, Mannedorf, Switzerland) at a wavelength of 492 nm. The glycogen concentration of the samples was calculated using the equation obtained from the linear regression of the standard curve.

3. Results and Discussion

3.1. Reconstruction of the Updated Genome-Scale Metabolic Network of A. platensis C1

Genome annotation showed that the rapid annotation of microbial genomes using Subsystems Technology (RAST) annotation servers [24] predicted a total of 2626 (1759 unique genes), whereas the KEGG Annotation Server (KAAS) [26] predicted a total of 1729 genes (1011 unique genes). Using the previously published genome-scale reconstruction for *A. platensis* C1 (*i*AK692) [23] as a reference, the annotation results overlapping among these three sources were analyzed. There were 513 conserved genes while 70 genes were shared between *i*AK692 [23] and KAAS [26]; 45 genes were shared between *i*AK692 [23] and RAST [24]; 275 genes were shared between RAST [24] and KAAS [26] (Figure S1). These large overlaps found between sources indicated good annotation quality. Genes found in all sources or two of the sources were considered to have the high reliability. Regarding to the transport gene annotation, approximately 116 new transport genes in the genome encoding for transporters or transport-related proteins were obtained from KAAS [26] and TCDB [28]. This is because the transportation mechanisms in *i*AK692 [23] were only diffusion reactions. Thus, in this reconstruction, ABC transport reaction and symport or antiport reactions were addressed based on annotation-based inference of transporter function to represent the transport machinery of *A. platensis* C1. Subsequently, the GPR of all candidate genes were then manually curated based upon various information such as physiological evidence in literature, gene expression [14], and biochemical databases. Importantly, the reactions involved in glycogen biosynthesis and degradation pathways were elaborated.

The reactions were balanced for charge and mass to prevent infeasible cycles. However, there were still mass- and charge-unbalanced reactions because either the associated metabolites contained an unspecified metabolite, R groups, or the correct reaction mechanism was unknown. During the curation, a metabolic gap was identified and filled through repeated cell growth simulations using FBA [22] until a positive flux on the biomass reaction was observed. These processes led to the modification of genes, reactions, and metabolites in *i*AK692 [23] and addition of new metabolic genes and their associated reactions in the updated model. For genes, 60 (6%) genes were removed, 268 (28%) genes were added, and 620 (66%) were refined. For reactions, 202 (15%) reactions were removed, 423 (33%) were added, and 673 (52%) were refined. For metabolites, 140 (12%) metabolites were removed, 297 (26%) were added, and 697 (62%) were refined (see Supplementary File S2 for details). Obviously, significant improvements in annotation resulted from not only re-annotation information collected from RAST [24] and KAAS [26] but also a manual effort to assess the reliability of such annotation. This is a crucial prior step to address the careful revision of the first genome-scale network reconstruction and derived constraint-based model of *A. platensis* C1 published in 2012 [23].

3.2. Characteristics of iAK888 and Comparison

*i*AK888 contains 1096 metabolic reactions, 994 metabolites, and 888 genes-representing 15% of total protein coding genes in the genome [11]. Of all reactions, 751 (68.5%) were gene associated enzymatic reactions whereas non-gene associated enzymatic reactions, transport reactions and exchange reactions were 66 (6%), 182 (16.6%) and 97 (8.9%), respectively (Table 2). The reactions revealed 9 major subsystems (Figure 1A) including 56 metabolic pathways, as defined by KEGG [29]. Vitamins and cofactors metabolism and transport represented the largest portions of the network. These likely represented *A. platensis* physiology, an excellent source of vitamins [9] and reflected the fact that approx. 6% of *A. platensis* C1 genome encoded for transporters [11]. The reactions distributed over six cellular compartments including carboxysome, thylakoid lumen, thylakoid membrane, cytoplasmic membrane, cytoplasm and periplasm, with the majority of reactions localized to the cytosol (Figure 1B). This observation agreed well with the known life cycle of *A. platensis* [9]. Figure 1C showed the non-gene associated and gene associated reaction involved in each pathway. Notably, these non-gene associated reactions were required to complete the metabolic network of *A. platensis* C1 and were also observed in other GEMs [54,55]. Besides, the metabolites localized in different compartments of the fully compartmentalized model are considered as distinct metabolites. Therefore, without considering subcellular sites, the model accounted for 796 unique metabolites. Additionally, the quality of model reconstruction was assessed using the confidence scores associated to each reaction (Figure S2). The overall confidence score was 3.38. Almost 89% of the internal reactions (936) have been either very well or well-studied where extensive physiological and sequence evidences are available, while 11% were primarily based on the genome annotation and modeling hypotheses. The new metabolic network reconstruction for *i*AK888 is provided in a spreadsheet format (Supplementary File S3) that includes curation notes and references.

Table 2. Comparison of network model characteristics of *A. platensis* species.

Arthrospira **Species**	*A. platensis* **C1**		*A. platensis* **NIE-39**
Genome Statistics	-		-
Genome size (bp)	6,089,210		6,788,435
Protein coding genes	6108		6630
Gene with enzymes	952		905
Transporter genes	345		NA
Model Name/Characteristics	*i*AK888	*i*AK692	-
Total genes in model	888 (15%)	692 (11%)	620 (9%)
- Metabolic genes	767	692	579
- Transporter genes	121	0	41
Total biochemical reactions	1096	875	746
- Metabolic reactions	817	699	652
- Transport reactions	182	88	60
- Exchange reactions	97	88	34
Metabolites	994	837	673
Compartments	6	2	2
Reference	This study	[23]	[55]

The properties of *i*AK888 were compared with the properties of two published GEM of *A. platensis* NIE-39 [55] and the first GEM of *A. platensis* C1 [23]. The *i*AK888 appeared to contain the largest number of genes, reactions, metabolites, and sub-cellular compartments (Table 2). It was evidently clear that the model reported in this research was the largest in terms of gene coverage. A comparison between *i*AK888 and *i*AK692 [23] specifically showed that *i*AK888 represented an increase in number of genes, reactions, and metabolites over *i*AK692 [23], by 196, 221, and 157, respectively (Figure 2A). Of the increased genes, 75 (38%) genes were the updated metabolic genes whereas 121 (62%) genes were the additional transport genes. The reactions associated to these genes covered a wide array of key metabolic functions mainly relevant to carbohydrate metabolism, lipid metabolism, vitamins and cofactors metabolism, and transports (Figure 2B). Furthermore, this effort was made to reconstruct some of the pathways which were either incomplete or not considered in the previous reconstruction such as tricarboxylic acid cycle (TCA) cycle, photosynthesis and oxidative phosphorylation, carbon concentrating mechanism (CCM), fatty acid biosynthesis, glycogen metabolism, polyhydroxyalkanoates biosynthesis, and hydrogen biosynthesis (Figure 2C). The incomplete TCA cycle in *i*AK692 [23] was improved based on the latest evidence reported in cyanobacteria [31,56]. Subsequently, the description of photosynthesis and oxidative phosphorylation was significantly improved according to a model organism, photosynthetic *Synechocystis* sp. PCC 6803 [35]. These included the photosynthetic linear electron flow (LEF) pathway [57], including photosystem I and II, alternate electron flow (AEF) pathways [58], and photorespiration [59]. In addition, the molecular components involved in CCM of *A. platensis* C1 were annotated and incorporated for a more precise understanding of the primary carbon metabolic route. There are 18 genes/proteins associated to the CCM in *i*AK888. Finally, reactions associated with polyhydroxyalkanoates and hydrogen biosynthesis were formulated to complement the physiological ability of *A. platensis* C1. In summary, *i*AK888 was considered to be the most comprehensive *A. platensis* model to date.

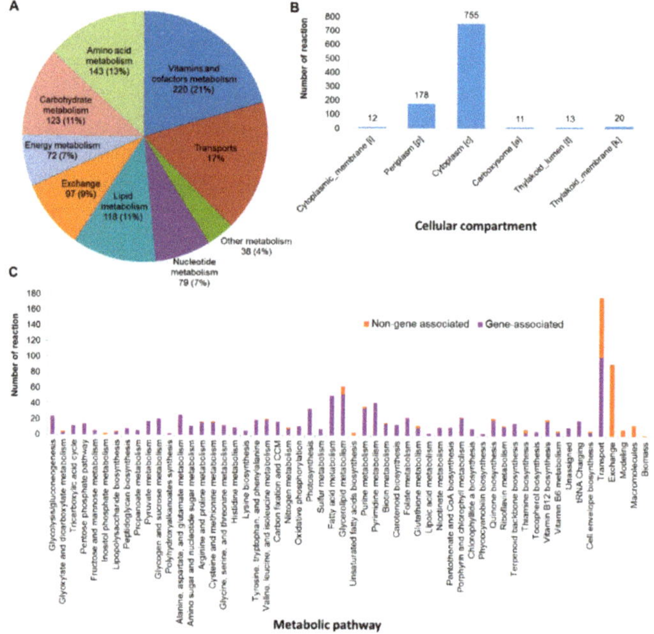

Figure 1. Properties of the updated genome-scale metabolic network of *A. platensis* C1. Distribution of reactions in each metabolism (**A**), Distribution of reactions in each cellular compartment (**B**), Gene associated and non-gene associated reaction in each metabolic pathway (**C**).

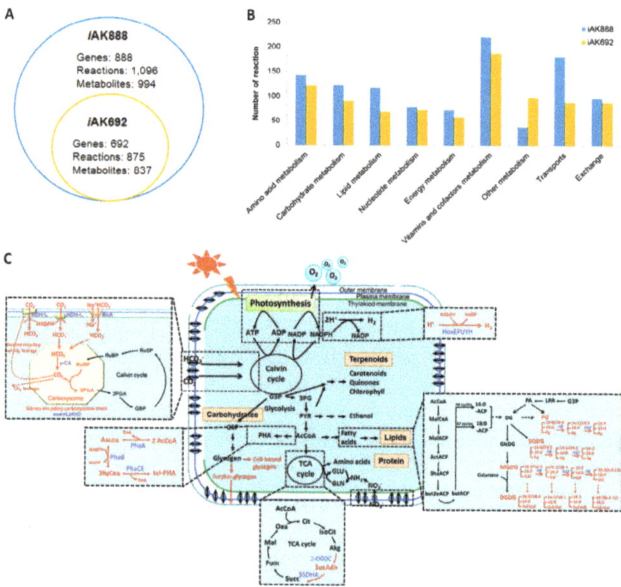

Figure 2. Comparison between *i*AK888 and *i*AK692. Overview features of *i*AK888 compared to *i*AK692 (**A**). Number of reactions in each metabolism of *i*AK888 compared to *i*AK692 (**B**). Schematic representation of *i*AK888 and the example of the filled pathways compared to *i*AK692 (**C**). Zoomed out sections are pathways that were completed in *i*AK888. Red texts and arrows indicate missing pathways in *i*AK692.

3.3. Validation of iAK888

For the growth verification, results showed that the model accurately predicted the growth rates with the error less than 5% in all culture conditions (Table 3). Moreover, the model generated oxygen under autotrophic and mixotrophic growth conditions. The released oxygen content under autotroph was higher than mixotroph. On the other hand, the model consumed oxygen and produced CO_2 under heterotrophic simulations. These results suggested that *iAK888* was able to represent the basic behavior of *A. platensis* C1. The models in SBML format for all four growth conditions are provided in Supplementary File S4–7. Besides, flux distributions of these three growth conditions are presented in Supplementary File S8.

Table 3. Comparison of in silico and experimental growth.

Growth Condition	Constraints of Consumed Metabolites			Maximal Specific Growth Rate (1/h)		% Error
	Photon Flux (µmol photons $m^{-2}\,s^{-1}$)	HCO_3^- Uptake Flux (mmol/g DCW/h)	Glucose Uptake Flux (mmol/g DCW/h)	Experiment	In Silico	
Autotroph	100	0.2	0	0.0255 [23]	0.0252	1.2
Autotroph	200	0.25	0	0.0331 [48]	0.0334	0.9
Heterotroph	0	0	0–0.017	0 [23]	0	0
Mixotroph	100	0.2	0–0.017	0.0262 [23]	0.0260	0.8

Furthermore, an effort was made to investigate the total carbohydrate production flux under NO_3^- depletion. The simulation was performed to imitate the experimental conditions [49] with the following parameters: HCO_3^- uptake rate, 0.2 mmol/gDCW/h; photon uptake rate, 100 µmol photons/m^2/s; no growth and NO_3^- uptake rate. The objective function was set to maximize the total carbohydrate production flux. Result showed that the *in silico* total carbohydrate production flux (0.0868 mmol/gDCW/h) was consistent with the experimental production (0.081 mmol/gDCW/h) [49]. The models in SBML format and flux distributions can be found in Supplementary File S9 and S10, respectively. These results suggested that *iAK888* could predict the growth rates and carbohydrate production of *A. platensis* C1 reasonably well.

3.4. Prediction of Glycogen Overproduction Using iAK888

Herein, the impacts of nutrient starvation on growth and glycogen production were simulated to predict cultivation strategies for enhancing glycogen content in *A. platensis* C1. *iAK888* was simulated by varying the uptake flux of NO_3^-, PO_4^{3-}, and SO_4^{2-} under autotrophic conditions with the following parameters: HCO_3^- uptake flux, 1.6 mmol/gDCW/h; photon uptake flux, 100 µmol photons/m^2/s. In overall, limitation of NO_3^-, PO_4^{3-}, and SO_4^{2-} uptakes and the excess HCO_3^- resulted in the negative effect on growth, while revealed positive effect on glycogen production (Figure 3A–C). Obviously, the specific growth rate rapidly decreased when limited levels of NO_3^-, PO_4^{3-}, and SO_4^{2-} were introduced. Apparently, *iAK888* expressed ability to rapidly accumulate glycogen when the uptake flux of each nutrient was lower than the optimal uptake flux for biomass production. On the other hand, when the uptake flux of each nutrient was higher than the optimal uptake flux for biomass production, the glycogen production flux exhibited no flux values.

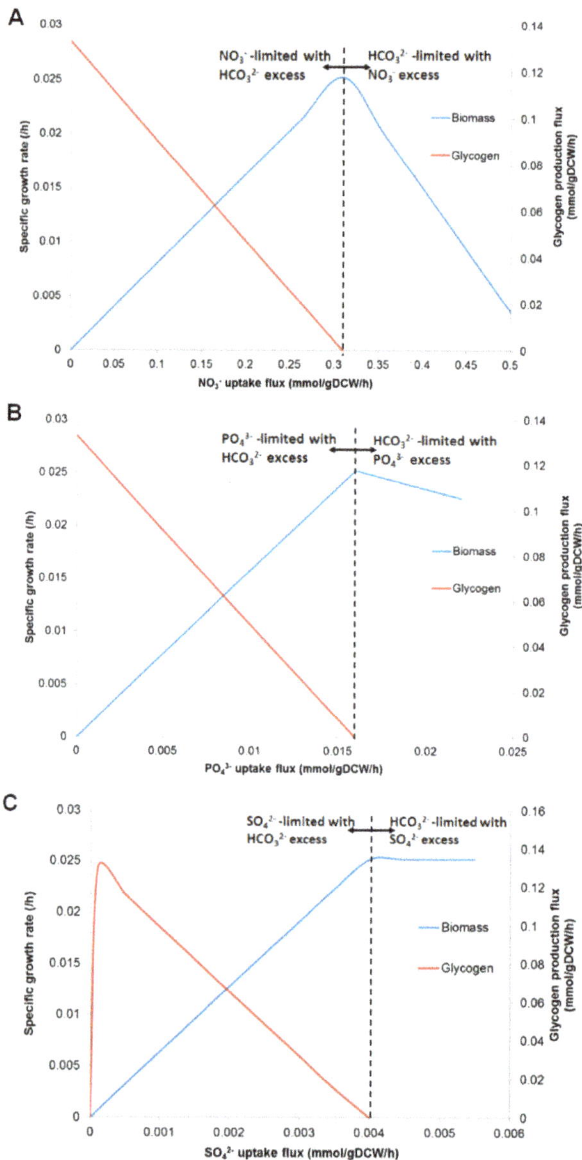

Figure 3. Prediction of nutrient effect on growth and glycogen production. The biomass and glycogen production as the functions of uptake fluxes of NO_3^- (**A**), PO_4^{3-} (**B**), and SO_4^{2-} (**C**). Dashed line indicates the optimal uptake rate for growth.

The internal metabolic fluxes throughout the central metabolism were determined using geometric FBA [52]. HCO_3^- uptake rate was constrained to 1.6 mmol/gDCW/h whereas, NO_3^-, PO_4^{3-}, and SO_4^{2-} were constrained to half of the optimum uptake rate for growth. Flux map (Figure 4) showed that all reactions in the central metabolism were activated to provide maximum biomass. The excess carbon was secreted as glycogen under NO_3^-, PO_4^{3-}, and SO_4^{2-}-insufficient growth. Fluxes comparison under different perturbations provided information on how central metabolic reactions respond to the altered growth condition, although each perturbation resulted in the similar overall phenotype. They exhibited different flux patterns. All fluxes in the central metabolisms were higher under NO_3^--limited growth than the normal growth considered as NO_3^--sufficient condition. All fluxes in the TCA cycle and some reactions of the glycolysis and non-oxidative pentose phosphate pathway were lower under PO_4^{3-}, and SO_4^{2-}-limited growth than the normal growth. A consequence of increased fluxes and metabolites in the central carbon pathway under NO_3^- starvation led to the enhanced synthesis of glycogen in *i*AK888. This showed some agreements with recent metabolomic observations during the glycogen production phase in *A. platensis* NIE-39 cultured under nitrate-free (Society of Toxicology) SOT medium [60]. The time-course analysis of the primary metabolites content revealed a transient increase of glucose-1-phosphate, glucose-6-phosphate, fructose-6-phosphate, 2-ketoglutarate, succinate, and malate. Besides, simulation of each nutrient limitation with excess carbon showed secretion of pyruvate, acetate, and lactate (see Supplementary File S11–13). In agreement with the experimental results, it was found that *A. platensis* NIES-39 cultured under nitrate-limited condition produced pyruvate, acetate, and lactate in the culture medium [55]. Regarding to PO_4^{3-}, and SO_4^{2-} starvation, the simulation results showed that the changes in flux patterns of both nutrients were very similar. Obviously, limitation of these nutrients greatly affected the fluxes through phosphoglycerate mutase, enolase, pyruvate kinase, and malate dehydrogenase reactions in comparison to the normal growth and NO_3^--limited conditions. These much lower fluxes occurred at the branch points where the precursor metabolites were drained for the synthesis of amino acid. However, future inclusion of such experimental data could be expected to verify the prediction of flux distribution under each nutrient limitation.

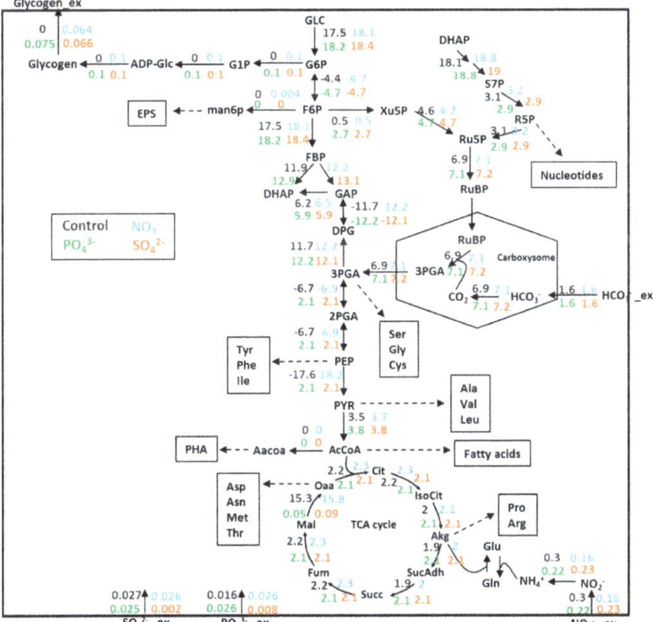

Figure 4. Illustrative fluxes in central metabolic predicted by geometric FBA for each growth condition: Control (black), NO_3^--limited (NO_3^-, blue), PO_4^{3-}-limited (PO_4^{3-}, green) and SO_4^{2-}-limited (SO_4^{2-}, orange). The diagram included key reactions including the glycolysis, the carbon fixation through CO_2-concentrating mechanism, the nonoxidative pentose phosphate pathway, the TCA cycle, and biosynthetic pathways of glycogen. Single and double-headed arrows indicate reactions assumed to be irreversible and reversible, respectively. Numbers labeled at the corresponding arrows show flux values in mmol/gDCW/h.

3.5. Experimental Validation of iAK888 Prediction for Glycogen Overproduction

Since *i*AK888 suggested that NO_3^-, PO_4^{3-}, and SO_4^{2-} starvation enhance glycogen content in *A. platensis* C1, -growth rate and glycogen production under NO_3^-, PO_4^{3-}, and SO_4^{2-} depletion was experimentally determined. *A. platensis* C1 was cultured with standard Zarrouk's medium [50] and cells were transferred to nitrogen, phosphorus, or sulfur depleted Zarrouk's medium [50]. Changes in cell growth and glycogen content of *A. platensis* C1 were investigated for 96 h. Regarding to growth in term of dry weight (Figure 5A) and growth rate (Figure 6A), the results showed no significant differences between control and experiments in the first 18 h. However, further incubation for 96 h, the dry weight and the growth rate under NO_3^- and SO_4^{2-} starvation decreased significantly ($p < 0.05$) whereas the dry weight under PO_4^{3-} starvation gradually increased, even though the growth rate was lower than that of the control. An increase in growth observed during the first 18 h might be due to the results from availability of endogenous nitrogen, phosphorus, and sulfur. In terms of glycogen content (Figure 5B) and glycogen production flux (Figure 6B), cells subjected to PO_4^{3-} and SO_4^{2-} depletion showed a significant increase in glycogen content and production flux ($p < 0.01$) in the first 18 h. In contrast, the significant increase in glycogen content and production flux were observed in the NO_3^--free medium after 24 h ($p < 0.01$).

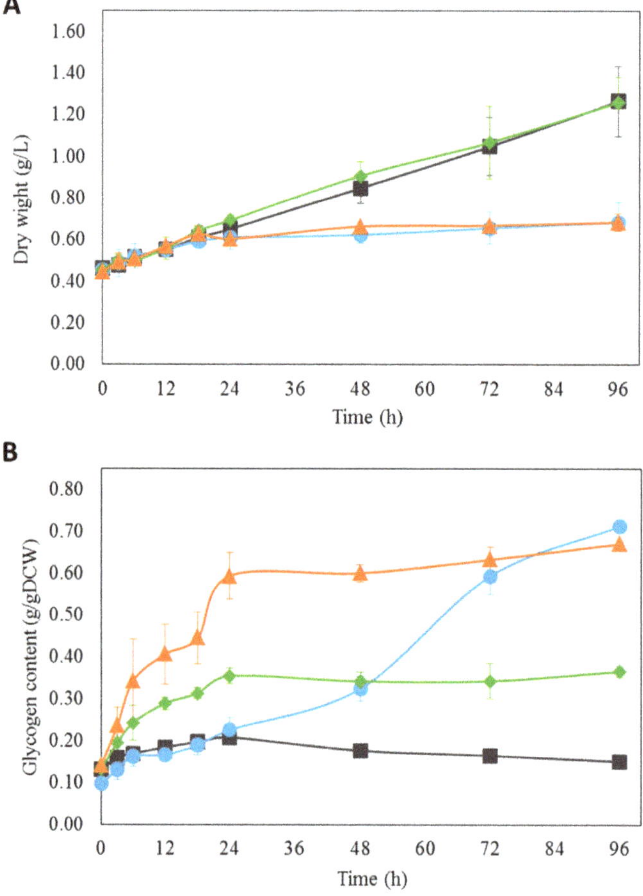

Figure 5. Growth (**A**) and glycogen content (**B**) of *A. platensis* C1 under nutrient starvation at 0–96 h. Culture conditions of Control (■), NO_3^- (●), PO_4^{3-} (◆) and SO_4^{2-} (▲).

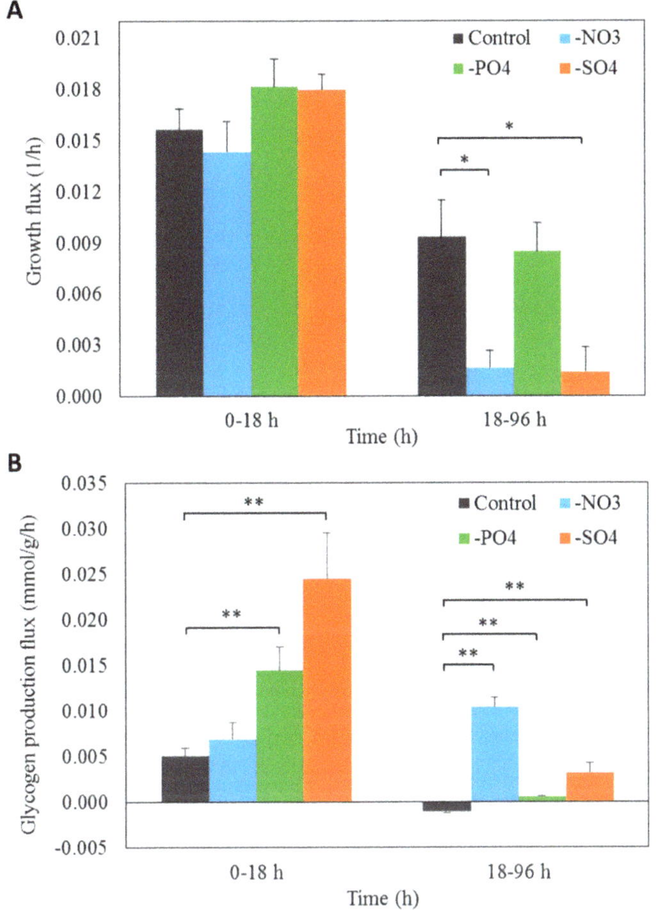

Figure 6. The specific growth rate growth (**A**) and the glycogen production flux (**B**) under the control, NO_3^--depleted, PO_4^{3-}-depleted and SO_4^{2-}-depleted conditions. * and ** represent significant difference at the p-value < 0.05 and the p-value < 0.01, respectively.

Compared to the results predicted by FBA simulations, glycogen flux was considered to be consistent with those obtained by experiments, suggesting that NO_3^-, PO_4^{3-}, and SO_4^{2-} depletion induced an increase of glycogen production in *A. platensis* C1. Notably, only the growth rate under PO_4^{3-} starvation between the experimental and *i*AK888-simulated fluxes had no significant difference. It should be noted that despite FBA [22] being widely used approach for simulating a large-scale cellular metabolism under a certain condition, this method could only be used to predict the steady state snapshot of flux distribution. Thus, using the FBA approach [22] was unable to directly analyze the transient states of cell metabolism such as a concentration of metabolite and dynamic change in the flux with time. These reasons might cause the discrepancy between the FBA simulation and the experiment.

4. Conclusions

In this study, an improved genome-scale model for *A. platensis* C1, *i*AK888, was performed. The *i*AK888 model displayed a highly detailed reconstruction capturing the fundamental knowledge and the significantly biotechnological capabilities as compared to the previous model, *i*AK692.

The *i*AK888 model was demonstrated to be a suitable model for prediction of growth and carbohydrate production flux of *A. platensis* C1 under various conditions. Moreover, it was demonstrated that the *i*AK888 model could suggest rational cultivation strategies for overproduction of glycogen in *A. platensis* C1. In the future, this new model shall be a versatile platform for further studies towards glycogen-enriched *A. platensis* C1 as an alternative feedstock for bioethanol production.

Supplementary Materials: The following are available online at http://www.mdpi.com/2218-1989/8/4/84/s1. Figure S1: The Venn diagrams illustrate the overlapped genes among RAST [24], KAAS [26], and *i*AK692 [23]. Figure S2: Heat map of the confidence score of each pathway in *i*AK888. The columns represent the level of confidence score (ranging from 1 to 4) and the rows represent the metabolic pathway. The various colors correspond to the percentage of pathway reactions that have the corresponding confidence score (orange = 100%, blue = 0%). Based on the confidence score system [37], 4 represents physiological and genetic evidence, 3 represents direct and indirect evidence for gene function, 2 represents indirect evidence from physiological data or only sequence-based evidence, and 1 represents modeling evidence. Supplementary File S1: Biomass composition of *A. platensis* C1. Supplementary File S2: Genome annotation. Supplementary File S3: Detailed GPR information of model *i*AK888. Supplementary File S4: SBML of autotrophic growth photon uptake 100 μmol photons/m^2/s. Supplementary File S5: SBML of autotrophic growth photon uptake 200 μmol photons/m^2/s. Supplementary File S6: SBML of heterotrophic growth. Supplementary File S7: SBML of mixotrophic growth. Supplementary File S8: Flux distribution of autotrophic, heterotrophic, and mixotrophic growth. Supplementary File S9: SBML of carbohydrate production under NO_3^- depletion. Supplementary File S10: Flux distribution of effect NO_3^- depletion on carbohydrate production. Supplementary File S11: Flux distribution of effect NO_3^- depletion on growth and glycogen. Supplementary File S12: Flux distribution of effect PO_4^{3-} depletion on growth and glycogen. Supplementary File S13: Flux distribution of effect SO_4^{2-} depletion on growth and glycogen.

Author Contributions: Reconstructed the genome-scale metabolic model, A.K.; performed the analyses, A.K.; wrote the manuscript, A.K.; conducted experiments for model validation, S.D. and K.C.; conceived and designed the overall study, P.P. and A.M.; revised the manuscript, P.P. and A.M.; approved the final manuscript, A.K., S.D., K.C., S.C., P.P. and A.M.

Funding: This work was supported by grants, P-11-01089, from the National Center for Genetic Engineering and Biotechnology (BIOTEC), NSTDA, Thailand and the research subsidy fund of fiscal year 2017 from King Mongkut's University of Technology Thonburi. Moreover, the authors acknowledge the financial support provided by King Mongkut's University of Technology Thonburi through the KMUTT 55th Anniversary Commemorative Fund.

Acknowledgments: The authors would like to specially thank the Algal Biotechnology Laboratory and the Biomedical Engineering Laboratory at KMUTT for allowing us to use their experimental facilities.

Conflicts of Interest: The authors declare no conflicts of interest.

References

1. Markou, G.; Angelidaki, I.; Georgakakis, D. Microalgal carbohydrates: An overview of the factors influencing carbohydrates production, and of main bioconversion technologies for production of biofuels. *Appl. Microbiol. Biotechnol.* **2012**, *96*, 631–645. [CrossRef] [PubMed]

2. Ho, S.-H.; Huang, S.-W.; Chen, C.-Y.; Hasunuma, T.; Chang, J.-S. Bioethanol production using carbohydrate-rich microalgae biomass as feedstock. *Bioresour. Technol.* **2013**, *135*, 191–198. [CrossRef] [PubMed]

3. Aikawa, S.; Joseph, A.; Yamad, R.; Izumi, Y.; Yamagishi, T.; Matsuda, F.; Kawai, H.; Chang, J.-S.; Hasunuma, T.; Kondo, A. Direct conversion of *Spirulina* to ethanol without pretreatment or enzymatic hydrolysis processes. *Energy Environ. Sci.* **2013**, *6*, 1844–1849. [CrossRef]

4. Möllers, K.B.; Cannella, D.; Jørgensen, H.; Frigaard, N.-U. Cyanobacterial biomass as carbohydrate and nutrient feedstock for bioethanol production by yeast fermentation. *Biotechnol. Biofuels* **2014**, *7*, 64. [CrossRef] [PubMed]

5. Gao, Z.X.; Zhao, H.; Li, Z.M.; Tan, X.M.; Lu, X. F Photosynthetic production of ethanol from carbon dioxide in genetically engineered cyanobacteria. *Energy Environ. Sci.* **2012**, *5*, 9857–9865. [CrossRef]

6. Wijffels, R.H.; Kruse, O.; Hellingwerf, K.J. Potential of industrial biotechnology with cyanobacteria and eukaryotic microalgae. *Curr. Opin. Biotechnol.* **2013**, *24*, 405–413. [CrossRef] [PubMed]

7. Taton, A.; Unglaub, F.; Wright, N.E.; Zeng, W.Y.; Paz-Yepes, J.; Brahamsha, B.; Palenik, B.; Peterson, T.C.; Haerizadeh, F.; Golden, S.S.; et al. Broad-host-range vector system for synthetic biology and biotechnology in cyanobacteria. *Nucleic Acids Res.* **2014**, *42*, e136. [CrossRef] [PubMed]

8. Aikawa, S.; Izumi, Y.; Matsuda, F.; Hasunuma, T.; Chang, J.S.; Kondo, A. Synergistic Enhancement of Glycogen Production in *Arthrospira platensis* by Optimization of Light Intensity and Nitrate Supply. *Bioresour. Technol.* **2012**, *108*, 211–215. [CrossRef] [PubMed]

9. Vonshak, A. *Spirulina platensis* (*Arthrospira*): Physiology, Cell Biology and Biotechnology. *J. Appl. Psychol.* **1997**, *9*, 295–296.

10. Markou, G.; Chatzipavlidis, I.; Georgakakis, D. Carbohydrates production and bio-flocculation characteristics in cultures of Arthrospira Spirulina platensis: Improvements through phosphorus limitation process. *BioEnergy Res.* **2012**, *5*, 915–925. [CrossRef]

11. Cheevadhanarak, S.; Paithoonrangsarid, K.; Prommeenate, P.; Kaewngam, W.; Musigkain, A.; Tragoonrung, S.; Tabata, S.; Kaneko, T.; Chaijaruwanich, J.; Sangsrakru, D.; et al. Draft Genome Sequence of *Arthrospira platensis* C1 (PCC 9438). *Stand. Genom. Sci.* **2012**, *6*, 43–53. [CrossRef] [PubMed]

12. Jeamton, W.; Dulsawat, S.; Tanticharoen, M.; Vonshak, A.; Cheevadhanarak, S. Overcoming Intrinsic Restriction Enzyme Barriers Enhances Transformation Efficiency in *Arthrospira platensis* C1. *Plant Cell Physiol.* **2017**, *58*, 822–830. [CrossRef] [PubMed]

13. Jeamton, W.; Mungpakdee, S.; Sirijuntarut, M.; Prommeenate, P.; Cheevadhanarak, S.; Tanticharoen, M.; Apiradee Hongsthong, A. A combined stress response analysis of *Spirulina platensis* in terms of global differentially expressed proteins, and mRNA levels and stability of fatty acid biosynthesis genes. *FEMS Microbiol. Lett.* **2008**, *281*, 121–131. [CrossRef] [PubMed]

14. Panyakampol, J.; Cheevadhanarak, S.; Sutheeworapong, S.; Chaijaruwanich, J.; Senachak, J.; Siangdung, W.; Jeamton, W.; Tanticharoen, M.; Paithoonrangsarid, K. Physiological and transcriptional responses to high temperature in *Arthrospira* (*Spirulina*) *platensis* C1. *Plant Cell Physiol.* **2014**, *56*, 481–496. [CrossRef] [PubMed]

15. Hongsthong, A.; Sirijuntarut, M.; Prommeenate, P.; Lertladaluck, K.; Porkaew, K.; Cheevadhanarak, S.; Tanticharoen, M. Proteome analysis at the subcellular level of the cyanobacterium *Spirulina platensis* in response to low-temperature stress conditions. *FEMS Microbiol. Lett.* **2008**, *288*, 92–101. [CrossRef] [PubMed]

16. Hongsthong, A.; Sirijuntarut, M.; Yutthanasirikul, R.; Senachak, J.; Kurdrid, P.; Cheevadhanarak, S.; Tanticharoen, M. Subcellular proteomic characterization of the high-temperature stress response of the cyanobacterium *Spirulina platensis*. *Proteome Sci.* **2009**, *7*, 33. [CrossRef] [PubMed]

17. Kurdrid, P.; Senachak, J.; Sirijuntarut, M.; Yutthanasirikul, R.; Phuengcharoen, P.; Jeamton, W.; Roytrakul, S.; Cheevadhanarak, S.; Hongsthong, A. Comparative analysis of the Spirulina platensis subcellular proteome in response to low- and high-temperature stresses: Uncovering cross-talk of signaling components. *Proteome Sci.* **2011**, *9*, 39. [CrossRef] [PubMed]

18. Senachak, J.; Cheevadhanarak, S.; Hongsthong, A. SpirPro: A *Spirulina* proteome database and web-based tools for the analysis of protein-protein interactions at the metabolic level in *Spirulina* (*Arthrospira*) *platensis* C1. *BMC Bioinform.* **2015**, *16*, 233. [CrossRef] [PubMed]

19. Likic, V.A.; McConville, M.J.; Lithgow, T.; Bacic, A. Systems Biology: The Next Frontier for Bioinformatics. *Adv. Bioinform.* **2010**, *2010*, 1–10. [CrossRef] [PubMed]

20. Monk, J.; Nogales, J.; Palsson, B.O. Optimizing genome-scale network reconstructions. *Nat. Biotechnol.* **2014**, *32*, 447–452. [CrossRef] [PubMed]

21. Kim, W.J.; Kim, H.U.; Lee, S.Y. Current State and Applications of Microbial Genome-scale Metabolic Models. *Curr. Opin. Syst. Biol.* **2017**, *2*, 10–18. [CrossRef]

22. Jeffrey, D.O.; Thiele, I.; Palsson, B.Ø. What Is Flux Balance Analysis? *Nat. Biotechnol.* **2010**, *28*, 245–248.

23. Klanchui, A.; Khannapho, C.; Phodee, A.; Cheevadhanarak, S.; Meechai, A. *i*AK692: A genome-scale metabolic model of *Spirulina platensis* C1. *BMC Syst. Boil.* **2012**, *6*, 71. [CrossRef] [PubMed]

24. Aziz, R.K.; Bartels, D.; Best, A.A.; DeJongh, M.; Disz, T.; Edwards, R.A.; Formsma, K.; Gerdes, S.; Glass, E.M.; Kubal, M.; et al. The RAST Server: Rapid annotations using subsystems technology. *BMC Genom.* **2008**, *9*, 75. [CrossRef] [PubMed]

25. Overbeek, R.; Olson, R.; Pusch, G.D.; Olsen, G.J.; Davis, J.J.; Disz, T.; Edwards, R.A.; Gerdes, S.; Parrello, B.; Shukla, M.; et al. The SEED and the Rapid Annotation of microbial genomes using Subsystems Technology (RAST). *Nucleic Acids Res.* **2014**, *42*, D206–D214. [CrossRef] [PubMed]

26. Moriya, Y.; Itoh, M.; Okuda, S.; Yoshizawa, A.C.; Kanehisa, M. KAAS: An Automatic Genome Annotation and Pathway Reconstruction Server. *Nucleic Acids Res.* **2007**, *35*, W182–W185. [CrossRef] [PubMed]

27. McGinnis, S.; Madden, T.L. Blast: At the Core of a Powerful and Diverse Set of Sequence Analysis Tools. *Nucleic Acids Res.* **2004**, *32*, W20–W25. [CrossRef] [PubMed]

28. Saier, M.H., Jr.; Reddy, V.S.; Tsu, B.V.; Ahmed, M.S.; Li, C.; Moreno-Hagelsieb, G. The Transporter Classification Database (TCDB): Recent Advances. *Nucleic Acids Res.* **2016**, *44*, D372–D379. [CrossRef] [PubMed]

29. Kanehisa, M.; Sato, Y.; Kawashima, M.; Furumichi, M.; Tanabe, M. KEGG as a reference resource for gene and protein annotation. *Nucleic Acids Res.* **2016**, *44*, D457–D462. [CrossRef] [PubMed]

30. Caspi, R.; Billington, R.; Ferrer, L.; Foerster, H.; Fulcher, C.A.; Keseler, I.M.; Kothari, A.; Krummenacker, M.; Latendresse, M.; Mueller, L.A.; et al. The MetaCyc database of metabolic pathways and enzymes and the BioCyc collection of pathway/genome databases. *Nucleic Acids Res.* **2016**, *44*, D471–D480. [CrossRef] [PubMed]

31. Steinhauser, D.; Fernie, A.R.; Araujo, W.L. Unusual cyanobacterial TCA cycles: Not broken just different. *Trends Plant Sci.* **2012**, *17*, 503–509. [CrossRef] [PubMed]

32. Klanchui, A.; Cheevadhanarak, S.; Prommeenate, P.; Meechai, A. Exploring Components of the CO_2-Concentrating Mechanism in Alkaliphilic Cyanobacteria through Genome-Based Analysis. *Comput. Struct. Biotechnol.* **2017**, *15*, 340–350. [CrossRef] [PubMed]

33. Satora, P.; Barwińska-Sendra, A.; Duda-Chodak, A.; Wajda, Ł. Strain-dependent production of selected bioactive compounds by Cyanobacteria belonging to the *Arthrospira* genus. *J. Appl. Microbiol.* **2015**, *119*, 736–743. [CrossRef] [PubMed]

34. Badri, H.; Monsieurs, P.; Coninx, I.; Nauts, R.; Wattiez, R.; Leys, N. Temporal Gene Expression of the Cyanobacterium *Arthrospira* in Response to Gamma Rays. *PLoS ONE* **2015**, *10*, e0135565. [CrossRef] [PubMed]

35. Knoop, H.; Grundel, M.; Zilliges, Y.; Lehmann, R.; Hoffmann, S.; Lockau, W.; Steuer, R. Flux balance analysis of cyanobacterial metabolism: The metabolic network of *Synechocystis* sp. PCC 6803. *PLoS Comput. Biol.* **2013**, *9*, e1003081. [CrossRef] [PubMed]

36. Orth, J.D.; Conrad, T.M.; Na, J.; Lerman, J.A.; Nam, H.; Feist, A.M.; Palsson, B.O. A comprehensive genome-scale reconstruction of *Escherichia coli* metabolism—2011. *Mol. Syst Biol.* **2011**, *7*, 535. [CrossRef] [PubMed]

37. Thiele, I.; Palsson, B.O. A protocol for generating a high-quality genome-scale metabolic reconstruction. *Nat. Protoc.* **2010**, *5*, 93–121. [CrossRef] [PubMed]

38. Kim, S.; Thiessen, P.A.; Bolton, E.E.; Chen, J.; Fu, G.; Gindulyte, A.; Han, L.; He, J.; He, S.; Shoemaker, B.A.; et al. PubChem Substance and Compound databases. *Nucleic Acids Res.* **2016**, *44*, D1202–D1213. [CrossRef] [PubMed]

39. Finn, R.D.; Bateman, A.; Clements, J.; Coggill, P.; Eberhardt, R.Y.; Eddy, S.R.; Heger, A.; Hetherington, K.; Holm, L.; Mistry, J.; et al. The Pfam protein families database. *Nucleic Acids Res.* **2014**, *42*, D222–D230. [CrossRef] [PubMed]

40. Panyakampol, J. (King Mongkut's University of Technology Thonburi, Bangkok, Thailand). Unpublished work. 2015.

41. Cogne, G.; Gros, J.B.; Dussap, C.G. Identification of a metabolic network structure representative of *Arthrospira (spirulina) platensis* metabolism. *Biotechnol. Bioeng.* **2003**, *84*, 667–676. [CrossRef] [PubMed]

42. Paithoonrangsarid, K. (King Mongkut's University of Technology Thonburi, Bangkok, Thailand). Unpublished work, 1997.

43. Abd El-Baky, H.H.; El Baz, F.K.; El-Baroty, G.S. *Spirulina* Species as a Source of Carotenoids and α-tocopherol and its Anticarcinoma Factors. *Biotechnology* **2003**, *2*, 222–240.

44. Shimamatsu, H. Mass production of *Spirulina*, an edible microalga. *Hydrobiologia* **2004**, *512*, 39–44. [CrossRef]

45. Ali, S.K.; Saleh, A.M. *Spirulina*—An Overview. *Int. J. Pharm. Pharm. Sci.* **2012**, *4*, 9–15.

46. Ataman, M.; Hatzimanikatis, V. Heading in the right direction: Thermodynamics-based network analysis and pathway engineering. *Curr. Opin. Biotechnol.* **2015**, *36*, 176–182. [CrossRef] [PubMed]

47. Schellenberger, J.; Que, R.; Fleming, R.M.; Thiele, I.; Orth, J.D.; Feist, A.M.; Zielinski, D.C.; Bordbar, A.; Lewis, N.E.; Rahmanian, S.; et al. Quantitative Prediction of Cellular Metabolism with Constraint-based Models: The COBRA Toolbox V2.0. *Nat. Protoc.* **2011**, *6*, 1290–1307. [CrossRef] [PubMed]

48. Phodee, A. (King Mongkut's University of Technology Thonburi, Bangkok, Thailand). Unpublished work. 2009.

49. Panyakampol, J.; Cheevadhanarak, S.; Senachak, J.; Dulsawat, S.; Siangdung, W.; Tanticharoen, M.; Paithoonrangsarid, K. Different Effects of the Combined Stress of Nitrogen Depletion and High Temperature Than an Individual Stress on the Synthesis of Biochemical Compounds in *Arthrospira platensis* C1 (PCC 9438). *J. Appl. Phycol.* **2016**, *28*, 2177–2186. [CrossRef]

50. Zarrouk, C. Contribution à l'étude d'une cyanophycéeInfluence de divers' facteurs physiques et chimiques sur la croissance et la photosynthèse de *Spirulina maxima*. PhD Thesis, Université de Paris, Paris, France, 1996. (In French)

51. Mahadevan, R.; Schilling, C.H. The effects of alternate optimal solutions in constraint-based genome-scale metabolic models. *Metab. Eng.* **2003**, *5*, 264–276. [CrossRef] [PubMed]

52. Smallbone, K.; Simeonidis, E. Flux balance analysis: A geometric perspective. *J. Theor. Biol.* **2009**, *258*, 311–315. [CrossRef] [PubMed]

53. Iwase, T.; Okai, C.; Kamata, Y.; Tajima, A.; Mizunoe, Y. A straightforward assay for measuring glycogen levels and RpoS. *J. Microbiol. Meth.* **2018**, *145*, 93–97. [CrossRef] [PubMed]

54. Karp, P.D. Call for an enzyme genomics initiative. *Genome Biol.* **2004**, *5*, 401. [CrossRef] [PubMed]

55. Yoshikawa, K.; Aikawa, S.; Kojima, Y.; Toya, Y.; Furusawa, C.; Kondo, A.; Shimizu, H. Construction of a Genome-Scale Metabolic Model of *Arthrospira platensis* NIES-39 and Metabolic Design for *Cyanobacterial* Bioproduction. *PLoS ONE* **2015**, *10*, e0144430. [CrossRef] [PubMed]

56. Zhang, S.; Bryant, D.A. The tricarboxylic acid cycle in cyanobacteria. *Science* **2011**, *334*, 1551–1553. [CrossRef] [PubMed]

57. Kramer, D.M.; Evans, J.R. The Importance of Energy Balance in Improving Photosynthetic Productivity. *Plant Physiol.* **2011**, *155*, 70–78. [CrossRef] [PubMed]

58. Nogales, J.; Gudmundsson, S.; Knight, E.M.; Palsson, B.O.; Thiele, I. Detailing the Optimality of Photosynthesis in *Cyanobacteria* through Systems Biology Analysis. *Proc. Natl. Acad. Sci. USA* **2012**, *109*, 2678–2683. [CrossRef] [PubMed]

59. Cooley, J.W.; Vermaas, W.F. Succinate Dehydrogenase and Other Respiratory Pathways in Thylakoid Membranes of *Synechocystis* sp. strain PCC 6803: Capacity Comparisons and Physiological Function. *J. Bacteriol.* **2001**, *183*, 4251–4258. [CrossRef] [PubMed]

60. Hasunuma, T.; Kikuyama, F.; Matsuda, M.; Aikawa, S.; Izumi, Y.; Kondo, A. Dynamic metabolic profiling of cyanobacterial glycogen biosynthesis under conditions of nitrate depletion. *J. Exp. Bot.* **2013**, *64*, 2943–2954. [CrossRef] [PubMed]

 *metabolites*

Article

Modulation of Polar Lipid Profiles in *Chlorella* sp. in Response to Nutrient Limitation

Daniel A. White [1],*,†, Paul A. Rooks [1], Susan Kimmance [1], Karen Tait [1], Mark Jones [1], Glen A. Tarran [1], Charlotte Cook [1] and Carole A. Llewellyn [2],*

[1] Plymouth Marine Laboratory, Prospect Place, The Hoe, Plymouth, Devon PL1 3DH, UK; paok@pml.ac.uk (P.A.R.); SUKIM@pml.ac.uk (S.K.); KTAIT@pml.ac.uk (K.T.); mark2@pml.ac.uk (M.J.); gat@pml.ac.uk (G.A.T.); chc@pml.ac.uk (C.C.)

[2] Department of Biosciences, Singleton Park, Swansea University, Swansea, Wales SA2 8PP, UK

* Correspondence: whited13@cardiff.ac.uk (D.A.W.); c.a.llewellyn@swansea.ac.uk (C.A.L.); Tel.: +44-2920-688921 (D.A.W.); +44-1792-606168 (C.A.L.)

† Current address: Division of Infection and Immunology, School of Medicine, Wales Heart Institute, Heath Park, Cardiff University, Cardiff, Wales CF14 4XN, UK.

Received: 16 January 2019; Accepted: 22 February 2019; Published: 28 February 2019

Abstract: We evaluate the effects of nutrient limitation on cellular composition of polar lipid classes/species in *Chlorella* sp. using modern polar lipidomic profiling methods (liquid chromatography–tandem mass spectrometry; LC-MS/MS). Total polar lipid concentration was highest in nutrient-replete (HN) cultures with a significant reduction in monogalactosyldiacylglycerol (MGDG), phosphatidylglycerol (PG), phosphatidylcholine (PC), and phosphatidylethanolamine (PE) class concentrations for nutrient-deplete (LN) cultures. Moreover, reductions in the abundance of MGDG relative to total polar lipids versus an increase in the relative abundance of digalactosyldiacylglycerol (DGDG) were recorded in LN cultures. In HN cultures, polar lipid species composition remained relatively constant throughout culture with high degrees of unsaturation associated with acyl moieties. Conversely, in LN cultures lipid species composition shifted towards greater saturation of acyl moieties. Multivariate analyses revealed that changes in the abundance of a number of species contributed to the dissimilarity between LN and HN cultures but with dominant effects from certain species, e.g., reduction in MGDG 34:7 (18:3/16:4). Results demonstrate that *Chlorella* sp. significantly alters its polar lipidome in response to nutrient limitation, and this is discussed in terms of physiological significance and polar lipids production for applied microalgal production systems.

Keywords: polar lipids; *Chlorella* sp.; LC-MS; nutrient limitation

1. Introduction

Lipids (e.g., triacylglycerols, wax esters, sterol esters, and polar diacylglycerols) are important constituents of microalgal cells with a broad range of cellular functions including energy storage, membrane structure and integrity, photosynthesis, metabolism, and cell–cell signaling [1–3].

In actively growing microalgae, the majority of the fatty acids produced are constituents of polar diacylglycerols (polar lipids), including different classes of glycolipids (monogalactosyldiacylglycerol (MGDG); digalactosyldiacylglycerol (DGDG); sulfoquinovosyldiacylglycerol (SQDG)), phospholipids (phosphatidylcholine (PC), phosphatidylglycerol (PG), phosphatidylethanolamine (PE)), and betaine lipids (e.g., diacylglyceryl trimethyl homoserine (DGTS); see Figure S1 for different polar lipid structures). In green algae, the glycolipids are predominantly composed of 16- and 18-carbon fatty acyl groups which are often polyunsaturated (PUFA), e.g., the omega-3 fatty acid α-linolenic acid (18:3), and are located in thylakoid and chloroplast membranes where they have pivotal roles in photosynthesis, signaling, and regulation [4]. MGDG is usually the most abundant glycolipid (40–55% of plant lipids) compared to DGDG (15–35%) and SQDG (2–40%). Unlike MGDG and DGDG, evidence suggests that SQDG has no specific role in photosynthesis but can act as a substituent for phospholipid under conditions of phosphate limitation [5]. Of the phospholipids, PG is essential for growth and for the photosynthetic transport of electrons and the development of chloroplasts [6]. PE is a nitrogen-containing non-bilayer forming lipid, constituting up to 50% of mitochondrial membrane lipids [7], and has been shown to have an activate role in the xanthophyll cycle [8]. Other than PG, the other lipid classes (PC, PE, betaine lipids) are mainly found in extra chloroplastid membranes within cells. The role of betaine lipids is largely unknown although evidence suggests they can act as a substituent for PC (zwitterionic membrane lipid) in non-PC-producing algae [9]. The composition of fatty acids (different chain lengths and degrees of unsaturation) associated with microalgal lipids varies with growth conditions, including nutrient limitation [10–16], giving rise to the potential for variation in the number and combination of molecular species contained within each polar lipid class. Therefore, assessment of polar lipid species regulation is essential in elucidating the fundamental lipidomic response and subsequent metabolic regulation within microalgae exposed to different conditions.

In recent years, there has been a growing interest in the applications and bioactivity of polar lipids (especially glycolipids) from algae for use in a range of biotechnological industries [17]. For example, demonstrated glycolipid (MGDG, DGDG, and SQDG) activities include induced apoptosis of human colon carcinoma Caco-2 cells [18], growth inhibition of human melanoma cells and human hepatocellular carcinoma cell lines [19,20], anti-inflammatory activity [21], cholinesterase inhibitory activity [22], activity against protozoans [23], and antibacterial, antiviral, and fungal growth inhibition [24–27]. The bioactivity of glycolipids has been demonstrated to be related to chemical structure where the sugar molecule and its anomeric configuration, and the carbon length of the acyl chains, are important [28]. The application of phospholipids (irrespective of source) is well recognised, especially in nutraceutical industries, which often take advantage of the amphiphilic nature of these chemicals (see [29]). Since a wealth of information suggests significant biotechnological potential of polar lipids extracted from algae including microalgae, and that the activity of these compounds is influenced by their chemical structure, research is also required to understand the effect of culture conditions on the yield and chemical nature of these compound classes and the species they contain with a view to commercial realisation of microalgal-polar lipid production systems.

The microalga *Chlorella* sp. (Chlorophyceae-green alga) is widely recognised as a useful strain for biotechnology [30], and has been extensively studied under a range of growth conditions, especially nutrient limitation, for the production of triacylglycerol lipids, with a focus on biofuel production [31–35]. Much less is known of the effects of nutrient limitation on the polar lipidome of microalgae, which is important for understanding how microalgae respond to stress by way of lipid metabolism regulation. To better understand these changes, the identification and quantification of polar lipid species can be achieved by using modern liquid chromatography-tandem mass spectrometry (LC-MS/MS) techniques. Indeed, using this approach, polar lipid species profiling has been performed on other microalgae species under different culture conditions, including nutrient stress [36–43]. However, detailed studies concerning the identification and quantification of polar lipid species and their modulation in *Chlorella* sp. under nutrient stress, are surprisingly limited.

In this study, a *Chlorella* sp. isolate was maintained and batch-cultured under different nutrient conditions (nutrient-replete and nutrient-deplete) and was comprehensively evaluated for polar lipid composition and production using a modern mass spectrometry-based approach (LC-MS/MS), to understand how the microalga alters the polar lipidome in response to these environmental conditions. It is envisaged that this fundamental information will be important for understanding the microalgal polar lipidomic response to nutrient deprivation, and for the optimisation of future applied microalgae-polar lipid production technologies and platforms. Findings are discussed in both these contexts.

2. Materials and Methods

2.1. Microalgal Strain and Culture Conditions

Stock cultures of the *Chlorella* sp. were maintained in the respective experimental media (see below) for a minimum of 6 months with weekly sub-culturing (10% v/v) in fresh media to ensure full physiological adaptation to media nutrient concentrations. All cultures were maintained at 25 °C under 100 µmols photons m^2 s^{-1} irradiance and 16:8 h light/dark cycle. The culture was originally isolated from an open pond textile factory wastewater in Chennai, India (provided by Dr. Sivasubramanian, Phycospectrum Environment Research Centre, Chennai, India). Species confirmation was provided both taxonomically and by molecular sequencing (GenBank: MF692949.1).

Experimentally, *Chlorella* sp. stock cultures were inoculated (1% v/v) into replicate ($n = 2$) aerated culture flaks (2 L) containing 1 L of either high nutrient-replete (HN) media where nutrients (nitrogen and phosphorous) were known to be replete throughout the culture period, or nutrient-limited media (LN media) where nutrients were known to be exhausted during the culture period (see Table S1 for composition of the media) and grown in batch cultures for up to 16 days. Replicate samples of cultures (typically 10 mL) were taken daily for nutrients, biomass (particulate organic carbon), and cell (*Chlorella* and bacteria) enumeration (see below). Although axenic practices were employed, the *Chlorella* sp. strain used was non-axenic, so bacteria numbers were assessed to determine contribution (based on cellular carbon estimates) to culture biomass and lipid composition. Replicate samples for polar lipid and total fatty acid analyses were taken at different growth stages of the batch cultures: exponential growth phase (Day 4), linear growth phase (Day 9), and late linear or stationary phase (depending on treatment, Days 14–15; see below).

2.2. Bacterial Cell Enumeration (Flow Cytometry)

Samples of culture (1 mL) were taken and fixed with 50 µl of 50% glutaraldehyde and stored at −80 °C until analysis. After thawing, the samples were stained with the DNA stain SYBR green (Fisher Scientific, Leicestershire, UK) for 1 h and then analysed using a FACSort flow cytometer (Becton Dickinson, Oxford, UK). Flow cytometer flow rate was calibrated (ca. 11 µL min^{-1}) and samples were diluted if required to maintain counts below 1000 events s^{-1}.

2.3. Particulate Organic Carbon (POC) and Nitrogen Analyses

Cellular material was harvested from cultures by filtration of accurately measured culture volume (typically 5–10 mL) onto ashed glass fiber filters (Whatman GF/F, 25 mm), dried at 60 °C and acidified by fuming with hydrochloric acid prior to analysis. All carbon and nitrogen analyses were carried out on a Thermoquest FlashEA 1112 elemental analyser. Lipid concentrations in cultures (mg mL^{-1}; see below) were subsequently normalised to culture biomass (mg C mL^{-1}) based on the culture volume from which lipid extracts were generated.

2.4. Nutrient Analyses

For media nutrient analyses, 10 mL samples of culture were membrane-filtered (0.2 µm) and the filtrate stored (−20 °C) in an acid-washed bottle to await analysis. After thawing, samples were analysed for nitrate and phosphate concentrations using a nutrient autoanalyser (Branne and Luebbe, AAIII, SPX Flow Technology Ltd., Brixworth, Northampton, UK) using standard methods [44,45].

2.5. Lipid Extraction

Culture samples (50 mL) were filtered under light vacuum onto ashed glass fiber filters (Whatman GF/F; 47 mm; $n = 3$) and then stored at −80 °C prior to analyses. Lipid extracts were generated using chloroform/methanol (2:1) and sonication to disrupt cellular material [46]. Samples were further extracted with 100% chloroform to ensure complete lipid extraction. Lipid-containing layers were pooled, dried under vacuum, and stored at −80 °C in 1 mL of chloroform/methanol (2:1).

2.6. Polar Lipid Analyses (Liquid Chromatography–Electrospray Ionisation Mass Spectrometry (LC-ESI-MS/MS))

Synthetic phospholipid (dipalmitoyl-PG, -PE, -PC) and betaine lipid (DGTS) standards were purchased from Avanti Lipids Inc. (Alabaster, AL, USA). Purified natural glycolipid standards (MGDG, DGDG, and SQDG) were purchased from Lipid Products (South Nutfield, UK).

Polar lipid standards or aliquots (400 µL) of dried (under N_2) lipid extract (see above) were re-suspended in dichloromethane/methanol (9:1 v/v; 100 µL) and analysed by LC-ESI-MS as previously described [47]. Samples were injected (2 µL) into an Agilent 1200 LC system (Agilent Technologies UK Ltd., Stockport, Cheshire, UK), and polar lipid classes were chromatographically separated on a 150 × 2.1 mm i.d., 5 µm diol particle size column (PrincetonSpher: Princeton Chromatography Inc., Cranbury, NJ, USA) using a gradient of 100% Solvent A to 52% Solvent B over a period of 20 min and to 71.5% B over 5 min, and this was then held for 10 min, prior to 10 min equilibration time at original solvent conditions. Flow rate was 0.4 mL min^{-1} for 40 min and then increased to 1 mL min^{-1} for rapid column equilibration. Solvents were as follows: Solvent A = 800:200:1.0:0.4 n-hexane/isopropanol/formic acid/25% aqueous ammonium hydroxide; Solvent B = 900:100:1.0:0.4 isopropanol/water/formic acid/25% aqueous ammonium hydroxide. The LC system was coupled to an ion trap mass spectrometer (Agilent 6330: Agilent Technologies UK Ltd., Stockport, Cheshire, UK) with an electrospray ionisation (ESI) source operated in both positive and negative ionisation modes (all samples run separately in each mode). Nitrogen was used as both the nebuliser and drying gas. The MS detector settings were as follows: nebuliser: 35 psi; dry gas flow: 10 L min^{-1}; drying temperature: 200 °C; capillary voltage: 4500V; MS/MS fragmentation amplitude: 1.5 V; isolation width: 4 m/z units. Other trap parameters were optimised by infusing standard solutions in isocratic mixtures of Solvents A and B.

Using these conditions under positive ionisation, protonated ions ($[M+H]^+$; PE, PC, DGTS) or ammonium adducts ($[M+NH_4]^+$; PG, MGDG, DGDG, SQDG, DGTS) of lipid species were formed, and characteristic class-specific neutral loss or product ions were generated in MS/MS data ([47]; see Table S2). Standard curves were generated from the extracted ion chromatogram of the molecular ion (or sum of molecular ions in the case of the glycolipid standards) in positive ionisation mode. All chromatograms (including standards) were smoothed (3 s cycle width) prior to integration. Linear detection ranges were 5–60 pg on column for the glycerophospholipids, SQDG, DGDG, and DGTS and 5–20 pg for MGDG. Zero values reported for polar lipid concentrations were below these linear limits of detection. Standard polar lipid concentrations were reproducible within 10% error between repeat injections, and a concentration range of standard mixtures was injected every 10 samples to monitor instrument response and ensure linearity.

For lipid extracts, determination of the major molecular ions within each lipid class was based on parent molecular ion, comparative retention times with standards, and confirmation of characteristic neutral loss and product fragmentation ions (headgroup) in MS/MS spectra in positive ionisation mode (see Table S2). A subset of samples was used to determine prominent acyl ion combinations in individual species by confirming acyl fragments in MS/MS spectra in either positive (MGDG, DGDG, SQDG; $[RCOO+C_3H_5O]^+$ fragments) or negative (PG, PE, PC, DGTS: $[RCOO]^-$ fragments) ionisation modes (see Figure S2 for examples). For quantification, extracted ion chromatograms (positive ionisation mode) were generated for each species (molecular ion) and peak areas integrated and quantified based on generated standard curves in the absence of appropriate internal standards (see above). Reported total lipid class concentrations (the sum of polar lipid species concentrations within the respective class) in biomass were subsequently calculated based on the total volume of lipid extract derived for known culture biomass (see particulate organic carbon and lipid extraction sections above).

2.7. Fatty Acid Analyses (Gas Chromatography-Mass Spectrometry: GC-MS)

Fatty acid composition was also determined in lipid extracts (no prior separation of lipid classes). 400 μL of lipid extract was dried under a gentle stream of nitrogen before adding nonadecanoic acid (C19:0; 20 μL, 1 mg mL^{-1}) as an internal standard. Cellular fatty acids were converted directly to fatty acid methyl esters (FAMEs) by adding 1 mL of transesterification mix (95:5 v/v 3N methanolic hydrochloric acid/2,2-dimethoxypropane) followed by incubation at 90 °C for 2 h [48]. After cooling, FAMEs were recovered by addition of a 1% w/v NaCl solution (1 mL) and n-hexane (1 mL) followed by vortexing. The upper hexane layer was injected directly into the GC-MS system and FAMEs were separated on a fused silica capillary column (30 m × 0.25 mm × 0.25 μm: Omegawax™ 250, Supelco, Sigma-Aldrich, Gillingham, Dorset, UK) using an oven temperature gradient of 75 °C to 240 °C at 4 °C min^{-1} followed by 15 min of hold time. Helium was used as the carrier gas (1 mL min^{-1}) and the injector and detector inlet temperatures were maintained at 280 °C and 230 °C, respectively. FAMEs were identified using retention times and qualifier ion response, and quantified using respective target ion responses. All parameters were derived from calibration curves generated from a FAME standard mix (Supelco, Sigma-Aldrich, Gillingham, Dorset, UK).

2.8. Data Analyses

Mean values from replicate cultures within treatment are reported ± variance. Comparison of biomass and lipid concentrations between treatments was conducted across the whole batch culture period using t-tests using the software package Minitab 17 (Minitab Ltd., Brandon Court, Coventry, UK). Linear Pearson's correlation coefficients between biomass and lipid were carried out using the same software. Figures were generated using Sigma Plot 12.0 (Systat Software Inc., San Jose, CA, USA).

Polar lipid species data was further analysed using non-parametric multivariate methods in Primer v6 [49,50]. Resemblances among samples were calculated using the Bray–Curtis resemblance measure [50]. Subsequently, the significance of dissimilarity in lipid data between samples was confirmed using ANOSIM (analysis of similarities; [51]) where large positive R values (range from 0 to 1) indicated distinct separation of lipid composition between the treatments. Subsequently, the polar lipid species contributing to the dissimilarity between samples were explored using the SIMPER routine in the same software package [52].

3. Results

Biomass measured as POC in both HN and LN media followed typical growth patterns with an initial lag phase (up to ca. 4 days) followed by linear growth in the HN cultures, but with early onset of the stationary phase at ca. 11 days in the LN cultures (Figure 1a).

Biomass in HN cultures at the end of the investigation was more than double (ca. 450 mg C L^{-1}) those grown in LN media (ca. 200 mg C L^{-1}). Based on 60% carbon composition of dry weight (dw), this equates to approximately 750 mg L^{-1} dw and 333 mg L^{-1} dw in HN and LN cultures, respectively. Biomass productivity within the linear phase of growth across both treatments (up to 9 days of batch culture) was similar at 15.6 ± 1.1 mg C L^{-1} day^{-1} and 13.2 ± 0.5 mg C L^{-1} day^{-1} for the HN and LN treatments, respectively. Biomass contribution from bacteria in cultures (carbon units) was assessed based on monitoring bacterial cell numbers by flow cytometry and estimates of carbon contribution based on 10 fg C cell^{-1} [53]. Across all treatments at any sampling point maximum biomass contribution of bacterial carbon was calculated as <1.6% w/w and was therefore deemed to have insignificant influence on lipid results (data not shown).

Media nitrate concentrations in the HN cultures were not fully depleted during the batch culture and were still high (ca. 5 mM) at the end of the experiment (16 days; Figure 1b). Likewise, phosphate levels were reduced to 0.01 mM but were not fully depleted. Nitrate concentrations in the LN cultures were reduced to 0.0003 mM, and phosphate concentrations were reduced to non-detectable concentrations (<0.0001 mM) at the end of the culture period. The carbon/nitrogen ratio of the biomass stayed consistent at a value of around 5 throughout the culture period in the HN cultures (Figure 1c). In the LN cultures, the carbon/nitrogen ratio increased linearly from 5 near the beginning of the culture (Day 4) to a value of 50 at the end of the culture period.

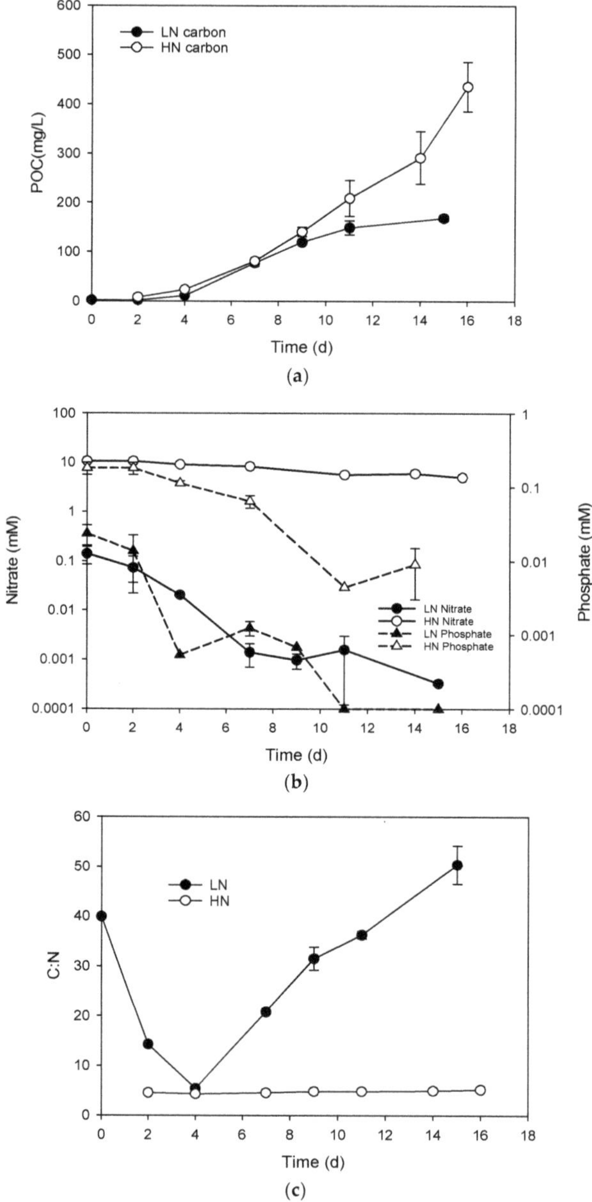

Figure 1. (**a**) Biomass (particulate organic carbon, POC) production, (**b**) media nutrient concentrations, and (**c**) particulate carbon/nitrogen ratio in *Chlorella* sp. grown in batch cultures containing nutrient-limited (LN) or nutrient-replete (HN) media (legends inset).

The biomass increase in cultures was reflected in the volumetric concentration of total polar lipids, which was highly correlated with POC ($R^2 = 0.99$, $p < 0.001$) in the HN cultures. The relationship between polar lipids and biomass was lower in the LN cultures and was just significant ($R^2 = 0.82$, $p = 0.048$). Final concentrations of total polar lipids (Figure 2a) was ca. 10 times higher in the HN cultures compared to the LN cultures (388 ± 73 mg L^{-1} and 42 ± 7 mg L^{-1}, respectively) and based on the linear phase of microalgal growth across treatments (up to 9 days) had a significantly higher productivity (24.0 mg L^{-1} day^{-1} and 6.0 ± 0.4 mg L^{-1} day^{-1}, respectively). The polar lipid/carbon ratio (Figure 2b) showed an increasing trend in the HN cultures from a value of 0.8 to 1.3 and a decreasing trend in the LN culture from 1.2 to 0.25 at the end of the batch culture and mean values over the course of the experiment were significantly different ($p = 0.04$).

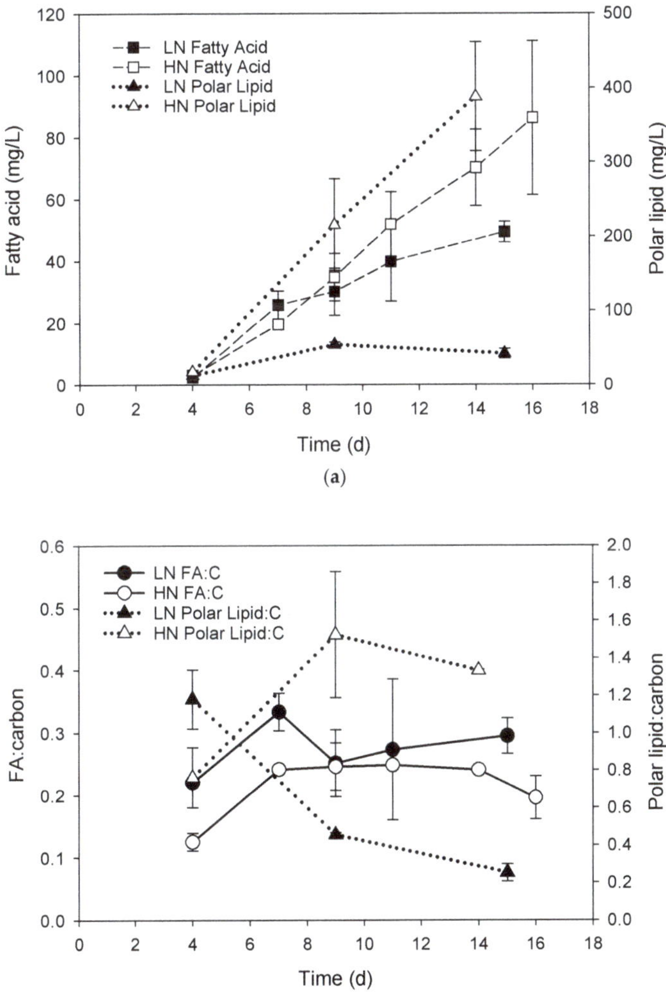

Figure 2. (**a**) Volumetric fatty acid and polar lipid concentrations and (**b**) fatty acid and polar lipid/carbon ratios (mg:mg) in *Chlorella* sp. grown in batch cultures containing LN or HN media (legends inset).

The total fatty acid concentrations of the HN and LN cultures were significantly correlated with biomass (R^2 = 0.98, $p < 0.001$ and 0.98, $p = 0.002$, respectively). Final concentrations of total fatty acids in the HN and LN cultures were 86.2 $\pm$ 24.9 and 49.5 $\pm$ 3.4 mg L^{-1}, respectively, but with similar productivities (over linear growth phase) of 3.9 $\pm$ 0.9 mg L^{-1} day^{-1} and 3.3 $\pm$ 0.8 mg L^{-1} day^{-1}, respectively. The fatty acid/carbon ratios in both the HN and LN cultures were similar and remained relatively steady throughout the course of the culture period (mean value of 0.22 $\pm$ 0.05 and 0.27 $\pm$ 0.04, respectively).

Closer examination of the polar lipid class concentration normalised for carbon (microalgal biomass; Figure 3) showed that irrespective of culture conditions the glycolipids (MGDG, DGDG and SQDG) were generally more abundant in *Chlorella* lipid extracts than the phospholipids (PG, PE, and PC) and betaine lipid (DGTS). Of these lipid classes, MGDG was the most abundant glycolipid constituting 35–43% consistently in the HN cultures throughout the batch culture but decreasing from 36 to 19% in the LN cultures. Concentrations of MGDG were overall significantly ($p = 0.025$) higher in *Chlorella*-consortia biomass grown in HN media with mean levels of 0.5 $\pm$ 0.22 mg mg C^{-1} compared to 0.2 $\pm$ 0.2 mg mg C^{-1} in LN cultures. Concentrations of DGDG increased in HN cultures biomass to a maximum of 0.15 $\pm$ 0.06 mg mg C^{-1} but relative contribution to total polar lipid remained constant (6–10%). The concentrations of DGDG in LN cultures biomass decreased from a maximum of 0.27 $\pm$ 0.01 mg mg C^{-1} to a minimum of 0.10 $\pm$ 0.03 mg mg C^{-1} at the end of culture. However, contribution to total polar lipid actually increased from 20 to 37%. Overall mean concentrations of DGDG were not significantly different ($p = 0.07$) between HN and LN cultures. Concentrations of SQDG increased in HN cultures biomass to maximum levels of 0.4 $\pm$ 0.1 mg mg C^{-1} and accounted for 11–31% of the total polar lipids. Conversely, concentrations of SQDG fell in LN cultures biomass from 0.3 $\pm$ 0.05 to 0.1 $\pm$ 0.02 mg mg C^{-1}. SQDG accounted for 22 to 34% of total polar lipid in LN cultures. Overall mean concentrations of SQDG across the culture period were not significantly different ($p = 0.109$). The concentrations of PG were relatively constant in the HN cultures biomass (0.08–0.1 mg mg C^{-1} and accounted for 7–11% of polar lipids) but decreased in LN cultures biomass from 0.1 to 0.002 mg mg C^{-1} (dropping from 9 to 1% of total polar lipid). Overall mean concentrations of PG were significantly higher ($p = 0.043$) in the HN cultures. Concentrations of PC constituted between 4 and 7% of polar lipid classes in HN cultures biomass (maximum concentration 0.05 $\pm$ 0.03 mg mg C^{-1}) and similarly between 2 and 4% in LN cultures (maximum concentration 0.04 mg/mg C) and mean concentrations were overall significantly higher ($p = 0.047$) in the HN cultures. Over the batch culture period, concentrations of PE fell in HN cultures biomass from 0.22 $\pm$ 0.16 to 0.07 $\pm$ 0.03 mg mg C^{-1} and PE concentrations in LN cultures fell from 0.03 to 0.002 mg mg C^{-1}. Overall mean concentrations of PE were significantly higher ($p = 0.015$) in the HN cultures. DGTS concentrations increased throughout the culture in HN culture biomass from 0.01 to 0.04 mg mg C^{-1}, but relative contribution to total polar lipids was consistent (ca. 2–4%). However, DGTS concentrations fell in LN cultures biomass (from 0.06 to 0.02 mg mg C^{-1}) but again retained consistent levels relative to other polar lipid classes (7–8%). Overall mean concentrations were not significantly different between LN and HN cultures ($p = 0.59$).

A total of 50 different polar lipid species (6 MGDG, 10 DGDG, 4 SQDG, 6 PG, 9 PE, 8 PC, and 7 DGTS) were identified in *Chlorella* lipid extracts using LC-MS/MS in this study (Table 1). The acyl groups associated with each species mainly constituted combinations of 16 and 18 carbon chains with different degrees of unsaturation (up to 4 double bonds) in agreement with the fatty acid composition of the same lipid extracts (Figure 4). In the HN treatments, the two major (>80% total) MGDG species (acyl carbon:double bonds) were 34:6 (18:3/16:3 and 18:2/16:4) and 34:7 (18:3/16:4). In the LN treatments, the composition of MGDG species varied with the day of the batch culture with notable decreases in MGDG 34:7 (18:3/16:4; from 78% to 2%) and increases in MGDG 34:6 (18:3/16:3; 15–42%), MGDG 34:5 (18:3/16:2; 1% to 28%), and MGDG 34:4 (18:1/16:3; 1–14%) from the start to the end of the batch culture. This is in agreement with the reduction in the fatty acids 16:4 and 18:3 (main acyl groups in MGDG 34:7).

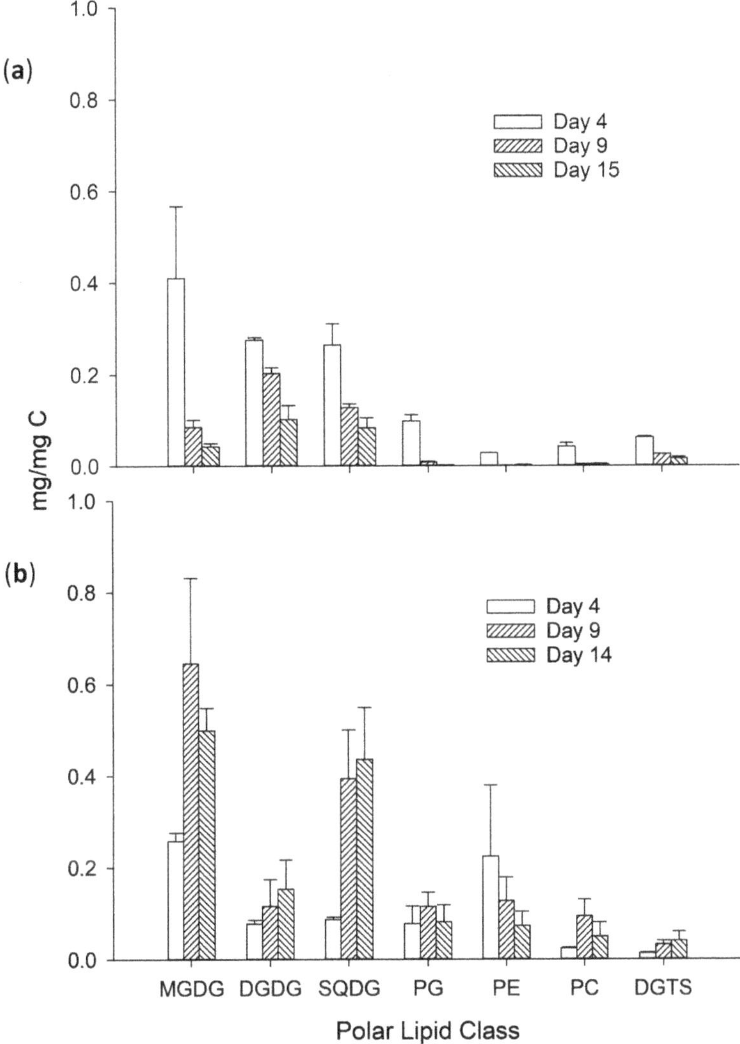

Figure 3. Polar lipid class concentrations in *Chlorella* sp. Biomass grown in batch cultures containing either (**a**) LN or (**b**) HN media (sampling day legend inset). Lipid classes were as follows: monogalactosyldiacylglycerol (MGDG); digalactosyldiacylglycerol (DGDG); sulfoquinovosyldiacylglycerol (SQDG); phosphatidylcholine (PC); phosphatidylglycerol (PG); phosphatidylethanolamine (PE), diacylglyceryl trimethyl homoserine (DGTS).

Table 1. Relative abundance (%) of polar lipid species (as a function of total species concentration within class) in *Chlorella* sp. grown in batch cultures containing either LN or HN media. Lipid classes were monogalactosyldiacylglycerol (MGDG); digalactosyldiacylglycerol (DGDG); sulfoquinovosyldiacylglycerol (SQDG); phosphatidylcholine (PC); phosphatidylglycerol (PG); phosphatidylethanolamine (PE), diacylglyceryl trimethyl homoserine (DGTS). The carbon:double bond column refers to the total number of carbon atoms and double bonds of acyl groups attached to the diglyceride moiety and the acyl groups column refers to the dominant fatty acid combinations as detected in MS/MS spectra (see Materials and Methods).

Lipid Class	Carbon: Double Bonds	Acyl Groups	LN			HN		
			Day 4	Day 9	Day 15	Day 4	Day 9	Day 14
MGDG	34:7	18:3/16:4	78.4 ± 0.5	11.1 ± 2.0	1.6 ± 0.5	78.5 ± 0.8	62.5 ± 1.3	58.6 ± 3.7
MGDG	34:6	18:3/16:3, 18:2/16:4	14.6 ± 0.2	39.9 ± 1.7	41.5 ± 2.8	15.0 ± 0.4	24.4 ± 1.1	20.8 ± 1.0
MGDG	34:5	18:3/16:2	0.6 ± 0.9	29.9 ± 3.3	28.2 ± 3.0	1.4 ± 0.1	5.6 ± 0.6	6.4 ± 1.0
MGDG	34:4	18:1/16:3	0.6 ± 0.8	12.0 ± 0.7	13.7 ± 0.8	1.6 ± 0.3	2.3 ± 0.3	3.8 ± 0.5
MGDG	34:3	18:1/16:2	0.0 ± 0.0	6.3 ± 0.8	10.1 ± 0.7	0.0 ± 0.0	0.8 ± 0.0	1.6 ± 0.4
MGDG	36:6	18:3/18:3	5.1 ± 0.7	0.7 ± 1.0	4.2 ± 0.2	3.5 ± 0.7	3.4 ± 0.5	5.8 ± 0.0
DGDG	34:7	18:3/16:4	14.6 ± 1.1	1.4 ± 0.6	0.0 ± 0.0	12.5 ± 0.7	20.1 ± 2.4	12.7 ± 0.9
DGDG	34:6	18:3/16:3	33.4 ± 2.4	22.3 ± 3.2	21.3 ± 0.4	31.3 ± 1.0	17.0 ± 1.4	14.5 ± 0.6
DGDG	34:5	18:2/16:3	10.2 ± 0.0	22.4 ± 1.0	21.7 ± 2.0	11.8 ± 0.5	10.2 ± 0.1	9.4 ± 0.8
DGDG	34:4	18:2/16:2	5.3 ± 0.0	9.0 ± 0.9	10.2 ± 0.2	9.5 ± 1.7	12.4 ± 1.0	10.9 ± 0.8
DGDG	34:3	18:3/16:0	16.1 ± 2.1	22.0 ± 1.4	21.4 ± 0.9	13.7 ± 0.5	11.4 ± 0.6	11.5 ± 1.1
DGDG	34:2	18:2/16:0	7.5 ± 0.8	14.6 ± 0.7	15.2 ± 0.5	8.2 ± 0.3	12.0 ± 0.7	14.9 ± 0.1
DGDG	34:1	18:1/16:0	5.6 ± 0.2	8.4 ± 0.2	10.3 ± 1.4	5.8 ± 0.7	8.7 ± 0.2	8.8 ± 0.5
DGDG	36:6	18:3/18:3	6.0 ± 0.2	0.0 ± 0.0	0.0 ± 0.0	5.4 ± 1.3	5.6 ± 1.3	8.9 ± 0.6
DGDG	36:5	18:3/18:2	1.3 ± 0.1	0.0 ± 0.0	0.0 ± 0.0	1.8 ± 0.5	1.8 ± 0.3	4.7 ± 0.3
DGDG	36:4	18:1/18:3	0.0 ± 0.0	0.0 ± 0.0	0.0 ± 0.0	0.0 ± 0.0	0.8 ± 1.1	3.5 ± 0.2
SQDG	32:0	16:0/16:0	37.5 ± 1.0	38.9 ± 2.5	37.9 ± 2.4	36.2 ± 1.3	32.5 ± 0.9	30.0 ± 3.7
SQDG	34:3	18:3/16:0	45.4 ± 0.2	27.1 ± 2.1	26.5 ± 0.7	46.2 ± 0.4	46.9 ± 3.3	45.5 ± 0.9
SQDG	34:2	18:2/16:2	12.0 ± 0.0	18.2 ± 0.1	17.0 ± 0.1	12.1 ± 0.2	14.5 ± 2.3	17.4 ± 2.0
SQDG	34:1	18:1/16:0	5.1 ± 1.2	15.8 ± 0.3	18.6 ± 1.8	5.5 ± 1.9	6.1 ± 0.2	7.1 ± 0.8
PG	32:1	16:0/16:1	0.0 ± 0.0	0.0 ± 0.0	0.0 ± 0.0	10.3 ± 0.2	3.2 ± 0.2	1.3 ± 0.1
PG	34:4	18:3/16:1	60.7 ± 0.0	31.3 ± 5.0	0.0 ± 0.0	10.8 ± 3.4	31.2 ± 2.5	33.2 ± 0.9
PG	34:3	18:3/16:0, 18:2/16:1	23.4 ± 0.7	23.5 ± 0.6	0.0 ± 0.0	1.8 ± 2.5	22.3 ± 0.5	24.0 ± 3.7
PG	34:2	18:1/16:1, 18:2/16:0	8.1 ± 1.0	16.5 ± 0.7	23.2 ± 8.0	43.9 ± 0.7	23.2 ± 2.4	21.6 ± 1.0
PG	34:1	18:1/16:0	7.8 ± 0.2	28.7 ± 5.0	76.8 ± 8.0	28.0 ± 0.1	14.8 ± 0.1	13.2 ± 1.4
PG	36:2	18:1/18:1	0.0 ± 0.0	0.0 ± 0.0	0.0 ± 0.0	5.2 ± 0.3	5.3 ± 0.6	6.7 ± 2.2

Table 1. *Cont.*

Lipid Class	Carbon: Double Bonds		LN			HN		
		Acyl Groups	Day 4	Day 9	Day 15	Day 4	Day 9	Day 14
PE	32:1	16:0/16:1	0.0 ± 0.0	0.0 ± 0.0	0.0 ± 0.0	14.2 ± 0.0	10.8 ± 1.1	10.1 ± 1.7
PE	33:1	16:0/17:1	0.0 ± 0.0	0.0 ± 0.0	0.0 ± 0.0	0.4 ± 0.2	0.8 ± 0.3	1.1 ± 0.0
PE	34:2	18:1/16:1	2.3 ± 0.3	13.3 ± 2.3	3.7 ± 5.2	50.7 ± 2.1	48.3 ± 0.3	47.5 ± 1.0
PE	34:1	18:1/16:0	0.0 ± 0.0	0.0 ± 0.0	0.0 ± 0.0	28.4 ± 0.1	22.5 ± 2.3	21.3 ± 1.6
PE	34:0	18:0/16:0	11.2 ± 1.6	0.0 ± 0.0	17.5 ± 11.0	1.4 ± 2.0	2.5 ± 0.6	2.5 ± 0.0
PE	36:5	18:2/18:3	46.2 ± 0.6	22.0 ± 1.4	6.6 ± 9.3	0.7 ± 0.2	2.9 ± 0.0	1.7 ± 1.1
PE	36:4	18:1/18:3, 18:2/18:2	29.7 ± 1.9	36.6 ± 0.4	27.2 ± 2.6	0.0 ± 0.0	2.2 ± 0.0	1.4 ± 2.0
PE	36:3	18:2/18:1	10.6 ± 0.6	28.0 ± 3.3	29.6 ± 1.1	0.0 ± 0.0	0.4 ± 0.5	0.7 ± 1.0
PE	36:2	18:1/18:1	0.0 ± 0.0	0.0 ± 0.0	15.5 ± 10.6	2.3 ± 0.4	3.3 ± 0.0	5.5 ± 0.1
PC	34:4	18:3/16:1	5.3 ± 0.1	1.8 ± 0.8	1.8 ± 0.0	4.4 ± 0.2	6.5 ± 1.2	5.1 ± 0.0
PC	34:3	18:3/16:0, 18:2/16:1	15.1 ± 0.2	6.7 ± 1.0	6.2 ± 1.7	13.5 ±1.8	9.6 ± 0.2	10.4 ± 0.7
PC	34:2	18:2/16:0, 18:1/16:1	3.5 ± 0.0	10.4 ± 1.7	10.3 ± 0.2	5.3 ± 0.3	3.3 ± 0.3	4.7 ± 0.9
PC	36:6	18:3/18:3	36.1 ± 1.2	15.1 ± 4.3	14.3 ± 2.6	36.4 ± 0.1	36.4 ± 0.1	28.4 ± 6.8
PC	36:5	18:3/18:2	22.8 ± 0.2	26.6 ± 1.3	23.5 ± 4.9	20.7 ± 0.2	20.9 ± 1.5	18.7 ± 0.2
PC	36:4	18:3/18:1	11.6 ± 0.7	18.7 ± 0.1	20.0 ±2.1	11.3 ± 0.5	11.7 ± 0.6	17.0 ± 1.9
PC	36:3	18:2/18:1	4.1 ± 0.2	10.9 ± 2.3	13.1 ± 0.5	4.0 ± 0.6	4.9 ± 0.1	8.9 ± 1.3
PC	36:2	18:1/18:1	1.7 ± 0.2	9.8 ± 0.3	11.1 ± 3.2	4.4 ± 0.8	3.3 ± 0.6	6.7 ± 2.1
DGTS	34:4	18:4/16:0	45.0 ± 0.1	25.7 ± 1.7	24.3 ± 0.8	44.6 ± 2.7	33.1 ± 1.2	31.6 ± 0.5
DGTS	34:3	18:3/16:0	16.3 ± 0.6	23.1 ± 1.0	28.8 ± 1.0	14.4 ± 0.9	11.1 ± 0.2	12.1 ± 1.5
DGTS	34:2	18:2/16:0	1.8 ± 0.0	12.0 ± 2.1	15.0 ± 0.3	0.0 ± 0.0	1.4 ± 1.9	3.5 ± 1.8
DGTS	36:7	18:3/18:4	14.8 ± 0.9	3.8 ± 0.4	2.3 ± 0.1	15.6 ± 0.8	23.7 ± 1.7	19.6 ± 4.6
DGTS	36:6	18:3/18:3	14.1 ± 0.2	9.1 ± 0.8	6.3 ± 0.7	14.8 ± 1.6	16.9 ± 0.3	15.2 ± 0.7
DGTS	36:5	18:3/18:2	6.3 ± 0.6	14.5 ± 0.8	11.3 ± 0.1	7.8 ± 1.0	9.2 ± 0.5	11.0 ± 1.0
DGTS	36:4	18:3/18:1	1.7 ± 0.1	11.4 ± 0.5	12.0 ± 0.2	2.9 ± 0.2	4.6 ± 0.5	6.9 ± 1.5

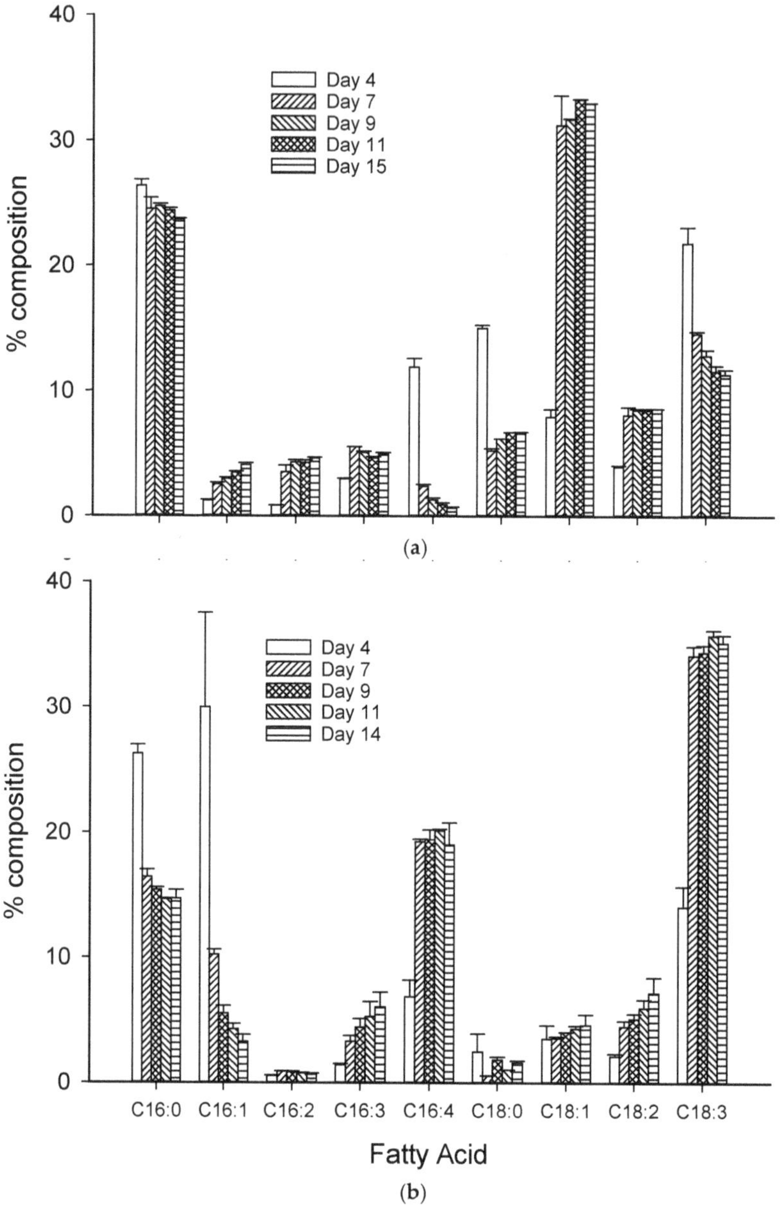

Figure 4. Fatty acid composition (% of total fatty acids) in *Chlorella* sp. in batch cultures containing either (**a**) LN media or (**b**) HN media (sampling day legend inset).

Similarly, DGDG species abundance was comparatively consistent in HN samples taken throughout the batch culture, but in LN cultures there were notable reductions in DGDG 34:7 (18:3/16:4; from 15 to 0%) and subsequent increases in the relative abundance of other DGDG species. The composition of SQDG species was again consistent in the HN cultures irrespective of sampling day but with notable reductions in SQDG 34:3 (18:3/16:0; 45–27%) and increases in SQDG 34:2 (18:2/16:2; 12–17%) and SQDG 34:1 (18:1/16:0; 5–19%) in LN cultures. This observed consistency in the relative composition of species noted for glycolipids in lipid extracts derived from HN *Chlorella* cultures was largely reflected in PC and DGTS classes, as was the decrease in PC and DGTS species, and indeed PG species, containing highly unsaturated acyl moieties in lipid extracts from the LN treatment. However, the relative abundance of PG species in HN extracts was more variable with notable decreases in PG 32:1 (16:0/16:1; from 10 to ca. 3%), PG 34:2 (18:1/16:1 and 18:2/16:0; from 44 to ca. 22%), and PG 34:1 (18:1/16:0; from 28 to ca. 14%) and increases in PG 34:4 (18:3/16:1; from 11 to ca. 32%) and PG 34:3 (18:3/16:0 and 18:2/16:1; from 2 to ca. 23%) between Day 4 and the rest of the batch culture sampling points, although overall this was viewed as an increase in the level of unsaturation of PG species.

Principal component analyses (PCAs) of the samples from different treatments based on the polar species composition (Figure 5) revealed separation along the PC1 axis (explaining 73.3% of the variation) of samples in the LN treatments based on the sample collection day. Notably, LN samples at Day 4 were clustered with HN samples taken at Days 9 and 14, which as a group were distinctive from LN samples taken at later stages in the batch culture. Some separation of samples along the PC2 axis to a lesser extent (explaining 17.3% of the variation), e.g., Day 4 HN samples, were distinctive from Day 9 and 14 HN samples.

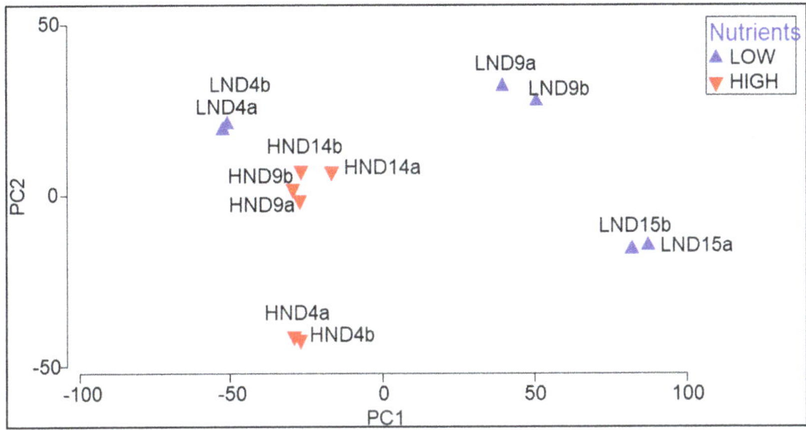

Figure 5. Principal component analyses (PCAs) of samples at different time points (Day = D) of *Chlorella* sp. grown in batch cultures (*n* = 2; a, b) containing LN or HN media based on the polar lipid species profiles (legends inset). PC1 and PC2 explain 73.3 and 17.3% of variation, respectively (cumulative 90.6% of variation explained by the model).

SIMPER analyses of polar lipid species contributing up to 70% of the dissimilarity between samples (see Table S3) revealed that the relative abundance of MGDG 34:7 contributed the most dissimilarity between LN samples taken at the beginning (Day 4) and middle (Day 9; 44% contribution) and at the end (Day 14; 29% contribution) with relative abundance being greater at Day 4 in both comparisons.

4. Discussion

Microalgae contain a diverse array of polar lipid structures that have several biological functions, including, importantly, maintaining photosynthesis. Recently there has been growing interest in utilising these polar lipid compounds as novel high-value phytochemical bioactives [54,55], as they often contain high concentrations of omega-3 fatty acids and express a range of activities that make them useful chemicals for the cosmetic, pharmaceutical, and functional food industries. Therefore, detailed research is required to understand the effect of culture conditions on the microalgal polar lipidome, to provide a fundamental understanding of lipid metabolism and regulation and with a view to biorefinery optimisation and commercial realisation of microalgal polar lipid production.

In this study, the productivity of polar lipids in cultures on a titre basis was higher when cultures were nutrient-replete, which was linked with the higher concentrations of biomass production. This is consistent with the fact that cultures initially supplied with lower concentrations of nutrients became nutrient-limited (as demonstrated by the concentrations of nitrate in the media and the carbon/nitrogen ratio in biomass) and as a result produced less total polar lipids for photosynthesis and growth. Coupled with the consistent total concentrations of fatty acids in biomass across the batch culture periods, this strongly supports the theory of physiological diversion of synthesised fatty acids from polar lipids into other neutral lipids, e.g., triacyclglycerols (not analysed here), which is a general metabolic feature of microalgae [56,57]. Indeed, authors have reported an increase in the activity of membrane galactolipid-specific acyl hydrolase in nitrogen-deficient cells of the alga *Dunaliella salina* [58] and *Chlamydomonas reinhardtii* [59], suggesting that liberated fatty acids are available for incorporation into TAGs. This would support the notion that the optimal commercial production of polar lipids in microalgal biomass will require nutrient-intensive culture systems. Therefore, consideration should be given to the cost requirements of nutrient supply versus the yield of bioactive polar lipids from microalgal biomass on this basis, and high nutrient wastewater systems, e.g., municipal and agricultural wastewater, should be considered where applicable [60]. Furthermore, semi-continuous or continuous production systems should be strongly considered over batch culture production, since the former permits a near constant supply of nutrients avoiding nutrient limitation scenarios.

The relative composition of polar lipid classes in lipid extracts from *Chlorella* cultures under nutrient-replete conditions was consistent with the literature where glycolipids dominated over other polar lipid classes in terms of abundance [57]. Glycolipids are principally found in the chloroplasts of eukaryotic algae, notably in thylakoid membranes (especially MGDG and DGDG) where they play critical roles in photosynthesis [4]. Other than PG, the other lipid classes identified are found mainly in extra-chloroplastid membranes.

The notable reduction in the concentrations of MGDG under conditions of nutrient limitation are consistent with the role of this polar lipid class as the major constituent thylakoid lipid and its role in the xanthophyll cycle and photosynthetic pigment-protein complexes [61]. Indeed, MGDG is enriched in chloroplast thylakoids compared to whole cells or envelope membranes in *Chlamydomonas* [9]. DGDG has a role in the structural integrity of Photosystem II, the assembly of the light harvesting complex, and the stability of Photosystem I [61], and it would be reasonable to expect a consistent relative level of reduction in line with MGDG. Interestingly, in this study the significant relative increase in the concentrations of DGDG as a function of total polar lipid (which was not generally consistent with the other polar lipid classes) suggests a physiological stress response adaptation to nutrient stress. These findings are consistent with previous works demonstrating a decrease in the concentrations of MGDG and corresponding increases in the concentrations of DGDG in nitrate-starved *Chlamydomonas nivalis* [62]. This was explained as a biological requirement for the stabilisation of thylakoid membranes in a bi-layer state where fewer proteins and pigments are available under nutrient stress, i.e., an increase of bilayer-forming DGDG and a reduction of hexagonal structures of MGDG. However, it is not possible here to say whether the relative increase in DGDG was a direct result of the substitution of phospholipids or as a result of a greater reduction of other lipid classes generally. It is unclear whether these stress mechanisms can be exploited to produce higher concentrations of DGDG should the bioactivity or biotechnological potential of this glycolipid be more desirable than MGDG. The relative levels of the glycolipid SQDG were consistent between high-nutrient and low-nutrient treatments. Unlike MGDG and DGDG, SQDG is not restricted to photosynthetic membranes.

The phospholipid PG showed a significant reduction in abundance as a function of biomass in nutrient-limited cultures. Since studies have demonstrated that PG is essential for growth of higher plants and cyanobacteria and for the photosynthetic transport of electrons and the development of chloroplasts [6], a reduction under nutrient-limited conditions may be anticipated. However, this is in contrast to the findings of Martin et al. [57], who demonstrated elevated concentrations of PG in both *Chlorella* sp. and *Nannochloropsis* sp. under N-limited conditions, although the absolute amounts remained relatively low. These authors explained an increase in PG with N-deprivation as a means to compensate for the loss of chloroplast glycolipids in order to maintain photosynthetic activity, since PG is found in plastid photosynthetic membranes [63] and has a vital role in photosynthesis. One possible explanation for the differences in these studies may be the substitution of PG with SQDG; both PG and SQDG are both anionic in nature at neutral pH, which has been demonstrated under conditions of phosphate limitation in cyanobacteria and plants [64,65]. In this study, the SQDG/PG molar ratio increased from 2.5 at the beginning of culture to a value of 28 at the end of the culture period in the nutrient-limited cultures. In comparison, the SQDG/PG ratio remained comparatively constant (1.0–4.9) in the nutrient-replete cultures. Phosphate was clearly depleted in the nutrient-limited cultures, and this ratio may suggest a substitution mechanism of SQDG for PG. However, given that a co-limitation of nitrate occurred, it may be that photosynthesis and growth was limited to an extent where substitution of PG with SQDG has no physiological significance. Owing to differences in experimentation and in the absence of comparable data, e.g., media nutrient levels, it is difficult to account for the differences in PG results in this study and those of Martin et al. [57]. Clearly the biosynthesis of PG under nutrient limitation in microalgae should be explored further, especially at the molecular level. Furthermore, since SQDG has been shown to have a range of important bioactivities [20], SQDG vs. PG production in commercially relevant microalgae should be studied, notably where phosphate availability is controlled as a single limiting nutrient.

The phospholipids PE and PC are nitrogen-containing lipids and the recorded reduction in their abundance was consistent with N-limitation in the nutrient-limited cultures [57]. Typically, green algae such as *Chlorella* contain DGTS and PC in interchangeable amounts, notably in response to phosphate limitation, presumably as these two lipids have an equivalent membrane function [9]. Overall, our results did not support the substitution of PC with DGTS under nutrient limitation conditions applied in this study.

Although overall there were reductions of polar lipid classes in biomass under nutrient limitation, there were apparent differences in the magnitude to which individual polar lipid classes were modulated as a function of biomass produced. This suggests that control of supplied nutrients, as a stressor to growth, could be controlled to optimise the relative composition of polar lipid classes depending on the desired class required balanced again reduced overall productivity. Obviously, this necessitates further research regarding the comparative bioactivity of specific classes.

Concerning the acyl composition of polar lipids, in general, MGDG tends to have high contents of polyunsaturated fatty acids, while PG and SQDG are generally more saturated with a high level of palmitate SQDG [9]. Green algae such as *Chlorella* and the Euglenophyta contain MGDG and, to a lesser extent, DGDG, with high amounts of 16:4 and 16:3 acyl groups [66]. This is consistent with the acyl distribution of these polar lipid classes in our cultures grown under nutrient-replete conditions. Under nutrient-limited conditions, the relative composition of FA acyl groups in polar lipids was affected similarly across most polar lipid classes, i.e., increase in the saturation of carbon chains. This was particularly noted in the reduction of the highly abundant MGDG 34:7 species (principally contained 18:3/16:4). The abundance of PG PUFAs (18:3) that was increased in the nutrient-replete cultures increased after 24 h, arguably as a physiological response to improve the incorporation of D1 proteins in the thylakoid membranes for maximum photosynthetic capacity. The reduction in the levels of fatty acid unsaturation generally in microalgae in response to nutrient limitation is well documented and is usually associated with greater deposition in neutral lipid classes [10,16,67]. This study supports the notion that the reduction in the levels of unsaturation is common to both neutral and polar lipid classes when nutrient limitation conditions are present. Since the bioactivity of polar lipids likely depends on chemical structure, which is largely affected by acyl combinations, this may offer the potential for modulating activity of polar lipids produced from microalgal based on nutrient control regimes at the expense of overall productivity. However, in-depth research of structure–activity relationships of individual polar lipid species from different classes is required.

In summary, through culture investigations and in-depth analysis using modern polar lipid species profiling (via LC-MS/MS), we have demonstrated the significant diversity of polar lipids species across all polar lipid classes in *Chlorella* sp. cultures and that the polar lipid productivity, class, and species composition can be modulated in response to nutrient availability. It is envisaged that similar studies exploring culture condition effects (particularly growth stressors) on polar lipid composition will inform our fundamental understanding of how microalgae lipid metabolism functions under changing environmental conditions. These findings, coupled with necessary novel research concerning bioactivity analyses of polar lipid classes and species, will help to realise the potential of a microalgal-polar lipid production platform for various future biotechnology applications.

Supplementary Materials: The following are available online at http://www.mdpi.com/2218-1989/9/3/39/s1. Figure S1: Chemical structures of different polar lipid classes analysed in this study; Figure S2: MS/MS fragmentation spectra for confirmation of acyl groups in polar lipid species using positive ionisation; Table S1: Nutrient composition in HN and LN media used in *Chlorella* sp. batch cultures; Table S2: Diagnostic neutral loss or product ions in MS/MS spectra used to determine class for lipid species analysed using LC-ESI-MS/MS in positive ionisation mode; Table S3: SIMPER analyses of polar lipid species contributing to the dissimilarity between lipid samples taken from *Chlorella* sp. grown in LN media at day 4, day 9, and days 4 and 15 of batch cultures.

Author Contributions: Authors of this study contributed in the following areas: conceptualisation, D.A.W. and C.A.L.; methodology, all authors; formal analyses, D.A.W., P.A.R., S.K., K.T., M.J., G.A.T., and C.C.; original draft preparation review and editing, D.A.W. and C.A.L.; funding acquisition, D.A.W., C.A.L., and K.T.; supervision, D.A.W., C.A.L.

Funding: This research was funded by the Biotechnology and Biological Sciences Research Council (BBSRC), UK, grant number BB/K020617/3. 'Using flow cytometry and genomics to characterise and optimise microalgal-bacterial consortia cultivated on wastewater to produce biomass for Biofuel'.

Conflicts of Interest: The authors declare no conflict of interest.

References

1. Guschina, I.; Harwood, J. Algal lipids and effect of the environment on their biochemistry. In *Lipids in Aquatic Ecosystems*; Kainz, M., Brett, M.T., Arts, M.T., Eds.; Springer: New York, NY, USA, 2009; pp. 1–24. [CrossRef]
2. Murata, N.; Siegenthaler, P.-A. Lipids in Photosynthesis: An Overview. In *Lipids in Photosynthesis: Structure, Function and Genetics*; Paul-André, S., Norio, M., Eds.; Springer: Berlin/Heidelberg, Germany, 1998; Volume 6, pp. 1–20.
3. Van Meer, G.; Voelker, D.R.; Feigenson, G.W. Membrane lipids: Where they are and how they behave. *Nat. Rev. Mol. Cell Biol.* **2008**, *9*, 112–124. [CrossRef] [PubMed]
4. Hölzl, G.; Dörmann, P. Structure and function of glycoglycerolipids in plants and bacteria. *Prog. Lipid Res.* **2007**, *46*, 225–243. [CrossRef] [PubMed]
5. Benning, C. Biosynthesis and function of the sulfolipid sulfoquinovosyl diacylglycerol. *Annu. Rev. Plant. Physiol. Plant. Mol. Biol.* **1998**, *49*, 53–75. [CrossRef] [PubMed]
6. Wada, H.; Mizusawa, N. The role of phosphatidylglycerol in photosynthesis. In *Lipids in Photosynthesis: Essential and Regulatory Functions*; Wada, H., Murata, N., Eds.; Springer: Berlin/Heidelberg, Germany, 2009; pp. 243–263.
7. Boudière, L.; Michaud, M.; Petroutsos, D.; Rébeillé, F.; Falconet, D.; Bastien, O.; Roy, S.; Finazzi, G.; Rolland, N.; Jouhet, J.; et al. Glycerolipids in photosynthesis: Composition, synthesis and trafficking. *Biochim. Biophys. Acta (BBA)* **2014**, *1837*, 470–480. [CrossRef] [PubMed]
8. Latowski, D.; Åkerlund, H.-E.; Strzałka, K. Violaxanthin De-Epoxidase, the Xanthophyll Cycle Enzyme, Requires Lipid Inverted Hexagonal Structures for Its Activity. *Biochemistry* **2004**, *43*, 4417–4420. [CrossRef] [PubMed]
9. Harwood, J.L. Membrane lipids in algae. In *Lipids in Photosynthesis: Structure, Function and Genetics*; Siegenthaler, P.-A., Murata, N., Eds.; Kluwer Academic Publishers: Dordrecht, The Netherlands, 1998; pp. 53–64.

10. Alonso, D.L.; Belarbi, E.H.; Fernandez-Sevilla, J.M.; Rodriguez-Ruiz, J.; Grima, E.M. Acyl lipid composition variation related to culture age and nitrogen concentration in continuous culture of the microalga *Phaeodactylum Tricornutum*. *Phytochemistry* **2000**, *54*, 461–471. [CrossRef]

11. Fidalgo, J.P.; Cid, A.; Torres, E.; Sukenik, A.; Herrero, C. Effects of nitrogen source and growth phase on proximate biochemical composition, lipid classes and fatty acid profile of the marine microalga *Isochrysis galbana*. *Aquaculture* **1998**, *166*, 105–116. [CrossRef]

12. Guiheneuf, F.; Fouqueray, M.; Mimouni, V.; Ulmann, L.; Jacquette, B.; Tremblin, G. Effect of UV stress on the fatty acid and lipid class composition in two marine microalgae Pavlova lutheri (Pavlovophyceae) and Odontella aurita (Bacillariophyceae). *J. Appl. Phycol.* **2010**, *22*, 629–638. [CrossRef]

13. Guiheneuf, F.; Mimouni, V.; Ulmann, L.; Tremblin, G. Environmental factors affecting growth and omega 3 fatty acid composition in *Skeletonema costatum*. The influences of irradiance and carbon source. *Diatom Res.* **2008**, *23*, 93–103. [CrossRef]

14. Guiheneuf, F.; Mimouni, V.; Ulmann, L.; Tremblin, G. Combined effects of irradiance level and carbon source on fatty acid and lipid class composition in the microalga *Pavlova lutheri* commonly used in mariculture. *J. Exp. Mar. Biol. Ecol.* **2009**, *369*, 136–143. [CrossRef]

15. Liang, Y.; Beardall, J.; Heraud, P. Effects of nitrogen source and UV radiation on the growth, chlorophyll fluorescence and fatty acid composition of Phaeodactylum tricornutum and Chaetoceros muelleri (Bacillarlophyceae). *J. Photochem. Photobiol. B Biol.* **2006**, *82*, 161–172. [CrossRef] [PubMed]

16. Reitan, K.I.; Rainuzzo, J.R.; Olsen, Y. Effect of nutrient limitation on fatty-acid and lipid-content of marine microalgae. *J. Phycol.* **1994**, *30*, 972–979. [CrossRef]

17. Plouguerné, E.; da Gama, B.A.P.; Pereira, R.C.; Barreto-Bergter, E. Glycolipids from seaweeds and their potential biotechnological applications. *Front. Cell. Infect. Microbiol.* **2014**, *4*. [CrossRef] [PubMed]

18. Hossain, Z.; Kurihara, H.; Hosokawa, M.; Takahashi, K. Growth inhibition and induction of differentiation and apoptosis mediated by sodium butyrate in caco-2 cells with algal glycolipids. *In Vitro Cell. Dev. Biol. Anim.* **2005**, *41*, 154–159. [CrossRef] [PubMed]

19. Imbs, T.I.; Ermakova, S.P.; Fedoreyev, S.A.; Anastyuk, S.D.; Zvyagintseva, T.N. Isolation of Fucoxanthin and Highly Unsaturated Monogalactosyldiacylglycerol from Brown Alga Fucus evanescens C Agardh and In Vitro Investigation of Their Antitumor Activity. *Mar. Biotechnol.* **2013**, *15*, 606–612. [CrossRef] [PubMed]

20. Tsai, C.-J.; Sun Pan, B. Identification of Sulfoglycolipid Bioactivities and Characteristic Fatty Acids of Marine Macroalgae. *J. Agric. Food Chem.* **2012**, *60*, 8404–8410. [CrossRef] [PubMed]

21. Banskota, A.H.; Stefanova, R.; Sperker, S.; Lall, S.P.; Craigie, J.S.; Hafting, J.T.; Critchley, A.T. Polar lipids from the marine macroalga Palmaria palmata inhibit lipopolysaccharide-induced nitric oxide production in RAW264.7 macrophage cells. *Phytochemistry* **2014**, *101*, 101–108. [CrossRef] [PubMed]

22. Fang, Z.; Jeong, S.Y.; Jung, H.A.; Choi, J.S.; Min, B.S.; Woo, M.H. Capsofulvesins A-C, cholinesterase inhibitors from Capsosiphon fulvescens. *Chem. Pharm. Bull.* **2012**, *60*, 1351–1358. [CrossRef] [PubMed]

23. Cantillo-Ciau, Z.; Moo-Puc, R.; Quijano, L.; Freile-Pelegrín, Y. The Tropical Brown Alga Lobophora variegata: A Source of Antiprotozoal Compounds. *Mar. Drugs* **2010**, *8*, 1292–1304. [CrossRef] [PubMed]

24. Treyvaud Amiguet, V.; Jewell, L.E.; Mao, H.; Sharma, M.; Hudson, J.B.; Durst, T.; Allard, M.; Rochefort, G.; Arnason, J.T. Antibacterial properties of a glycolipid-rich extract and active principle from Nunavik collections of the macroalgae Fucus evanescens C. Agardh (Fucaceae). *Can. J. Microbiol.* **2011**, *57*, 745–749. [CrossRef] [PubMed]

25. Plouguerné, E.; Ioannou, E.; Georgantea, P.; Vagias, C.; Roussis, V.; Hellio, C.; Kraffe, E.; Stiger-Pouvreau, V. Anti-microfouling Activity of Lipidic Metabolites from the Invasive Brown Alga Sargassum muticum (Yendo) Fensholt. *Mar. Biotechnol.* **2010**, *12*, 52–61. [CrossRef] [PubMed]

26. El Baz, F.K.; El Baroty, G.S.; Abd El Baky, H.H.; Abd El-Salam, O.I.; Ibrahim, E.A. Structural characterization and Biological Activity of Sulfolipids from selected Marine Algae. *Grasas Aceites* **2013**, *64*. [CrossRef]

27. Al-Fadhli, A.; Wahidulla, S.; D'Souza, L. Glycolipids from the red alga Chondria armata (Kütz.) Okamura. *Glycobiology* **2006**, *16*, 902–915. [CrossRef] [PubMed]

28. Zhang, J.; Li, C.; Yu, G.; Guan, H. Total Synthesis and Structure-Activity Relationship of Glycoglycerolipids from Marine Organisms. *Mar. Drugs* **2014**, *12*, 3634–3659. [CrossRef] [PubMed]

29. Gunstone, F.D. *Phospholipid Technologies and Applications*; Woodhead Publishing: Sawston/Cambridge, UK, 2008.

30. Safi, C.; Zebib, B.; Merah, O.; Pontalier, P.-Y.; Vaca-Garcia, C. Morphology, composition, production, processing and applications of Chlorella vulgaris: A review. *Renew. Sustain. Energy Rev.* **2014**, *35*, 265–278. [CrossRef]

31. Guccione, A.; Biondi, N.; Sampietro, G.; Rodolfi, L.; Bassi, N.; Tredici, M.R. Chlorella for protein and biofuels: From strain selection to outdoor cultivation in a Green Wall Panel photobioreactor. *Biotechnol. Biofuels* **2014**, *7*, 1–12. [CrossRef] [PubMed]

32. Li, Y.; Chen, Y.-F.; Chen, P.; Min, M.; Zhou, W.; Martinez, B.; Zhu, J.; Ruan, R. Characterization of a microalga *Chlorella sp.* well adapted to highly concentrated municipal wastewater for nutrient removal and biodiesel production. *Bioresour. Technol.* **2011**, *102*, 5138–5144. [CrossRef] [PubMed]

33. Xu, H.; Miao, X.; Wu, Q. High quality biodiesel production from a microalga Chlorella protothecoides by heterotrophic growth in fermenters. *J. Biotechnol.* **2006**, *126*, 499–507. [CrossRef] [PubMed]

34. Converti, A.; Casazza, A.A.; Ortiz, E.Y.; Perego, P.; Del Borghi, M. Effect of temperature and nitrogen concentration on the growth and lipid content of Nannochloropsis oculata and Chlorella vulgaris for biodiesel production. *Chem. Eng. Process.* **2009**, *48*, 1146–1151. [CrossRef]

35. Liu, J.; Huang, J.; Sun, Z.; Zhong, Y.; Jiang, Y.; Chen, F. Differential lipid and fatty acid profiles of photoautotrophic and heterotrophic Chlorella zofingiensis: Assessment of algal oils for biodiesel production. *Bioresour. Technol.* **2011**, *102*, 106–110. [CrossRef] [PubMed]

36. Liang, Y.; Maeda, Y.; Yoshino, T.; Matsumoto, M.; Tanaka, T. Profiling of Polar Lipids in Marine Oleaginous Diatom Fistulifera solaris JPCC DA0580: Prediction of the Potential Mechanism for Eicosapentaenoic Acid-Incorporation into Triacylglycerol. *Mar. Drugs* **2014**, *12*, 3218–3230. [CrossRef] [PubMed]

37. Cutignano, A.; Luongo, E.; Nuzzo, G.; Pagano, D.; Manzo, E.; Sardo, A.; Fontana, A. Profiling of complex lipids in marine microalgae by UHPLC/tandem mass spectrometry. *Algal Res.* **2016**, *17*, 348–358. [CrossRef]

38. He, H.; Rodgers, R.P.; Marshall, A.G.; Hsu, C.S. Algae polar lipids characterized by online liquid chromatography coupled with hybrid linear quadrupole ion trap/fourier transform ion cyclotron resonance mass spectrometry. *Energy Fuels* **2011**, *25*, 4770–4775. [CrossRef]

39. Lu, N.; Wei, D.; Chen, F.; Yang, S.T. Lipidomic profiling and discovery of lipid biomarkers in snow alga Chlamydomonas nivalis under salt stress. *Eur. J. Lipid Sci. Technol.* **2012**, *114*, 253–265. [CrossRef]

40. Xu, J.L.; Chen, D.Y.; Yan, X.J.; Chen, J.J.; Zhou, C.X. Global characterization of the photosynthetic glycerolipids from a marine diatom Stephanodiscus sp by ultra performance liquid chromatography coupled with electrospray ionization-quadrupole-time of flight mass spectrometry. *Anal. Chim. Acta* **2010**, *663*, 60–68. [CrossRef] [PubMed]

41. Li, S.; Xu, J.; Jiang, Y.; Zhou, C.; Yu, X.; Zhong, Y.; Chen, J.; Yan, X. Lipidomic analysis can distinguish between two morphologically similar strains of Nannochloropsis oceanica. *J. Phycol.* **2015**, *51*, 264–276. [CrossRef] [PubMed]

42. Yan, X.; Chen, D.; Xu, J.; Zhou, C. Profiles of photosynthetic glycerolipids in three strains of Skeletonema determined by UPLC-Q-TOF-MS. *J. Appl. Phycol.* **2011**, *23*, 271–282. [CrossRef]

43. Vieler, A.; Wilhelm, C.; Goss, R.; Süß, R.; Schiller, J. The lipid composition of the unicellular green alga Chlamydomonas reinhardtii and the diatom Cyclotella meneghiniana investigated by MALDI-TOF MS and TLC. *Chem. Phys. Lipids* **2007**, *150*, 143–155. [CrossRef] [PubMed]

44. Brewer, P.G.; Riley, J.P. The automatic determination of nitrate in sea water. *Deep-Sea Res.* **1965**, *12*, 765–772. [CrossRef]

45. Zhang, J.Z.; Chi, J. Automated analysis of nanomolar concentrations of phosphate in natural waters, with liquid waveguide. *Environ. Sci. Technol.* **2002**, *36*, 1048–1053. [CrossRef] [PubMed]

46. Folch, J.; Lees, M.; Sloane Stanley, G.H. A simple method for the isolation and purification of total lipides from animal tissues. *J. Biol. Chem.* **1957**, *226*, 497–509. [PubMed]

47. Popendorf, K.J.; Fredricks, H.F.; Van Mooy, B.A.S. Molecular ion-independent quantification of polar glycerolipid classes in marine plankton using triple quadrupole MS. *Lipids* **2013**, *48*, 185–195. [CrossRef] [PubMed]

48. White, D.A.; Pagarette, A.; Rooks, P.; Ali, S.T. The effect of sodium bicarbonate supplementation on growth and biochemical composition of marine microalgae cultures. *J. Appl. Phycol.* **2013**, *25*, 153–165. [CrossRef]

49. Clarke, K.R.; Gorley, R.N.; Somerfied, P.J.; Warwick, R.M. *Changes in Marine Communities: An Approach to Statistical Analysis and Interpretation*, 3rd ed.; PRIMER-E: Plymouth, UK, 2014.

50. Clarke, K.R.; Somerfield, P.J.; Chapman, M.G. On resemblance measures for ecological studies, including taxonomic dissimilarities and a zero-adjusted Bray-Curtis coefficient for denuded assemblages. *J. Exp. Mar. Biol. Ecol.* **2006**, *330*, 55–80. [CrossRef]

51. Clarke, K.R.; Green, R.H. Statistical design and analysis for a biological effects study. *Mar. Ecol. Prog. Ser.* **1988**, *46*, 213–226. [CrossRef]

52. Clarke, K.R.; Ainsworth, M. A method of linking multivariate community structure to environmental variables. *Mar. Ecol. Prog. Ser.* **1993**, *92*, 205–219. [CrossRef]

53. Herndl, G.J.; Reinthaler, T.; Teira, E.; van Aken, H.; Veth, C.; Pernthaler, A.; Pernthaler, J. Contribution of Archaea to total prokaryotic production in the deep Atlantic Ocean. *Appl. Environ. Microbiol.* **2005**, *71*, 2303–2309. [CrossRef] [PubMed]

54. Da Costa, E.; Melo, T.; Moreira, A.S.P.; Alves, E.; Domingues, P.; Calado, R.; Abreu, M.H.; Domingues, M.R. Decoding bioactive polar lipid profile of the macroalgae Codium tomentosum from a sustainable IMTA system using a lipidomic approach. *Algal Res.* **2015**, *12*, 388–397. [CrossRef]

55. Da Costa, E.; Silva, J.; Mendonça, S.H.; Abreu, M.H.; Domingues, M.R. Lipidomic Approaches towards Deciphering Glycolipids from Microalgae as a Reservoir of Bioactive Lipids. *Mar. Drugs* **2016**, *14*, 101. [CrossRef] [PubMed]

56. Roessler, P. Environmental control of glycerolipid metabolism in microalgae: Commercial implications and future research directions. *J. Phycol.* **1990**, *26*, 393–399. [CrossRef]

57. Martin, G.J.O.; Hill, D.R.A.; Olmstead, I.L.D.; Bergamin, A.; Shears, M.J.; Dias, D.A.; Kentish, S.E.; Scales, P.J.; Botte, C.Y.; Callahan, D.L. Lipid profile remodeling in response to nitrogen deprivation in the microalgae *Chlorella sp.* (Trebouxiophyceae) and *Nannochloropsis sp.* (Eustigmatophyceae). *PLoS ONE* **2014**, *9*, e103389. [CrossRef] [PubMed]

58. Sung Ho, C.; Thompson, G.A. Properties of a fatty acyl hydrolase preferentially attacking monogalactosyldiacylglycerols in Dunaliella salina chloroplasts. *Biochim. Biophys. Acta* **1986**, *878*, 353–359. [CrossRef]

59. Li, X.; Moellering, E.R.; Liu, B.; Johnny, C.; Fedewa, M.; Sears, B.B.; Kuo, M.-H.; Benning, C. A Galactoglycerolipid Lipase Is Required for Triacylglycerol Accumulation and Survival Following Nitrogen Deprivation in Chlamydomonas reinhardtii. *Plant. Cell* **2012**, *24*, 4670–4686. [CrossRef] [PubMed]

60. Cai, T.; Park, S.Y.; Li, Y. Nutrient recovery from wastewater streams by microalgae: Status and prospects. *Renew. Sustain. Energy Rev.* **2013**, *19*, 360–369. [CrossRef]

61. Dormann, P.; Holzl, P. The role of glycolipids in photosynthesis. In *Lipids in Photosynthesis: Essential and Regulatory Functions*; Wada, H., Murata, N., Eds.; Springer Science+ Business Media B.V.: Berlin/Heidelberg, Germany, 2009; pp. 265–282.

62. Lu, N.; Wei, D.; Chen, F.; Yang, S.-T. Lipidomic profiling reveals lipid regulation in the snow alga *Chlamydomonas nivalis* in response to nitrate or phosphate deprivation. *Process. Biochem.* **2013**, *48*, 605–613. [CrossRef]

63. Jouhet, J.; Maréchal, E.; Block, M.A. Glycerolipid transfer for the building of membranes in plant cells. *Prog. Lipid Res.* **2007**, *46*, 37–55. [CrossRef] [PubMed]

64. Shimojima, M. Biosynthesis and functions of the plant sulfolipid. *Prog. Lipid Res.* **2011**, *50*, 234–239. [CrossRef] [PubMed]

65. Sato, N. Roles of the acidic lipids sulfoquinovosyl diacylglycerol and phosphatidylglycerol in photosynthesis: Their specificity and evolution. *J. Plant. Res.* **2004**, *117*, 495–505. [CrossRef] [PubMed]

66. Kates, M. Glycolipids of higher plants, algae, yeasts and fungi. In *Handb Lipid Res*; Kates, M., Ed.; Springer: Boston, MA, USA, 1990; Volume 6, pp. 235–320.

67. Suen, Y.; Hubbard, J.S.; Holzer, G.; Tornabene, T.G. Total lipid production of the green alga *Nannochloropsis sp.* QII under different nitrogen regimes. *J. Phycol.* **1987**, *23*, 289–296. [CrossRef]

Communication

The Bacterial Phytoene Desaturase-Encoding Gene (*CRTI*) is an Efficient Selectable Marker for the Genetic Transformation of Eukaryotic Microalgae

Ana Molina-Márquez [1], Marta Vila [1,2], Javier Vigara [1], Ana Borrero [1] and Rosa León [1,*]

[1] Laboratory of Biochemistry, Faculty of Experimental Sciences, Marine International Campus of Excellence (CEIMAR), University of Huelva, 2110 Huelva, Spain; ana.molina@dqcm.uhu.es (A.M.-M.); mvila@phycogenetics.com (M.V.); vigara@uhu.es (J.V.); anika_b92@hotmail.com (A.B.)

[2] PhycoGenetics SL, C/Joan Miró N°6, Aljaraque, 21110 Huelva, Spain

* Correspondence: rleon@uhu.es; Tel.: +34-959219951

Received: 7 February 2019; Accepted: 5 March 2019; Published: 12 March 2019

Abstract: Genetic manipulation shows great promise to further boost the productivity of microalgae-based compounds. However, selection of microalgal transformants depends mainly on the use of antibiotics, which have raised concerns about their potential impacts on human health and the environment. We propose the use of a synthetic phytoene desaturase-encoding gene (*CRTIop*) as a selectable marker and the bleaching herbicide norflurazon as a selective agent for the genetic transformation of microalgae. Bacterial phytoene desaturase (CRTI), which, unlike plant and algae phytoene desaturase (PDS), is not sensitive to norflurazon, catalyzes the conversion of the colorless carotenoid phytoene into lycopene. Although the expression of *CRTI* has been described to increase the carotenoid content in plant cells, its use as a selectable marker has never been testedin algae or in plants. In this study, a version of the *CRTI* gene adapted to the codon usage of *Chlamydomonas* has been synthesized, and its suitability to be used as selectable marker has been shown. The microalgae were transformed by the glass bead agitation method and selected in the presence of norflurazon. Average transformation efficiencies of 550 colonies μg^{-1} DNA were obtained. All the transformants tested had incorporated the *CRTIop* gene in their genomes and were able to synthesize colored carotenoids.

Keywords: microalgae; *Chlamydomonas reinhardtii*; genetic transformation; carotenoid; *CRTI*; phytoene desaturase

1. Introduction

Microalgae have attracted considerable interest for the production of a wide range of compounds of applied interest due to their easy growth, their ability to fix atmospheric CO_2, and the valuable metabolites that some species can produce, which include pigments, food supplements, vitamins, antioxidants, polysaccharides, lipids, and other bioactive products [1–3]. Moreover, in the last years, there has been an increasing interest in microalgae as a potential source of biofuels [4,5]. However, despite the expectations generated, the production of biofuels and other useful compounds from microalgae will not be economically feasible unless the cost of microalgae cultivation and harvesting is lowered and the productivity increased. Genetic manipulation and synthetic biology show great promise to further boost the productivity of microalgae-based compounds [6–9]. Significant advances have been achieved in the development of molecular tools for genetic manipulation of microalgae. However, important challenges remain. One important issue is the fact that the selection of microalgal transformants depends mainly on the use of antibiotics as selective agents [10]. Antibiotic resistance genes continue to be the most commonly used selectable markers for the genetic manipulation of algae and plants. However, the risk of horizontal gene transfer has raised concerns about their potential impacts on human health and the environment, and has encouraged the search for new non-antibiotic-based selection procedures [11].

Herbicide resistance genes are a good alternative for the selection of genetically modified plant cells. Examples of this are the glyphosate aminotransferase [12] or the acetolacetate synthase genes [13], which confer resistance to glyphosate and sulfomturon methyl herbicides, respectively, and have been used as reporter genes in the transformation of the unicellular chlorophyte *Chlamydomonas reinhardtii* and other microalgae, such as *Porphyridium* sp. [14] or *Parietochloris incisa* [15]. An interesting herbicide-based selective strategy is the use of mutated versions of the phytoene desaturase gene (*PDS*) resistant to bleaching herbicides such as norflurazon [16,17]. Phytoene desaturase (PDS) catalyzes the conversion of the colorless phytoene into ζ-carotene, which is converted to lycopene by ζ-carotene desaturase (ZDS) and carotene isomerase (CRTISO). PDS is a membrane-associated protein that uses the flavin adenine dinucleotide (FAD) as a redox cofactor, through which electrons are transferred to the plastoquinone, thereby connecting the desaturation of carotenoids with the photosynthetic electronic transport chain [1]. Treatment with norflurazon causes inhibition of phytoene desaturase by competition with its cofactors, resulting in suppression of carotenoid synthesis and cellular whitening [18]. By modifying key amino acids in the FAD binding domain, some authors have obtained mutated norflurazon-resistant versions of PDS and setup genetic selective procedures based on norflurazon as a selective agent [17,19].

In bacteria and fungi, however, the three reactions that convert phytoene into lycopene are carried out by a single enzyme, bacterial phytoene desaturase (CRTI), which presents a low degree of homology with the corresponding plant phytoene desaturase [20]. The *CRTI* gene seems to have emerged independently in evolution and, unlike plant and algae PDS, is not sensitive to norflurazon [21]. The *CRTI* gene was first identified and functionally analyzed by Misawa and coworkers [22] from the soil bacteria *Erwinia uredovora*, currently renamed as *Pantotea ananatis*. The pioneering work of Sandman's group showed that expression of this bacterial *CRTI* gene in tobacco plants enhanced the production of β-carotene. Furthermore, they observed higher resistance to the herbicide norflurazon in transgenic *CRTI*-expressing plants [23]. Several subsequent studies reported the expression of the bacterial carotenoid gene cluster from *Pantotea*, including the *CRTI* gene, to increase the carotenoid content in higher plants such as rice [24], tomato [25], or potato [26]. However, the possible use of *CRTI* as a selectable marker gene has never been investigated in algae or in plants. Poor expression of bacterial *CRTI* in microalgae, beside potential silencing and lack of stability of the transgene in algae, has withdrawn its use as a selectable marker. In the present study, we have synthesized a codon-adapted version of the bacterial *CRTI* gene and showed its suitability to be used as selectable marker gene in the genetic transformation of the model microalga *Chlamydomonas reinhardtii*.

2. Results and Discussion

2.1. Construction of Plasmid pSI06PLK-CRTIop

A synthetic *CRTI* gene with the codon usage adapted to *C. reinhardtii* was designed and synthesized by Genescript Co (Piscataway, NJ, USA). The mean difference between codon usage of the wild type *CRTI* gene and the genome of *C. reinhardtii* was 24%, as calculated using the graphical codon usage analyser v. 2.0 (http://gcua.schoedl.de/index.html). This difference was reduced to only 13% for the synthetic codon-optimized version. The *CRTIop* gene was fused to a DNA fragment which encoded the chloroplast transit peptide of the RuBisCo small subunit, and this tp*CRTIop* fusion product was cloned between the *BstBI*/*BamHI* restriction sites of the p106PLK plasmid, described in Section 3.2, generating the expression cassette outlined in Figure 1.

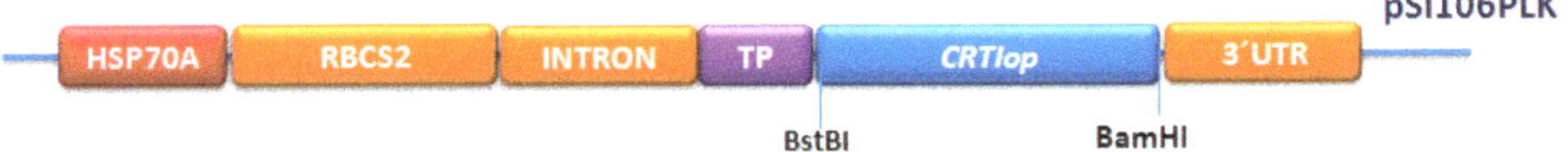

Figure 1. Expression cassette of the pSI106PLK-tp*CRTIop* plasmid.Abbreviations: HSP70A, heat shock protein 70A promoter; RBCS2, ribulose 1,5-biphosphate carboxylase small subunit promoter; TP, chloroplastic transit peptide; *CRTIop*, codon-optimized bacterial phytoene desaturase; 3′UTR terRBCS2, ribulose 1,5-biphosphate carboxylase small subunit terminator region.

2.2. Sensitivity of the Chlorophyte Microalgae to the Bleaching Herbicide Norflurazon

The ability of *CRTI* to act as a selectable marker for microalgae transformation is based on the sensitivity of the target microalgae to herbicides which inhibit phytoene desaturase (PDS). We have tested the sensitivity of the model chlorophyte *Chlamydomonas reinhardtii* to the bleaching herbicide norflurazon by culturing it with growing concentrations of the herbicide and determining the lethal doses (Figure 2). Samples of control non-transformed *C. reinhardtii* cultures were harvested at the exponential phase of growth and 100-fold concentrated by centrifugation, and 10 µL drops of the concentrated suspension were spotted on multi-well plates with growing concentrations (0–25 µg mL^{-1}) of the herbicide norflurazon. Concentration of the culture was done to mimic the conditions in which the transformation experiments are done (see Section 3.3). The minimal inhibitory norflurazon concentration for *C. reinhardtii* was 1.5 µg mL^{-1} after 15 days of incubation in the presence of norflurazon, as can be observed in Figure 2B. It is interesting to note that at shorter times, herbicide concentrations as low as 0.5 µg mL^{-1} seemed to inhibit *C. reinhardtii* growth. However, the microalgaeare able to survive and grow after an adaptation period. To establish the real lethal dose and avoid false negatives, in the subsequent transformation experiments it is necessary to follow the inhibitory effect of norflurazon for a long time period. All *C. reinardtii* transformants were selected at norflurazon concentrations ≥1.5 µg mL^{-1}.

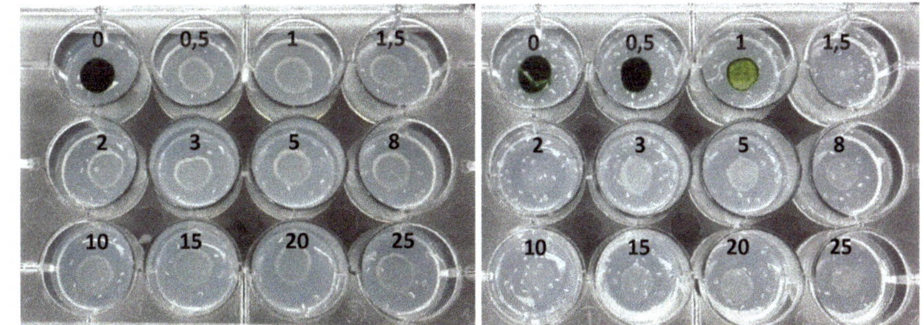

Figure 2. Norflurazon sensitivity test for the clorophyceae microalga *Chlamydomonas reinhardtii*. Drops (10 µL) of a 100-fold concentrated *C. reinhardtii* control untransformed culture were spotted on agar-solidified Tris-acetate phosphate (TAP) culture medium with increasing concentrations of the bleaching herbicide norflurazon, and their growth was evaluated 4 (**A**) and 15 (**B**) days after inoculation.

Furthermore, sensitivity experiments were carried out for other microalgal species, such as the freshwater trebouxiophyceae *Chlorella sorokiniana*, the marine prasinophyceae *Tetraselmis suecica*, and the halophilic chlorophyceae *Dunaliella salina* and *Dunaliella bardawill*. These studies revealed that the inhibitory norflurazon concentration dose was between 0.5 and 2 µg mL^{-1} for all the tested species, and that norflurazon can be an adequate selective agent for many different microalgal species, including marine microalgae, for which traditional selective antibiotic agents are usually inefficient due to interference with the saline concentration of the medium (Figure S1).

2.3. Transformation of Chlamydomonas reinhardtii with the Plasmid pSI106PLK-tpCRTIop and Selection of Norflurazon-Resistant Norf^R-Chlamydomonas Transformants

Chlamydomonas reinhardtii cells were transformed by the glass beads agitation method with the plasmid pSI106PLK-tp*CRTIop* and selected in Tris-acetate phosphate (TAP) medium with norflurazon (1.5 µg mL^{-1}). Transformation efficiencies of 550 colonies µg^{-1} DNA were obtained (Figure 3A). This transformation efficiency is of the same order as that usually obtained for transformations with paromomycin as the selective agent [27]. A randomly selected group of the obtained transformants were cultured in 2 mL of liquid TAP medium with norflurazon for 48h, and their cellular density was then adjusted to the same value. Drops of each transformed culture were spotted on TAP agar plates with a higher concentration of norflurazon (4 µg mL^{-1}), and those which grew at this concentration of the herbicide were further checked by PCR using the specific primers CRTIopFor (CAGCCGCGCCGTGTTCAAAGAG) and CRTIpoRev (CAGCAGGTCGCGGTAGGTGTG), as illustrated in Figure 3B. It is necessary to consider that nuclear transformations of microalgae take place by heterologous recombination. This means that the *CRTIop* gene is randomly inserted into the algal genome, and its expression level and stability largely depend on the insertion point. The two-round selection strategy with increasing concentrations of norflurazon allows the selection of the transformants with the highest resistance to the herbicide. Insertion of the *CRTIop* into the *C. reinhardtii* nuclear genome was confirmed in all the transformants checked (Figure 3C).

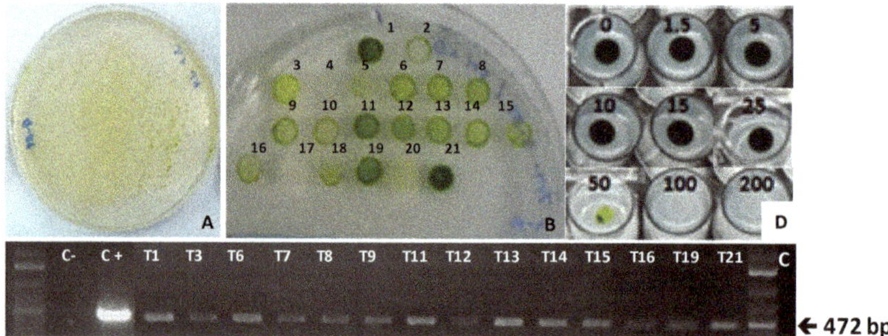

Figure 3. Molecular and phenotypic analysis of Norf^R-*Chlamydomonas* transformants. *Chlamydomonas* transformants selected in TAP with 1.5 µg mL^{-1} of norflurazon (**A**) were cultured in TAP agar plates with 4 µg mL^{-1} of norflurazon for 10 days (**B**). A band of the expected size (472 bp) corresponding to the *CRTIop* amplicon was shown in all the transformants tested (**C**). A norflurazon sensitivity test for the selected transformant T21 (**D**) was carried out as described in the Figure 2 legend with the indicated concentrations of norflurazon (from 0 to 200 µg mL^{-1}).

The transformants which grew more vigorously were subjected to a norflurazon sensitivity test as described in Figure 1. *C. reinhardtii* transformed with *CRTIop* showed a 33-fold increase in their tolerance to norflurazon, with an inhibitory dose of 50 µg mL^{-1}, equivalent to 160 µM, as shown for transformant T21 (Figure 3D) and for the other Norf^R-transformants selected (Figure S2). The synthetic tp*CRTIop* gene allowed for norflurazon-based selection of transformants with no background of spontaneous herbicide-resistant clones. The level of resistance to norflurazon acquired by the *C. reinhardtii* transformed with *CRTIop* is of the same order as the resistance reported by Suarez et al. [28] or Bruggeman and coworkers [12], who found 30- and 40-fold increases, respectively, in the tolerance to norflurazon for transgenic *Chlamydomonas* harbouring modified versions of its own *PDS* gene. Similar strategies using mutated PDS versions and norflurazon have been successfully used for the selection of other transformed microalgae species, such as *Haematococcuspluvialis* [29,30], *Chlorella zofingiensis* [31,32], or *Isocrhrysis* [33].

2.4. Carotenoid Composition of Norflurazon-Resistant Norf[R]-Chlamydomonas Transformants

The phenotypic characteristics of the selected transformants were further studied by chromatographic analysis. First, we compared the phytoene contents in the Norf[R]-transformants with that of the control cells grown in the presence and in the absence of norflurazon (Table 1). As expected, in *C. reinhardtii* cultures grown without the herbicide, only trace levels of phytoene were found. After 24h of growth in the presence of norflurazon (1.5 μg mL^{-1}), the phytoene contents in control untransformed cells reached 4.4 mg g^{-1} DW. However, the phytoene intracellular concentrations of Norf[R]-transformants grown with and without norflurazon ranged between 1.5 mg g^{-1} DW, for transformant T21, to 2.1 mg g^{-1} DW, for transformant T11. This is far from the intracellular level of phytoene in the control, which is between two and three times higher. This shows that the *CRTIop* gene is correctly expressed in the transformed microalgae and is able to convert phytoene into the downstream carotenoids. However, the presence of certain contents of phytoene in all the transformants tested indicates that the foreign CRTI is less efficient than the endogenous PDS in the absence of the herbicide.

Table 1. Concentration of the colorless carotenoid phytoene in control cells (C2) and Norf[R]-transformants (T), grown for 24h with norflurazon. Control cells non-treated with norflurazon (C1) have also been included as a reference. Values are the average of three biological replications. Standard deviation is indicated.

Strain	C1	C2	T1	T11	T19	T21
Norflurazon (μg mL^{-1})	-	1.5	1.5	1.5	1.5	1.5
Phytoene (μg mL^{-1})	0.1 ± 0.5	4.4 ± 0.4	1.8 ± 0.1	2.1 ± 0.2	2 ± 0.2	1.5 ± 0.1

A representative Norf[R]-transformant (T21) and the control untransformed strain were grown in TAP medium with norflurazon (1.5 μg mL^{-1}) for a complete analysis of their carotenoid profiles along the time. Samples were withdrawn every 24h, and pigments were extracted and analyzed (Figure 4). Typical chromatograms of the pigment extracts from Control (C2) and Norf[R]-transformant cells registered at 288 and 450 nm, and are shown in the Supplementary Material (Figure S3).

In control parental *Chlamydomonas* treated with norflurazon, there was a reduction of all the colored carotenoids, excepting zeaxanthin, and an important accumulation of phytoene, due to the inhibitory effect of norflurazon. By contrast, in transformant cells treated with norflurazon, the reduction of the colored carotenoids and the accumulation of phytoene was much lower. After 24 h of incubation with norflurazon, the content of lutein in the Norf[R]-transfomant was 50% higher than in the controls, while the content in β-carotene and violaxanthin was 2.5 times the level of control untransformed cells. These differences were even more acute for longer periods of incubation with norflurazon. The level of phytoene, on the contrary, was between three and five times higher in the norflurazon-treated control cells, reaching intracelullar levels of 4.4 mg g^{-1} DW at 24 h and 6.7 mg g^{-1} DW at 48h. This confirms that the *CRTIop* gene was working in the transformant, allowing the conversion of phytoene into lycopene and the subsequent carotenoids in the presence of norflurazon, which inhibits the endogenous PDS. Zeaxanthin was the only colored carotenoid which increased in the control untransformed cells, indicating a higher level of fotooxidative stress in these cells. This induces the xantophyll cycle, which catalyzes the conversion of violaxanthin into zeaxanthin. In the Norf[R]-transfomants, the levels of zeaxanthin were inappreciable during the first 48 h. Only after 72 h of growth, when cultures started to be nutrient-limited, there was certain synthesis of this xanthophyll.

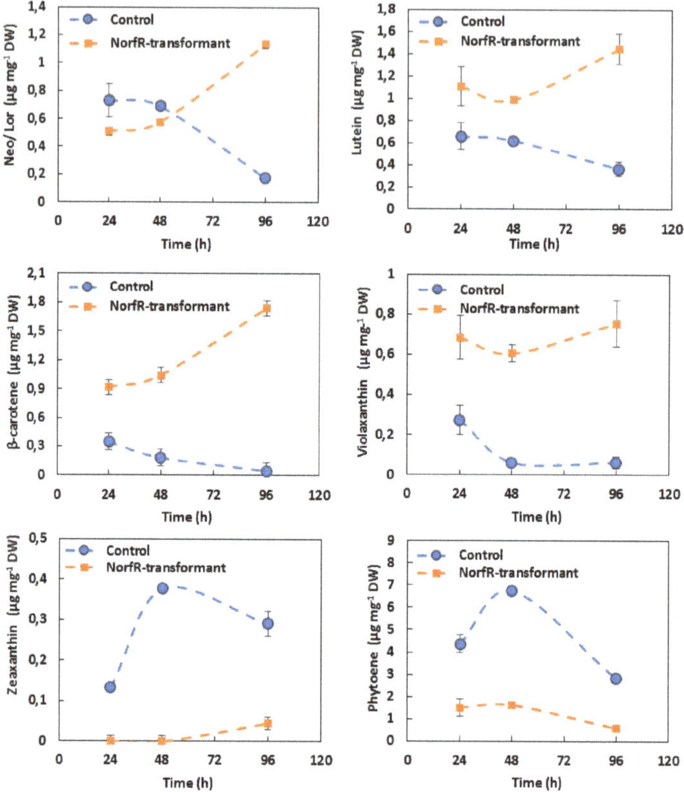

Figure 4. Time-course evolution of the main carotenoid pigments in *C. reinhardtii* control (●) and Norf[R]-transformant (■) cells incubated with norflurazon (1.5 µg mL^{-1}). Values are the average of three biological replicates and bars indicate standard deviation. Neoxanthin and Loroxanthin (Neo/Lor) are expressed as unique values since they are not resolved in the analytical conditions used.

In the presence of norflurazon, the growth of the parental strain was severely affected. Meanwhile, the growth rate of the Norf[R]-transformantswas similar with and without norflurazon, and was slightly higher than the growth rate of the control wild type without herbicide (Figure S4A,B). The carotenoid content of the Norf[R]-transformants grown without herbicide was also checked and resulted to be very similar to that of control cells (Figure S4C,D). This means that the transformants selected in the presence of norflurazon with *CRTI* as a marker gene can grow at a normal rate and have practically normal contents of carotenoids [19,20]. It could be expected that the expression of an exogenous phytoene desaturase caused an increase in the contents of carotenoids. The fact that the *CRTIop* transformants studied had intracellular levels similar to that of the control cells in the absence of norflurazon could be due to a limitation in the supply of precursors from the previous step catalyzed by the Phytoene synthase (PSY). However, further studies should be done to confirm this issue.

PDS is the second step in carotenoid biosynthesis and an important regulatory point of the pathway [19,20]. The inhibitory effect of norflurazon is well known and has been widely used to study the physiological consequences resulting from the lack of carotenoids in higher plants and microalgae [34]. We have corroborated that blocking PDS activity by chemical inhibition with norflurazon impedes the formation of downstream carotenoids in *Chlamydomonas* (Figure 4), which is in agreement with what Nigoyi's group observed by mutagenesis-induced PDS inactivation [34]. Furthermore, we demonstrated that the expression of the foreign *CRTIop*, which is not affected by the herbicide, allows bypass, at least partially, of the norflurazon-blocked step and enables the synthesis of colored carotenoids. Similar conclusions were reported by Liu and coworkers [35] or Steinbrenner and Sandmann [29], who used a modified *PDS* gene as a selectable marker for the genetic transformation of *Chlorella zofingiensis* and as *H. pluvialis*, respectively, and found that the transformants had the same or even higher carotenoid content than the untransformed controls.

3. Materials and Methods

3.1. Strains and Culture Conditions

Chlamydomonas reinhardtii 704 strain (Cw15, Arg7, mt+) was kindly donated by Dr. R. Loppes and cultured photomixotrophically in liquid or agar-solidified Tris-acetate phosphate (TAP) medium [36]. *Tetraselmissuecica*, kindly provided by IFAPA-Aguas del Pino station (Huelva, Spain), was cultured in F/2 medium in filtered sea water at pH 8, as reported by Guillard and Ryther [37]. *Dunaliella salina* (CCAP 19/18) was obtained from the culture collection of algae and protozoa (Scotland, UK) and grown in the culture medium described by Johnson and coworkers [38]. All were grown in a culture chamber at 25 °C under continuous white light irradiation (50 µE m^{-2} s^{-1} PAR).

3.2. Microalgal Expression of pSI106PLK Plasmid

Plasmid pSI106PLK is a renewed version of plasmid pSI104PLK [39]. It contains an expression cassette in which a multiple cloning site is preceded by the strong chimeric fusion promoter *HSP70A:RBCS2*, designed by Sizova [40], and the first intron of the *RBCS2* gene, and is terminated by the 3′ untranslated region of *RBCS2* (Figure 1).

3.3. Chlamydomonas Nuclear Transformation

Nuclear transformation of *C. reinhardtii* was carried out using the glass beads method [41] with minor modifications. *Chlamydomonas* cultures were grown as described in Section 3.1. to a cell density of 5 × 10^6 cells mL^{-1}, and resuspended to get a 100-fold concentrated cell suspension. 0.3 g of sterile glass beads (0.4–0.6mm Ø), were added to 0.6 mL of concentrated cell suspension with 0.2 mL of 20% PEG (MW8000) and about 1 µg of the desired plasmid (pSI106PLK-tpCRTIop). Negative controls, done in the same conditions with the empty plasmid (pSI106PLK), were included in all the transformation reactions. This mixture was agitated for 10s, resuspended in fresh TAP medium, and spread onto the selective solid medium with the indicated concentration of norflurazon. Transformed colonies were visible after 4 or 5 days.

3.4. Determination of Carotenoids

Samples from *C. reinhardtii* cultures, grown in TAP liquid medium supplemented with 1.5 µg mL of norflurazon in the same conditions described in Section 3.1., were used for the extraction of carotenoids with methanol as described by Linchtehaler [42]. The chromatographic analysis was performed in a Merck Hitachi HPLC equipped with a diode array detector as described by Young and coworkers [28] using an RP-18 column, a flow rate of 1mL min^{-1}, and a final injection volume of 100 µL. Two mobile phases were used: Solvent A (ethyl acetate 100%) and solvent B (acetonitrile:H$_2$O; 9:1 *v/v*). The gradients applied were: 0–16 min 0–60% A; 16–30 min 60% A; and 30–35 min 100% A. Standards were supplied by DHI (Hoersholm, Denmark). All experiments were done in triplicate, and the average values and standard deviation are represented. The significant differences have been analysed by a t-student test with a confidence level of 95%.

3.5. Dry Weight Determination

Dry weight was determined by filtering an exact volume of microalgae culture (30 mL) on pre-tared glass-fiber filters (GF/F Whatman). The filter was washed with a solution of ammonium formate (0.5 M) to remove salts and dried at 100 °C for 24 h. The dried filters were weighed in an analytical balance and the dry weight calculated by the difference. Values are the average of three measurements.

3.6. Herbicide Sensitivity Test

Norflurazon sensitivity was assayed on multi-well plates with the corresponding agar-solidified medium supplemented with the indicated concentrations of the herbicide.

4. Conclusions

The use of a synthetic codon-adapted *CRTIop* gene fused to a chloroplastic transit peptide as a selectable marker for the genetic transformation of *Chlamydomonas reinhardtii* has been shown to be a reliable and efficient approach for the selection of transformants, contributing to increase the non-antibiotic-dependent markers available for microalgae and plant cells. *C. reinhardtii* transformants selected on norflurazon have been shown to have the *CRTIop* gene correctly inserted into their genomes to acquire a 33-fold increased resistance to norflurazon and be stable for long periods of time. This is a good alternative to genetic markers based on resistance to antibiotics, which can be very useful to establish genetic transformations systems for new microalgal species which are recalcitrant to inhibition with traditional antibiotics, as usually happens with marine and halophyllic microalgae. It can also be an interesting alternative for the selection of higher plants transformants without using antibiotics.

Supplementary Materials: The following are available online at http://www.mdpi.com/2218-1989/9/3/49/s1. Figure S1: Norflurazon sensitivity test for several microalgae species. Drops (10 µL) of concentrated cultures of the microalgae *Tetraselmis suecica*, *Dunaliella salina*, *Dunaliella bardawill* and *Chlorella sorokiniana* were spotted on agar-solidified culture medium with increasing concentrations of the bleaching herbicide norflurazon (from 0 to 5 µg mL^{-1}) and their growth was evaluated 10 days after inoculation. The culture media used was that described in Materials and Methods for each species, excepting that the concentration of NaCl was reduced to half the salinity of the seawater for *T. suecica* and to 0.5M for *D. salina* and *D. bardawill*. Figure S2: Norflurazon sensitivity test for different Nor^R-transformants of the microalga *Chlamydomonas reinhardtii*. Drops (10 µL) of concentrated cultures of the transformants T1, T11 and T19 obtained as described in the legend of Figure 2, were spotted on agar-solidified TAP culture medium with increasing concentrations of the bleaching herbicide norflurazon (from 0 to 200 µg mL^{-1}) and their growth was evaluated 10 days after inoculation. Figure S3: Chromatographic analysis of pigments extracts from control and Norf^R-*Chlamydomonas* transfomants grown in the presence of norflurazon (1.5 µg mL^{-1}). Conditions for chromatographic separation are described in Materials and Methods. Peaks were identified as: neoxanthin/loroxanthin (1), violaxanthin (2), antheraxanthin (3), lutein (4), zeaxanthin (5), chlorophyll b (6), chlorophyll a (7) and β-carotene (8) and phytoene (9). Figure S4: Growth curves (A,B) and carotenoid content (C,D) of untransformed (Control) and transformed (Norf^R) strains of *Chlamydomonas reinhardtii* cultured without and with increasing concentrations of the herbicide norflurazon. Cultures of the control untransformed cells and of the Norf^R-transformant T21 were harvested at the beginning of exponential phase of growth and resuspended in fresh TAP medium without norflurazon or supplemented with 1.5, 8 and 20 µg mL^{-1} of norflurazon. Dry weight (DW) was determined every 24h. In addition, the carotenoid content of control and transformed (Norf^R) cells grown for 24h without or with 1.5 µg mL^{-1} of norflurazon were also determined. Neoxanthin and loroxanthin (Neo/Lor) are expressed a unique value since they are not resolved in the analytical conditions used. All the values are the average of three biological replicates and bars represent standard deviation.

Author Contributions: All authors have contributed to this research and agree to authorship, and submit this manuscript for its revision and publication. Conceptualization, R.L. and A.M.-M.; methodology, A.M.-M., A.B., and M.V.; software, A.M.-M.; investigation, A.M.-M. and A.B.; data curation, J.V. and M.V.; Writing—Original Draft preparation, R.L.; Writing—Review Andediting, A.M.-M., M.V., J.V., and R.L.; supervision, R.L. and J.V.; project administration, R.L. and J.V.; funding acquisition, R.L. and J.V.

Funding: This research was funded by the Spanish Agencia Estatal de Investigación [AGL2016-74866-C32R-AEI/FEDER] and the EU-INTERREG Program [INTERREG VA POCTEP-055 ALGARED_PLUS_5E].

Conflicts of Interest: The authors declare no conflict of interest.

References

1. Varela, J.C.; Pereira, H.; Vila, M.; León, R. Production of carotenoids by microalgae: Achievements and challenges. *Photosynth. Res.* **2015**, *125*, 423–436. [CrossRef] [PubMed]
2. Valverde, F.; Romero-Campero, F.J.; León, R.; Guerrero, M.G.; Serrano, A. New challenges in microalgae biotechnology. *Eur. J. Protistol.* **2016**, *55*, 95–101. [CrossRef]
3. Khan, M.I.; Shin, J.H.; Kim, J.D. The promising future of microalgae: Current status, challenges, and optimization of a sustainable and renewable industry for biofuels, feed, and other products. *Microb. Cell Factor.* **2018**, *17*, 36. [CrossRef] [PubMed]
4. Schulze, P.S.C.; Guerra, R.; Pereira, H.; Schüler, L.M.; Varela, J.C.S. Flashing LEDs for Microalgal Production. *Trends Biotechnol.* **2017**, *35*, 1088–1101. [CrossRef] [PubMed]
5. Mallick, N.; Bagchi, S.K.; Koley, S.; Singh, A.K. Progress and Challenges in Microalgal Biodiesel Production. *Front. Microbiol.* **2016**, *7*, 1019. [CrossRef]
6. Rengel, R.; Smith, R.T.; Haslam, R.P.; Sayanova, O.; Vila, M.; León, R. Overexpression of acetyl-CoA synthetase (ACS) enhances the biosynthesis of neutral lipids and starch in the green microalga Chlamydomonas reinhardtii. *Algal Res.* **2018**, *31*, 183–193. [CrossRef]
7. Bellou, S.; Triantaphyllidou, I.-E.; Aggeli, D.; Elazzazy, A.M.; Baeshen, M.N.; Aggelis, G. Microbial oils as food additives: Recent approaches for improving microbial oil production and its polyunsaturated fatty acid content. *Curr. Opin. Biotechnol.* **2016**, *37*, 24–35. [CrossRef]
8. Gimpel, J.A.; Henríquez, V.; Mayfield, S.P. In Metabolic Engineering of Eukaryotic Microalgae: Potential and Challenges Come with Great Diversity. *Front. Microbiol.* **2015**, *6*, 1376. [CrossRef]
9. Scranton, M.A.; Ostrand, J.T.; Fields, F.J.; Mayfield, S.P. *Chlamydomonas* as a model for biofuels and bio-products production. *Plant J.* **2015**, *82*, 523–531. [CrossRef]
10. León, R.; Fernández, E. Nuclear Transformation of Eukaryotic Microalgae. In *Transgenic Microalgae as Green Cell Factories*; Springer: New York, NY, USA, 2007; Volume 616, pp. 1–11.
11. Nielsen, K.M.; Bøhn, T.; Townsend, J.P. Detecting rare gene transfer events in bacterial populations. *Front. Microbiol.* **2014**, *4*, 415. [CrossRef]
12. Brueggeman, A.J.; Kuehler, D.; Weeks, D.P. Evaluation of three herbicide resistance genes for use in genetic transformations and for potential crop protection in algae production. *Plant Biotechnol. J.* **2014**, *12*, 894–902. [CrossRef]
13. Kovar, J.L.; Zhang, J.; Funke, R.P.; Weeks, D.P. Molecular analysis of the acetolactate synthase gene of Chlamydomonas reinhardtii and development of a genetically engineered gene as a dominant selectable marker for genetic transformation. *Plant J.* **2002**, *29*, 109–117. [CrossRef]
14. Lapidot, M.; Raveh, D.; Sivan, A.; Arad, S.M.; Shapira, M. Stable chloroplast transformation of the unicellular red alga Porphyridium species. *Plant Physiol.* **2002**, *129*, 7–12. [CrossRef] [PubMed]
15. Grundman, O.; Khozin-Goldberg, I.; Raveh, D.; Cohen, Z.; Vyazmensky, M.; Boussiba, S.; Shapira, M. Cloning, mutagenesis, and characterization of the microalga *Parietochloris incisa* acetohydroxyacid synthase, and its possible use as an endogenous selection marker. *Biotechnol. Bioeng.* **2012**, *109*, 2340–2348. [CrossRef]
16. Michel, A.; Arias, R.S.; Scheffler, B.E.; Duke, S.O.; Netherland, M.; Dayan, F.E. Somatic mutation-mediated evolution of herbicide resistance in the nonindigenous invasive plant hydrilla (*Hydrilla verticillata*). *Mol. Ecol.* **2004**, *13*, 3229–3237. [CrossRef] [PubMed]
17. Arias, R.S.; Dayan, F.E.; Michel, A.; Howell, J.; Scheffler, B.E. Characterization of a higher plant herbicide-resistant phytoene desaturase and its use as a selectable marker. *Plant Biotechnol. J.* **2006**, *4*, 263–273. [CrossRef]
18. Breitenbach, J.; Zhu, C.; Sandmann, G. Bleaching Herbicide Norflurazon Inhibits Phytoene Desaturase by Competition with the Cofactors. *J. Agric. Food Chem.* **2001**, *49*, 5270–5272. [CrossRef] [PubMed]
19. Chamovitz, D.; Sandmann, G.; Hirschberg, J. Molecular and biochemical characterization of herbicide-resistant mutants of cyanobacteria reveals that phytoene desaturation is a rate-limiting step in carotenoid biosynthesis. *J. Biol. Chem.* **1993**, *268*, 17348–17353.
20. Cunningham, F.X.; Gantt, E. Genes And Enzymes of Carotenoid Biosynthesis in Plants. *Annu. Rev. Plant Physiol. Plant Mol. Biol.* **1998**, *49*, 557–583. [CrossRef]
21. Sandmann, G.; Römer, S.; Fraser, P.D. Understanding carotenoid metabolism as a necessity for genetic engineering of crop plants. *Metab. Eng.* **2006**, *8*, 291–302. [CrossRef]

22. Misawa, N.; Nakagawa, M.; Kobayashi, K.; Yamano, S.; Izawa, Y.; Nakamura, K.; Harashima, K. Elucidation of the Erwinia uredovora carotenoid biosynthetic pathway by functional analysis of gene products expressed in *Escherichia coli*. *J. Bacteriol.* **1990**, *172*, 6704–6712. [CrossRef] [PubMed]

23. Misawa, N.; Yamano, S.; Linden, H.; de Felipe, M.R.; Lucas, M.; Ikenaga, H.; Sandmann, G. Functional expression of the *Erwinia uredovora* carotenoid biosynthesis gene crtI in transgenic plants showing an increase of beta-carotene biosynthesis activity and resistance to the bleaching herbicide norflurazon. *Plant J.* **1993**, *4*, 833–840. [CrossRef] [PubMed]

24. Paine, J.A.; Shipton, C.A.; Chaggar, S.; Howells, R.M.; Kennedy, M.J.; Vernon, G.; Wright, S.Y.; Hinchliffe, E.; Adams, J.L.; Silverstone, A.L.; et al. Improving the nutritional value of Golden Rice through increased pro-vitamin A content. *Nat. Biotechnol.* **2005**, *23*, 482–487. [CrossRef] [PubMed]

25. Fraser, P.D.; Romer, S.; Shipton, C.A.; Mills, P.B.; Kiano, J.W.; Misawa, N.; Drake, R.G.; Schuch, W.; Bramley, P.M. Evaluation of transgenic tomato plants expressing an additional phytoene synthase in a fruit-specific manner. *Proc. Natl. Acad. Sci. USA* **2002**, *99*, 1092–1097. [CrossRef] [PubMed]

26. Ducreux, L.J.M.; Morris, W.L.; Hedley, P.E.; Shepherd, T.; Davies, H.V.; Millam, S.; Taylor, M.A. Metabolic engineering of high carotenoid potato tubers containing enhanced levels of beta-carotene and lutein. *J. Exp. Bot.* **2005**, *56*, 81–89.

27. Díaz-Santos, E.; Vila, M.; Vigara, J.; León, R. A new approach to express transgenes in microalgae and its use to increase the flocculation ability of *Chlamydomonas reinhardtii*. *J. Appl. Phycol.* **2016**, *28*, 1611–1621. [CrossRef]

28. Suarez, J.V.; Banks, S.; Thomas, P.G.; Day, A. A new F131V mutation in *Chlamydomonas phytoene* desaturase locates a cluster of norflurazon resistance mutations near the FAD-binding site in 3D protein models. *PLoS ONE* **2014**, *9*, e99894. [CrossRef] [PubMed]

29. Steinbrenner, J.; Sandmann, G. Transformation of the green alga *Haematococcus pluvialis* with a phytoene desaturase for accelerated astaxanthin biosynthesis. *Appl. Environ. Microbiol.* **2006**, *72*, 7477–7484. [CrossRef]

30. Sharon-Gojman, R.; Maimon, E.; Leu, S.; Zarka, A.; Boussiba, S. Advanced methods for genetic engineering of *Haematococcus pluvialis* (Chlorophyceae, Volvocales). *Algal Res.* **2015**, *10*, 8–15. [CrossRef]

31. Huang, J.; Liu, J.; Li, Y.; Chen, F. Isolation and Characterization of the Phytoene Desaturase Gene As a Potential Selective Marker for Genetic Engineering of the Astaxanthin-Producing Green Alga *Chlorella zofingiensis* (Chlorophyta). *J. Phycol.* **2008**, *44*, 684–690. [CrossRef]

32. Xue, J.; Niu, Y.-F.; Huang, T.; Yang, W.-D.; Liu, J.-S.; Li, H.-Y. Genetic improvement of the microalga Phaeodactylum tricornutum for boosting neutral lipid accumulation. *Metab. Eng.* **2015**, *27*, 1–9. [CrossRef]

33. Prasad, B.; Vadakedath, N.; Jeong, H.-J.; General, T.; Cho, M.-G.; Lein, W. Agrobacterium tumefaciens-mediated genetic transformation of haptophytes (Isochrysis species). *Appl. Microbiol. Biotechnol.* **2014**, *98*, 8629–8639. [CrossRef] [PubMed]

34. Tran, P.T.; Sharifi, M.N.; Poddar, S.; Dent, R.M.; Niyogi, K.K. Intragenic Enhancers and Suppressors of Phytoene Desaturase Mutations in *Chlamydomonas reinhardtii*. *PLoS ONE* **2012**, *7*, e42196. [CrossRef] [PubMed]

35. Liu, J.; Sun, Z.; Gerken, H.; Huang, J.; Jiang, Y.; Chen, F. Genetic engineering of the green alga *Chlorella zofingiensis*: A modified norflurazon-resistant phytoene desaturase gene as a dominant selectable marker. *Appl. Microbiol. Biotechnol.* **2014**, *98*, 5069–5079. [CrossRef] [PubMed]

36. Loppes, R.; Radoux, M.; Ohresser, M.C.; Matagne, R.F. Transcriptional regulation of the Nia1 gene encoding nitrate reductase in *Chlamydomonas reinhardtii*: Effects of various environmental factors on the expression of a reporter gene under the control of the Nia1 promoter. *Plant Mol. Biol.* **1999**, *41*, 701–711. [CrossRef] [PubMed]

37. Guillard, R.R.; Ryther, J.H. Studies of marine planktonic diatoms. I. Cyclotella nana Hustedt, and *Detonula confervacea* (cleve) Gran. *Can. J. Microbiol.* **1962**, *8*, 229–239. [CrossRef] [PubMed]

38. Johnson, M.K.; Johnson, E.J.; MacElroy, R.D.; Speer, H.L.; Bruff, B.S. Effects of salts on the halophilic alga *Dunaliella viridis*. *J. Bacteriol.* **1968**, *95*, 1461–1468. [PubMed]

39. León, R.; Couso, I.; Fernández, E. Metabolic engineering of ketocarotenoids biosynthesis in the unicelullar microalga *Chlamydomonas reinhardtii*. *J. Biotechnol.* **2007**, *130*, 143–152. [CrossRef]

40. Sizova, I.; Fuhrmann, M.; Hegemann, P. A Streptomyces rimosus aphVIII gene coding for a new type phosphotransferase provides stable antibiotic resistance to *Chlamydomonas reinhardtii*. *Gene* **2001**, *277*, 221–229. [CrossRef]

41. Kindle, K.L. High-frequency nuclear transformation of Chlamydomonas reinhardtii. *Proc. Natl. Acad. Sci. USA* **1990**, *87*, 1228–1232. [CrossRef]
42. Lichtenthaler, H.K. Chlorophylls and carotenoids: Pigments of photosynthetic biomembranes. *Methods Enzymol.* **1987**, *148*, 350–382.

 *metabolites*

Article

Intracellular and Extracellular Metabolites from the Cyanobacterium *Chlorogloeopsis fritschii*, PCC 6912, During 48 Hours of UV-B Exposure

Bethan Kultschar [1,*], Ed Dudley [2], Steve Wilson [3] and Carole A. Llewellyn [1,*]

[1] Department of Biosciences, Swansea University, Singleton Park, Swansea SA2 8PP, UK
[2] Swansea University Medical School, Swansea University, Singleton Park, Swansea SA2 8PP, UK; E.Dudley@swansea.ac.uk
[3] Unilever Corporate Research, Colworth Park, Sharnbrook, Bedfordshire MK44 1LQ, UK; Steve.Wilson@Unilever.com
* Correspondence: 878620@swansea.ac.uk (B.K.); C.A.Llewellyn@swansea.ac.uk (C.A.L.); Tel.: +44-1792-606168 (C.A.L.)

Received: 12 March 2019; Accepted: 13 April 2019; Published: 16 April 2019

Abstract: Cyanobacteria have many defence strategies to overcome harmful ultraviolet (UV) stress including the production of secondary metabolites. Metabolomics can be used to investigate this altered metabolism via targeted and untargeted techniques. In this study we assessed the changes in the intra- and extracellular low molecular weight metabolite levels of *Chlorogloeopsis fritschii* (*C. fritschii*) during 48 h of photosynthetically active radiation (PAR) supplemented with UV-B (15 µmol m^{-2} s^{-1} of PAR plus 3 µmol m^{-2} s^{-1} of UV-B) and intracellular levels during 48 h of PAR only (15 µmol m^{-2} s^{-1}) with sampling points at 0, 2, 6, 12, 24 and 48 h. Gas chromatography–mass spectrometry (GC–MS) was used as a metabolite profiling tool to investigate the global changes in metabolite levels. The UV-B time series experiment showed an overall significant reduction in intracellular metabolites involved with carbon and nitrogen metabolism such as the amino acids tyrosine and phenylalanine which have a role in secondary metabolite production. Significant accumulation of proline was observed with a potential role in stress mitigation as seen in other photosynthetic organisms. 12 commonly identified metabolites were measured in both UV-B exposed (PAR + UV-B) and PAR only experiments with differences in significance observed. Extracellular metabolites (PAR + UV-B) showed accumulation of sugars as seen in other cyanobacterial species as a stress response to UV-B. In conclusion, a snapshot of the metabolome of *C. fritschii* was measured. Little work has been undertaken on *C. fritschii*, a novel candidate for use in industrial biotechnology, with, to our knowledge, no previous literature on combined intra- and extracellular analysis during a UV-B treatment time-series. This study is important to build on experimental data already available for cyanobacteria and other photosynthetic organisms exposed to UV-B.

Keywords: cyanobacteria; *C. fritschii*; UV-B; PAR; time-series; intracellular; extracellular; metabolites; GC–MS

---

1. Introduction

Cyanobacteria are gram-negative bacteria with the ability to photosynthesise, assimilating CO_2 into a variety of biochemical compounds through different metabolic pathways [1]. Cyanobacteria can thrive in a wide variety of extreme habitats such as high ultraviolet radiation (UVR) due to their adaptive capabilities such as the production of secondary metabolites [1]. Metabolomics can be used to determine changes at the metabolite level during varying environmental stimuli and is a useful tool in cyanobacterial research [2]. The metabolome provides information closely reflecting the interaction between an organism and its environment. Some metabolites produced by cyanobacteria under stress conditions are unique and are of increasing interest from a biotechnological perspective as sustainable sources of ingredients in a variety of industries [3,4].

The effect of UVR on cyanobacteria has been widely researched including the interaction with biomolecules, production of reactive oxygen species (ROS) which cause oxidative stress, impaired growth, partial inhibition of photosynthesis and decreased enzyme activity [5–7]. UVR also has a role as an activator of secondary metabolite production such as mycosporine-like amino acids (MAAs) [8] and other protective secondary metabolites [9]. Many studies have been conducted to identify these targeted intracellular metabolites during UV-B and UV-A exposure in *Lyngbya* sp. CU2555 [10], *Nostoc commune* [11], *Anabaena variabilis* PCC 7937 [12], *Calothrix* sp. [13] and *Chlorogloeopsis fritschii* (*C. fritschii*), PCC 6912, [14] to name a few. Other studies conducted have sought to evaluate changes at the protein level [15,16], targeted and untargeted metabolomic analysis using different intensities of UV-B [17] and combined metabolomic and proteomic analysis during UV-A exposure [18].

Cyanobacteria convert CO_2 into reduced carbon which forms the backbone of metabolites and are central to life. Like many other microorganisms, cyanobacteria release these carbon-based primary and secondary metabolites into their surrounding area which drives carbon cycling within microbial communities [19,20]. These released metabolites are by-products of metabolism within cells and make up a small proportion of the dissolved organic matter (DOM) pool within freshwater and marine ecosystems [19]. Consisting of a variety of chemical compositions such as; polysaccharides, proteins, lipids, organic compounds or inorganic molecules, they are released for communication, structural organisation, and defence against biotic and abiotic factors [20–22]. The uptake and release of metabolites change with varying environments; examples include the release of exopolysaccharides during high light and UVR [11,23].

Monitoring industrially relevant metabolites released by microorganisms into their surroundings is a widely used technique in the fermentation industry. It can be used in bioprocess monitoring, fermentation biomarker identification, for monitoring metabolite levels in fermentation processes and microbial contamination [24,25].

Combining intracellular and extracellular analysis is useful in the study of cyanobacteria providing a more holistic picture of metabolite production during growth and its response to different environmental conditions [25,26].

Little work has been undertaken on monitoring both intracellular and extracellular metabolites in cyanobacteria especially *C. fritschii,* a potential candidate for use in industrial biotechnology due to its scalability [27] and tolerance to different growth conditions [28–30]. In this study, we observe the changes in metabolites produced by *C. fritschii* during 48 h of UV-B exposure as detected by untargeted gas chromatography-mass spectrometry (GC–MS). We were able to identify metabolites with altered levels comparing UV-B treatment (PAR + UV-B) to cultures irradiated with PAR only.

2. Results

2.1. Intracellular and Extracellular Analysis of C. fritschii during UV-B Stress Response

The metabolite profiles of *C. fritschii* cultures ($n = 3$) were investigated during 48 h of UV-B exposure. At each time point (0, 2, 6, 12, 24 and 48 h) intracellular and extracellular metabolites were analysed by untargeted GC–MS to evaluate the global changes in metabolite production during UV-B stress.

A total of 300 and 412 peaks were detected from the intracellular and extracellular time-series data respectively (Tables S1 and S2). Using a match factor of 60% or above, 135 and 218 peaks were putatively identified within the intracellular and extracellular data respectively (Tables S1 and S2). The identified chemical structures belonged to a variety of classes such as; acids, alcohols, amino acids, aromatics, fatty acids, heterocycles and sugars.

A Principle Component Analysis (PCA) model was used as an unsupervised multivariate statistical tool to plot and visualise the variance between UV-B exposed samples over time. A total variance of

37.2% for intracellular (Figure 1A, PC1 = 17%, PC2 = 11.1%, PC3 = 9.1%) and 36.4% for extracellular (Figure 1B, PC1 = 18.9%, PC2 = 9.2%, PC3 = 8.3%) was observed.

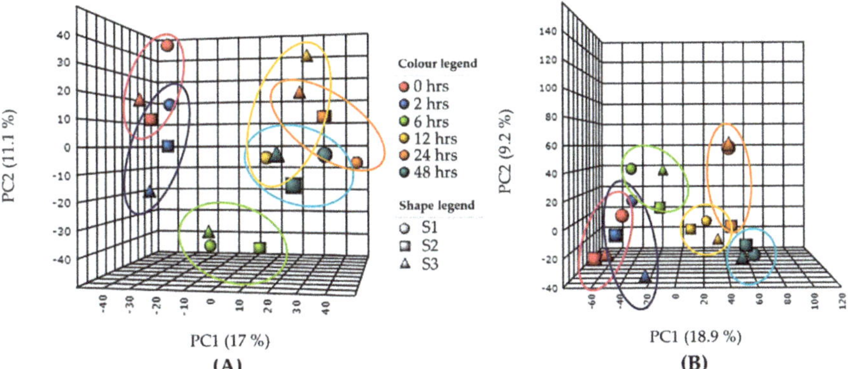

Figure 1. Principle component analysis (PCA) of (**A**) intracellular and (**B**) extracellular gas chromatography-mass spectrometry (GC–MS) data of UV-B exposed (PAR + UV-B) *Chlorogloeopsis fritschii (C. fritschii)* cultures showing PC1 vs PC2 only. Each ring represents distribution of biological replicates. S1 = replicate 1, S2 = replicate 2, S3 = replicate 3.

The results of the PCA for intracellular samples (Figure 1A) showed good separation over time between the control (0 h) and 6, 12, 24 and 48 h of UV-B. Less variation was observed between 0 and 2 h of UV-B with clustering seen between 12, 24 and 48 h of UV-B. This result was consistent with the two sample T-test results comparing control (0 h) with each time point where the number of significant features increases with length of UV-B exposure (Table S1). After a one-way analysis of variance (ANOVA) with repeated measures, 112 statistically significant peaks were observed with $p \leq 0.05$ (Figure S1A), 10 of which remained significant after Bonferroni correction.

From the extracellular data PCA (Figure 1B) a similar pattern was observed with increasing variance with increasing length of UV-B exposure. Statistically significant changes between control (0 h) and each time point, measured using a two-sample T-test also showed increasing significance ($p \leq 0.05$) with increasing length of UV-B up to 24 h (Table S2). A one-way ANOVA with repeated measures calculated 114 statistically significant peaks with $p \leq 0.05$ (Figure S1B).

2.1.1. Intracellular Metabolites

28 metabolites (13 represented for simplicity, Figure 2), selected as being involved in the central carbon and nitrogen metabolism within cyanobacteria, were identified within the intracellular GC–MS results Table S3). Many changes in metabolite levels were observed comparing between time points. Glucose, pyruvate and lactate all decreased in abundance after UV-B exposure with significant reduction after 2 h (pyruvate $p \leq 0.05$, 0 vs. 2 h), 6 h (glucose $p \leq 0.05$, 2 vs. 6 h) and 12 h (lactate $p \leq 0.05$, 0 vs. 12 h). Lactate was present during the whole time course whereas glucose and pyruvate were below detection limit after 6 h ($p \leq 0.05$) and 24 h ($p \leq 0.001$) respectively.

6 proteinogenic amino acids were detected; serine (ser), glycine (gly), glutamate (glu), proline (pro), tyrosine (tyr) and phenylalanine (phe). All detected amino acids decreased after 6 or 12 h of exposure with the exception of pro. A decrease in tyr, phe and gly was seen after 6 h (tyr $p \leq 0.05$, 0 vs. 6 h; phe $p \leq 0.01$, 2 vs. 6 h; gly $p \leq 0.05$, 2 vs. 6 h) of UV-B followed by no detection at 6, 12, 24, and 48 h (tyr $p \leq 0.05$; phe $p \leq 0.01$; gly $p \leq 0.05$). Ser and glu decreased significantly after 12 h of treatment (ser $p \leq 0.05$, 0 vs. 12 h; glu $p \leq 0.05$, 0 vs. 12 h), ser was below detection limit between 12 and 48 h ($p \leq 0.05$) whereas glu was detected throughout the time series. Proline showed no significant decrease after UV-B exposure with a significant increase observed after 24 h ($p \leq 0.05$, 6 vs. 24 h).

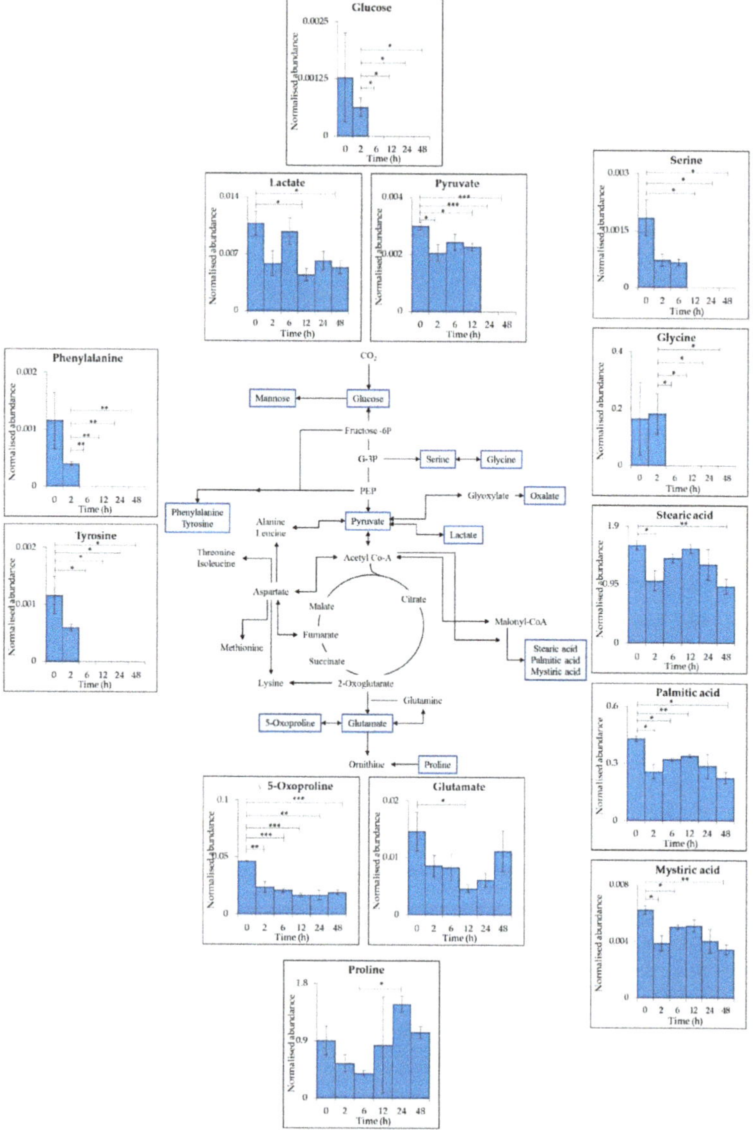

Figure 2. Schematic representation of a generalised reduced carbon metabolism in the cyanobacterium *C. fritschii* showing glycolysis, the citric acid (TCA) cycle, amino acid and fatty acid biosynthesis. Primary metabolites identified in intracellular samples using GC–MS are highlighted in blue with each insert presenting mean values of normalised abundance (normalised to internal standard and dry weight) ± standard error of each metabolite during supplemented UV-B exposure PAR + UV-B). Statistical significance between control (0 h) and UV-B exposure (2, 6, 12, 24 and 48 h) and between each treatment time point was measured using a two-sample T-test with equal variance; * = $0.05 \geq p \geq 0.01$, ** = $0.01 \geq p \geq 0.001$ and *** = $p \leq 0.001$.

The fatty acids stearic acid, palmitic acid and mystiric acid all decreased significantly after 2 h of treatment ($p \leq 0.05$) and their abundance remained lowered throughout the time series (stearic acid $p \leq 0.01$, 0 vs. 48 h; palmitic acid $p \leq 0.05$, 0 vs. 48 h; mystiric acid $p \leq 0.01$, 0 vs. 48 h).

Carotenoid and MAA Analysis

Carotenoid concentration (Figure 3A) and MAA content (Figure 3B) were analysed by UV-visible spectroscopy and high performance liquid chromatography (HPLC) respectively. Total carotenoid concentration decreased after 2 h ($p \leq 0.05$); with a steady significant increase up to 48 h with a final concentration of 2.59 µg/mg dry weight ($p \leq 0.05$).

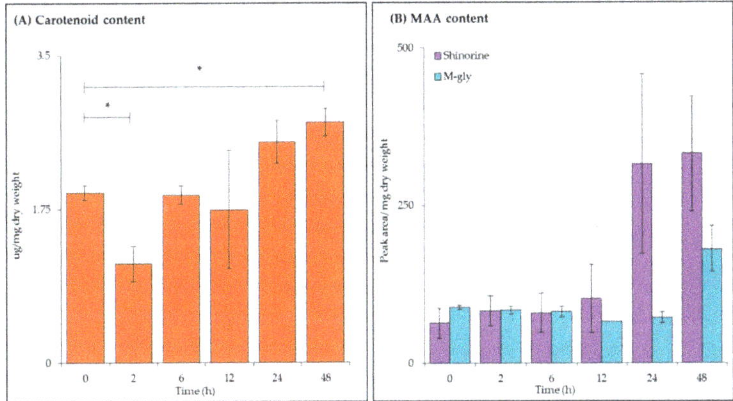

Figure 3. Carotenoid and mycosporine-like amino acid (MAA) analysis of *C. fritschii* extracts during UV-B exposure. **(A)** Total carotenoid concentration as measured by UV-visible spectroscopy and **(B)** shinorine and mycosporine-glycine (m-gly) content measured by high-performance liquid chromatography (HPLC) analysis. All values are the mean of three biological replicates (normalised to dry weight) ± standard error. Statistical significance was measured using a two-sample T-test with equal variance; * = $0.05 \geq p \geq 0.01$, ** = $0.01 \geq p \geq 0.001$ and *** = $p \leq 0.001$.

As described above, UV-B also induces the production of the photoprotective compounds, MAAs. The two forms found in *C. fritschii* are mycosporine-glycine (m-gly) and shinorine [29], both were detected during this experiment with peaks identified using their retention time and absorption maxima (λmax) values. As expected an increase in shinorine (retention time ~4.9 min, λmax = 334 nm) was observed with increasing length of UV-B exposure. No significance was observed with m-gly (retention time ~10.8 min, λmax = 310 nm) content during this experimental time series.

2.1.2. Extracellular Metabolites

29 biologically relevant dissolved metabolites (Table S3) were detected within the extracellular data set (13 represented for simplicity, Figure 4). Citrate, a component of BG-11 medium [31] and involved in the citric acid (TCA) cycle, was consistently detected throughout the time series (ANOVA, $p \leq 0.05$) along with succinate. Other TCA substrates; malate and fumarate were also detected at 0 h with decreasing abundance after 2 h of UV-B (malate, $p \leq 0.001$). Other metabolites detected at 0 h which decreased after UV-B exposure were leucine ($p \leq 0.01$, 0 vs. 6 h), putrescine ($p \leq 0.05$, 0 vs. 2 h) octanoic acid ($p \leq 0.001$, 0 vs. 24 h) mystiric acid ($p \leq 0.01$, 2 vs. 24 h) and fructose ($p \leq 0.01$, 0 vs. 2 h). Accumulation of the sugars galactose, xylose, lyxose and arabinose was seen after 6 h (galactose $p \leq 0.01$; arabinose $p \leq 0.01$) 12 h (arabinose $p \leq 0.05$), 24 h (arabinose $p \leq 0.05$; xylose $p \leq 0.05$; lyxose $p \leq 0.01$) and 48 h (galactose $p \leq 0.05$; arabinose $p \leq 0.001$; lyxose $p \leq 0.01$) of UV-B exposure. Trehalose was also identified throughout the time series with no significant changes.

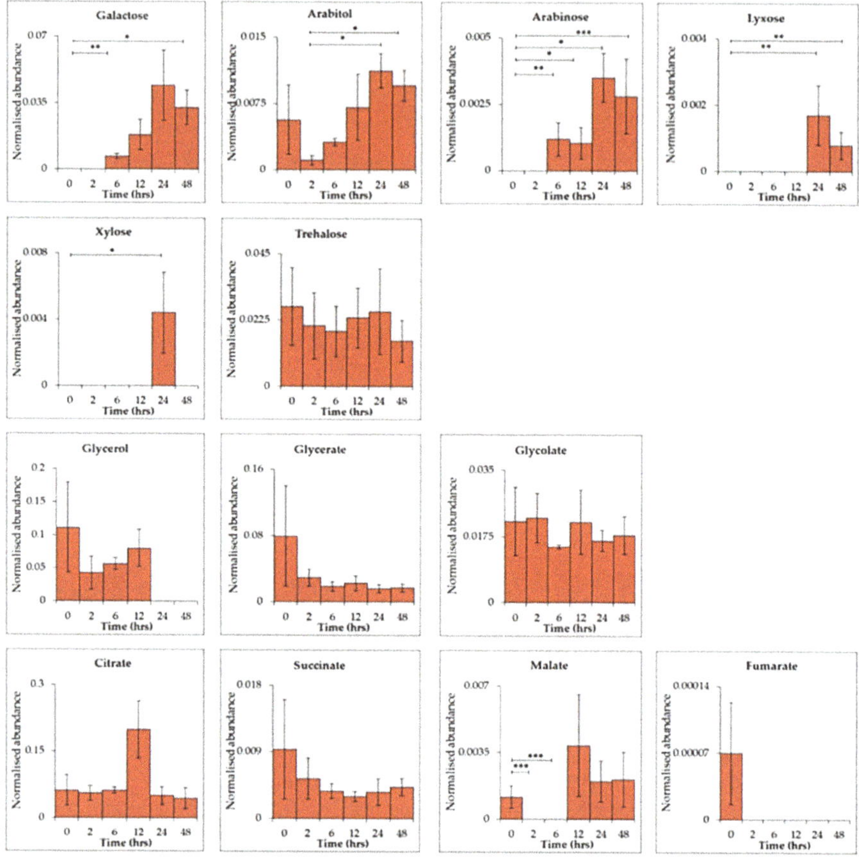

Figure 4. Time-series exometabolomics data of *C. fritschii* over 48 h of UV-B exposure showing primary metabolites found in extracellular samples only. Statistical significance was measured using a two-sample T-test comparing control (0 h) and UV-B exposure (2, 6, 12, 24 and 48 h) and between each treatment time point, * = $0.05 \geq p \geq 0.01$, ** = $0.01 \geq p \geq 0.001$ and *** = $p \leq 0.001$.

2.2. Intracellular Analysis of C. fritschii (PAR Only)

Intracellular Metabolites

A time-series analysis of PAR only (without UV-B supplementation) over 48 h (Table S4) revealed 35 key primary intracellular metabolites (Table S3). 12 were commonly identified between PAR only conditions and during UV-B supplementation (nine represented for simplicity in pathway schematic, Figure 5).

In general, comparing both supplement UV-B and PAR only, the nine common metabolites (Figure 6) showed differences in log2 fold change (FC). The metabolites detected during supplemented UV-B showed negative log2(FC) values which corresponds to reduced metabolite abundances compared to 0 h. Positive log2 (FC)) was generally observed for metabolites detected during PAR only indicating increased abundances compared to 0 h. The main exception is 5-oxoproline that reduced in both UV-B + PAR and PAR only experiments.

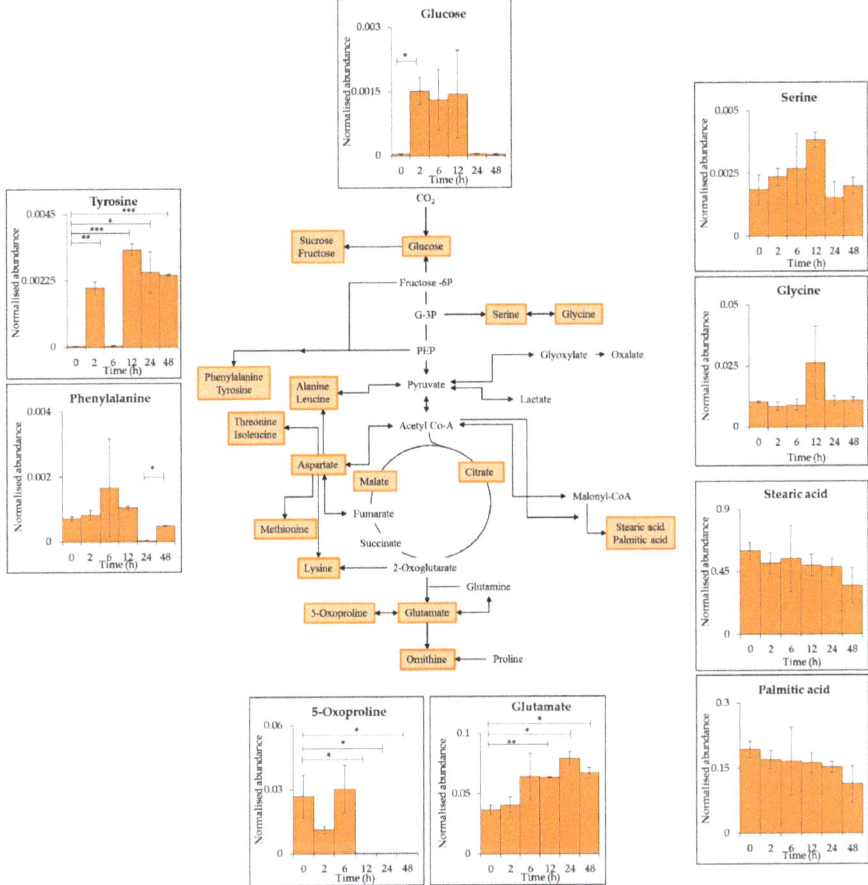

Figure 5. Schematic representation of a generalised reduced carbon metabolism in the cyanobacterium *C. fritschii* showing glycolysis, TCA cycle, amino acid and fatty acid biosynthesis. Primary metabolites identified in intracellular samples using GC–MS are highlighted in orange with each insert presenting mean values of normalised abundance (normalised to internal standard and dry weight) ± standard error of common metabolites found during PAR only conditions and supplemented UV-B (Figure 2). Statistical significance between 0 h and each time point as well as between time points were measured using a two-sample T-test with equal variance; * = $0.05 \geq p \geq 0.01$, ** = $0.01 \geq p \geq 0.001$ and *** = $p \leq 0.001$.

A significant accumulation of glu ($p \leq 0.01$) was observed after 12 h of PAR only whereas a significant decrease was seen after 12 h of UV-B exposure. Tyr accumulation ($p \leq 0.001$) was also seen after 2 and 12 h of PAR only conditions with a significant decrease during UV-B treatment. Palmitic and stearic acid remained relatively stable over time with a significant reduction during UV-B exposure observed. Gly and ser abundance remained consistent with significant decreases after 6 and 12 h respectively during UV-B exposure.

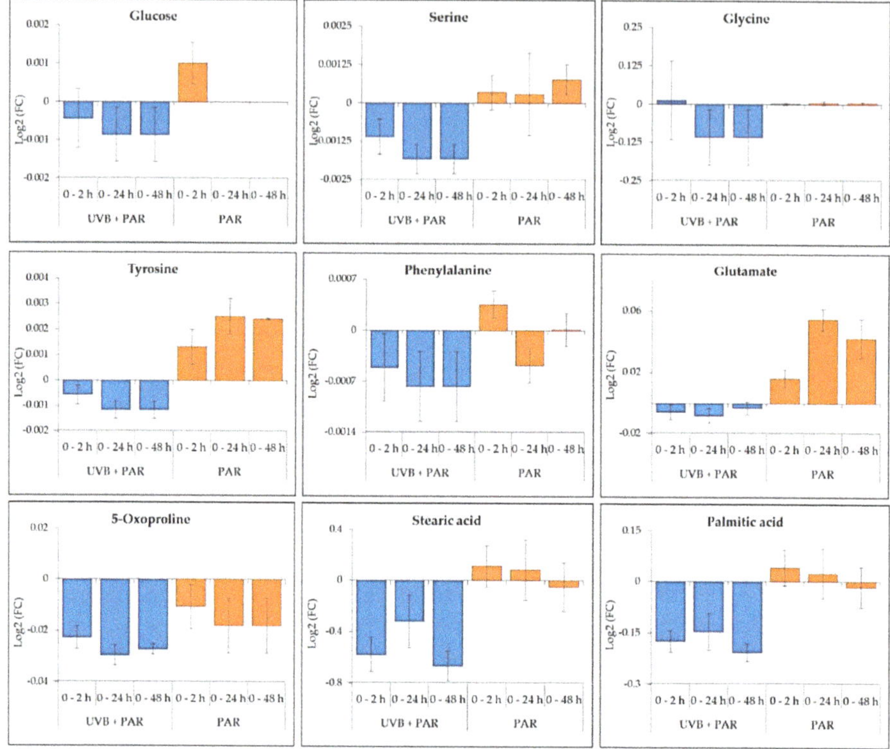

Figure 6. The changes in common metabolites identified in both UV-B + PAR and PAR only experiments comparing 0 h with 2, 24 and 48 h. Data are presented as Log2 Fold Change (FC) ± standard error.

3. Discussion

3.1. Intracellular Metabolite Changes and Pathway Analysis

UV-B exposure is known to reduce growth, photosynthesis and nitrogen fixation in cells to divert energy from key primary pathways to adaptive mechanisms such as; the production of secondary metabolites, MAAs; antioxidant production and DNA/protein repair [32]. A reduction in average dry weight was measured over 24 h of UV-B exposure (Figure S2) with the recovery of the initial biomass concentration and further growth measured between 24 and 48 h. These results showed no significance ($p \geq 0.05$) across the time series suggesting acclimation of cells to UV-B where damage to photosynthetic systems are counterbalanced by repair and mitigation strategies. A reduction in carotenoid concentration (Figure 3A) was observed after 2 h ($p \leq 0.05$) due to damage to photosystems caused by UV-B. Accumulation of total carotenoids could be indicative of antioxidant activity as a response to ROS production [6,33].

A reduction in glucose, pyruvate and lactate (Figure 2) could indicate a reduction in CO_2 fixation via photosynthesis and further biochemical processes. This could be due to the reduced production of ATP and $NADPH_2$ from photosynthesis [34].

The 13 selected intracellular metabolites (Figure 2) were reduced across the UV-B time series indicating the reduction of cellular processes. A decrease in phe and tyr could be due to their role as precursors to many secondary metabolites such as aromatic nitrogen-containing alkaloids [13]. M-gly and shinorine are produced via a combination of the shikimate/pentose phosphate pathway which also involves the addition of gly and ser to form the final MAAs [9,35]. The reduction of these amino acids coincides with an increase in MAA levels (Figure 3B).

5-oxoproline and glu are involved in glutathione metabolism. 5-oxoproline reduction ($p \leq 0.01$) could be due to its interconversion into glutamate which is further converted into the antioxidant glutathione [36]. Glu is also produced from the assimilation of nitrogen during nitrogen fixation which is reduced during UV-B exposure [34].

Pro has been studied in many UV-B experiments involving different photosynthetic organisms and its accumulation is thought to have a role in stress response by providing additional defence as a ROS scavenger and molecular chaperone [37–39]. Accumulation of proline has been observed in *Nostoc punctiforme* during 24 h of UV-A stress [18], in the model plant organism *Arabidopsis* after 24 h of UV-B treatment [40], and also in *C. fritschii* after 24 h of UV-B exposure within this study.

Overall less significant differences were observed during PAR only conditions (Figure 5) compared to PAR supplemented with UV-B (Figure 2).

3.2. Extracellular Metabolite Changes

The movement of metabolites and substrates between cells and their surrounding environment (or vice versa) can occur via passive and active uptake and efflux systems. Reactions can also occur on cell surface membranes and as transformations of media components [41]. Identification of extracellular metabolite uptake and release from cyanobacteria is, therefore, a complex process due to the high turnover rates of intracellular processes [25]. Extracellular metabolites can be released during stress and as by-products of intracellular reactions [19]. Sugars such as galactose, arabinose, lyxose and xylose are actively released during UV-B stress [5] as seen in this experiment (Figure 4).

7 metabolites from the identified biologically relevant pool were found in both intra- and extracellular metabolite samples (Figure S3). The fatty acid mystiric acid shows a similar pattern of reduced abundance with increasing length of UV-B in both samples (intracellular $p \leq 0.05$; extracellular $p \leq 0.05$). Ethanolamine, involved in glycerophospholipid metabolism, and 2-oxobutanoate, involved in amino acid biosynthesis show opposite patterns with intracellular levels decreasing ($p \leq 0.05$) and extracellular levels increasing (ethanolamine, $p \leq 0.01$, 2-oxobutanoate, $p \leq 0.001$) (S1: Figure 3). Further detailed analysis using a combination of -omic techniques and the application of isotopic labelling such as ^{13}C flux balance analysis would be required to better understand the uptake and release of extracellular metabolites and their possible use within cyanobacterial metabolite production [42].

4. Conclusions

In summary, an untargeted GC–MS workflow was used to evaluate intra- and extracellular metabolites under supplemented UV-B exposure (PAR + UV-B). Most significantly we found a reduction of intracellular metabolites such as the amino acids, tyr, phe, ser, gly and glu and the accumulation of pro, which to our knowledge has not been previously reported in *C. fritschii*. Compared to PAR only, intracellular metabolites showed less significant changes with amino acids tyr, phe and glu accumulation observed.

Although a time series analysis was conducted, this only represents a minuscule proportion of the true changes within the metabolome. This study is important to build on experimental data already available for cyanobacteria and other photosynthetic organisms exposed to UV-B. To understand the changes in primary metabolites and metabolic process with increasing length of UVR exposure to help further understand secondary metabolite production and adaptation of cyanobacteria to UV stress. Further studies are needed to understand and verify these processes within cyanobacteria to aid in the understanding of UV stress adaptation at the metabolite level.

5. Materials and Methods

5.1. Organism and Growth Conditions

The cyanobacterium *C. fritschii,* PCC 6912, was obtained from the Pasteur culture collection (PCC) and grown in autoclaved deionised water with filtered BG-11 growth medium (Sigma Aldrich). The strain was maintained in 50 mL BG-11 at a temperature of 27 ± 2 °C under continuous PAR illuminated at 15 µmol m^{-2} s^{-1} (measured using a PAR light sensor, Enviromonitors, West Sussex, UK). Experimental cultures were pre-grown in 300 mL BG-11 media under the same conditions with constant shaking at 80 rpm.

5.2. Experimental Setup

5.2.1. Supplemented UV-B Experiment (PAR + UV-B)

After 6 days of pre-growth, triplicate experimental *C. fritschii* cultures were transferred into three Quartz Erlenmeyer flasks (H.Baumbach & CO.LTD, Suffolk, UK) at an optical density at 750 nm (OD$_{750nm}$) of approx. 0.14 to allow even UV-B exposure. The cultures were exposed to a total of 48 h of UV-B radiation using a UVB broadband (290–315 nm, centered at 310 nm) fluorescent tube (Philips TL 20W/12 RS SLV/25, Proflamps, Eindhoven, The Netherlands) emitting 3 µmol m^{-2} s^{-1} of UV-B radiation (measured using a UVR light sensor, Enviromonitors, West Sussex, UK). The experiment was carried out under continuous PAR at 15 µmol m^{-2} s^{-1} and shaking at 100 rpm for even UVB exposure of cells. For time course analysis samples were collected at no UV-B (0 h), 2, 6, 12, 24 and 48 h for dry weight, pigment, MAA, and GC–MS analysis.

5.2.2. PAR Only Experiment (PAR only)

C. fritschii cultures were pre-grown for 6 days prior to experimental analysis. After 6 days of pre-growth, triplicate experimental *C. fritschii* cultures (OD$_{750nm}$ of approx. 0.13) were grown at a temperature of 27 ± 2 °C under continuous PAR illuminated at 15 µmol m^{-2} s^{-1} (measured using a PAR light sensor, Enviromonitors, West Sussex, UK) and continuous shaking at 100 rpm. For time course analysis samples were collected at 0, 2, 6, 12, 24 and 48 h for dry weight and GC–MS analysis.

5.3. Sample Harvest and Growth Analysis

Forty mL volumes of UV-B exposed ($n = 3$) and PAR only cultures ($n = 3$) were harvested at each time point by centrifugation at 4400 rpm for 20 min to produce a pellet and supernatant. The supernatants (40 mL) were collected and freeze-dried (Edwards, super modulyo) for 72 h. The remaining pellets were transferred into pre-weighed Eppendorf's and freeze-dried for 24 h (Scanvac, CoolSafe^{TM}, LaboGene^{TM}, Vassingerød, Denmark) for dry weight measurements. Both pellets (PAR + UV-B and PAR only) and dried supernatant (PAR + UV-B only) were stored at −20 °C until analysis. OD was monitored using absorbance at 750 nm using a UV-visible spectrophotometer (Shimadzu, UV-2550, Kyoto, Japan).

5.4. GC–MS Analysis

5.4.1. Sample Preparation

Polar and non-polar metabolites were extracted from UV-B exposed and PAR only dried cell pellets for GC–MS analysis. Briefly, approx. 0.5–1 mg (UV-B exposed) or 1.5–3 mg (PAR only) of dried biomass was re-suspended in 1 mL methanol:chloroform:water (2:2:1) and sonicated using a sonicator probe (Fisher Scientific, FB50) using 6 cycles of 20 s pulses at 40 Hz at 0 °C. After centrifugation (5 min at 12,000 rpm), 100 µL of each solvent layer (both methanolic and chloroform layers) were aliquoted into new Eppendorf's and evaporate to dryness using a rotary vacuum concentrator (Eppendorf concentrator 5301).

Dried supernatant (PAR + UV-B) was re-suspended in 1 mL of methanol and centrifuged (4000 rpm, 5 min). Two hundred µL was aliquoted into new Eppendorf's followed by evaporation to dryness and derivatisation as below.

5.4.2. Sample Derivatisation

To each 200 µL dried sample, 30 µL of methoxyamine hydrochloride (23 mg) in pyridine (1.5 mL) was added and samples were heated at 70 °C for 45 min. Once cooled to room temperature, 50 µL of MSTFA+TMCS (Thermo Scientific^{TM}, product no: TS-48915) was added and samples heated for an additional 90 min at 40 °C. Once cooled to room temperature, 10 µL of tetracosane dissolved in hexane (2 mg/mL) was added as an internal standard. Derivatised samples were transferred into auto-sample vials ready for analysis.

5.4.3. GC–MS Analysis

Derivatised sample (1 µL) was loaded onto an Agilent HP-5MS capillary column (30 m × 0.25 mm × 0.25 um) in splitless mode at 250 °C. The GC was operated at a constant flow of 1 mL min^{-1} helium. The temperature program started at 60 °C for 1 min, followed by temperature ramping at 10 °C min^{-1} temp of to a final 180 °C, this was followed by a second temperature ramping at 4 °C min^{-1} to a final temp of 300 °C and held constant at 300 °C for 15 min. Data acquisition included a mass range of 50 to 650 and resulted in .D data files for analysis.

Chromatograms were deconvoluted using AMDIS (Automated Mass Spectral Deconvolution and Identification System) followed by alignments using the online portal SpectConnect, http://spectconnect.mit.edu/ [43], before identifying peaks using Golm metabolome database, www.gmd.mpimp-golm.mpg.de/, and the NIST 05 (National Institute of Standards and Technology) library [44]. MetaboAnalyst, www.metaboanalyst.ca/, was used for statistical analysis [45,46].

5.4.4. GC–MS Data Processing

GC–MS data sets need deconvolution of co-eluting compounds, the freely available software AMDIS was used to process the chromatograms (.D) and produce .ELU files for alignment and conservative component identification using SpectConnect [43,47]. Settings for AMDIS were as followed [18]; Resolution = medium, sensitivity = medium, shape requirement = medium and component width was 10. The resulting .ELU files were uploaded to SpectConnect to produce matrices for further analysis and processing in Excel 2010 (Microsoft, USA). The integrated signal (IS) matrix generated was used for relative quantification of peaks. Triplicate missing data points (within time points) were assumed to be lower than the detection limit and replaced with half of the minimum integrated signal within each data set. Data normalisation was carried out using the peak area of the internal standard tetracosane and dry weight of each sample (for intracellular data only; approx. 0.5–1 mg of UV-B exposed cells and 1.5–3 mg of PAR only cells). All duplicate retention times were removed before further analysis.

5.4.5. Identification

Identification of peaks was carried out in AMDIS by analysing each chromatogram using the Golm databases as a target library, followed by searching the NIST 05 library with a match factor of 60% or above. Reports were exported in .xls format from AMDIS with the first hit only included for further processing using Excel 2010 (Microsoft, Redmond, WA, USA). A true hit was considered when two or more biological replicates (within the same time-point) contained the peak. If none of the time points contained 'true hits,' the peaks were removed before further analysis. Metabolites reported belonged to level 2 (putatively annotated compounds) and level 4 (unknown compounds) identifications in accordance with the Metabolomics Standards Initiative [48].

5.4.6. Statistical Analysis

MetaboAnalyst was used to statistically analyse the IS peak lists (in .csv format) of the time series data using the time-series/two-factor module. Multivariate analysis was carried out using each column as a different time point and each row representing a metabolite (data type = peak intensity table; study design = time-series only; data format = samples in columns) [46]. Missing data points were uploaded as blanks and replaced with half of the minimum integrated signal within each data set. Peaks were normalised to total sum of peaks, log-transformed and mean centered prior to statistical analysis. PCA, a one-way repeated ANOVA ($p \leq 0.05$) and hierarchical heat map clustering was used to evaluate the data. A two-sample T-test with equal variance was also used as a univariate statistical tool to evaluate data comparing 0 h with each treatment time point (2, 6, 12, 24 and 48 h) as well as between each time points in Excel.

5.5. MAA analysis

5.5.1. Sample Preparation

0.5–1 mg of UV-B exposed dried biomass was re-suspended in 100% HPLC grade methanol (1 mL) and left in the dark at 4 °C overnight (24 h). After centrifugation (5 min at 12000 rpm), the supernatant was removed and evaporated to dryness using a rotary vacuum concentrator. The dried extract was re-dissolved in 600 µL of deionised water and transferred to autosample vials for HPLC analysis [49].

5.5.2. HPLC Analysis

HPLC analysis was performed using an Agilent 1100 system equipped with a binary pump (G1312A), an autosampler injector (ALS, G1313A), thermostatted column compartment (G1316A) and diode array detector (DAD, G1315A) connected via an interface module to a computer running ChemStation software. The stationary phase was an Alltima™ Altech™ C18, 4.6 × 150 mm, 5 µm column heated to 35 °C. The mobile phases consisted of; Eluent A: Water (0.01% TFA, *v/v*) and Eluent B: 70% methanol (0.054% TFA, *v/v*) with a gradient of; 99% A for 10 min, to 80% A over 5 min, to 1% A over 5 min, held for 3 min and increased to 99% A over 2 min. The samples were injected at a volume of 100 µL and MAA's were monitored at wavelengths of; 310, 320 and 330 nm, absorption spectra between 200–400 nm were stored in each detected peak.

5.6. Pigment Analysis

Sample Preparation

To approx. 0.5–1 mg of dried biomass, 100% HPLC grade methanol (1 mL) was added and vortexed to re-suspend. Samples were sonicated under low light conditions using a sonicator probe for 6 cycles of 20 s pulses at 40 Hz at 0 °C. After centrifugation (5 min at 12,000 rpm), the supernatant was removed and absorbance spectra measured using a UV-visible spectrophotometer between 400–800 nm with 100% methanol as a blank. Carotenoid concentration was calculated using the equations as described in [50,51].

Supplementary Materials: The following are available online at http://www.mdpi.com/2218-1989/9/4/74/s1, Figure S1: Hierarchical heatmap visualisation of the significant (A) intracellular and (B) extracellular peak intensities ($p \leq 0.05$) during UV-B exposure using a one-way repeated measure ANOVA in MetaboAnalyst. Data is arranged in triplicate with increasing length of UV-B exposure from left to right (0–48 h). S1 = replicate 1, S2 = replicate 2, S3 = replicate 3; Figure S2: Dry weight measurements of *C. fritschii* cultures during 48 h of supplemented UV-B exposure. All values are the mean of three biological replicates ± standard error; Figure S3: Time-series metabolomics data of *C. fritschii* during 48 h of UV-B exposure showing primary metabolites found in both intra- and extracellular during GC–MS analysis. (A) Metabolites showing statistically significant ($p \leq 0.05$) changes over time in intra- and extracellular data; (B) metabolites showing statistically significant ($p \leq 0.05$) changes over time in intracellular samples only. Statistical significance was measured using a two-sample T-test comparing control (0 h) to each treatment time point (2, 6, 12, 24 and 48 h) and between treatment time points, * = $0.05 \geq p \geq 0.01$, ** = $0.01 \geq p \geq 0.001$ and *** = $p \leq 0.001$; Table S1: Intracellular compounds detected during UV-B exposure using GC–MS including; Retention time (RT), name, class, model (*m/z*), normalised mean abundance ($n = 3$) ± standard error (SE). T-test and ANOVA results showing statistical significant compounds, * = $0.05 \geq p \geq 0.01$, ** = $0.01 \geq p \geq 0.001$, *** = $p \leq 0.001$; Table S2: Extracellular compounds detected during UV-B exposure using GC–MS including; Retention time (RT), name, class, model (*m/z*), normalised mean abundance ($n = 3$) ± standard error (SE). T-test and ANOVA results showing statistical significant compounds, * = $0.05 \geq p \geq 0.01$, ** = $0.01 \geq p \geq 0.001$, *** = $p \leq 0.001$; Table S3: Biologically relevant metabolites detected during PAR + UV-B (intra- and extracellular metabolites) and PAR only (intracellular metabolites), including possible biosynthetic pathways; Table S4: Intracellular compounds detected during PAR only exposure using GC–MS including; Retention time (RT), name, class, model (*m/z*), normalised mean abundance ($n = 3$) ± standard error (SE). T-test and ANOVA results showing statistical significant compounds, * = $0.05 \geq p \geq 0.01$, ** = $0.01 \geq p \geq 0.001$, *** = $p \leq 0.001$.

Author Contributions: Authors in this study contributed in the following areas: B.K. was responsible for the design of the experiment, execution of experimental work and data analysis. E.D. supported experimental and data acquisition. B.K. wrote the manuscript with input from C.A.L., E.D. and S.W. Supervision by C.A.L., E.D. and S.W.

Funding: This research was funded by the Biotechnology and Biological Sciences Research Council (BBSRC iCASE studentship), UK, grant number BB/N503630/1.

Conflicts of Interest: The authors declare no conflict of interest.

References

1. Gupta, V.; Ratha, S.K.; Sood, A.; Chaudhary, V.; Prasanna, R. New Insights into the Biodiversity and Applications of Cyanobacteria (Blue-Green Algae)—Prospects and Challenges. *Algal Res.* **2013**, *2*, 79–97. [CrossRef]
2. Schwarz, D.; Orf, I.; Kopka, J.; Hagemann, M. Recent Applications of Metabolomics Toward Cyanobacteria. *Metabolites* **2013**, *3*, 72–100. [CrossRef] [PubMed]
3. Wijffels, R.H.; Kruse, O.; Hellingwerf, K.J. Potential of Industrial Biotechnology with Cyanobacteria and Eukaryotic Microalgae. *Curr. Opin. Biotechnol.* **2013**, *24*, 405–413. [CrossRef] [PubMed]
4. Rastogi, R.P.; Sinha, R.P. Biotechnological and Industrial Significance of Cyanobacterial Secondary Metabolites. *Biotechnol. Adv.* **2009**, *27*, 521–539. [CrossRef] [PubMed]
5. Rastogi, R.P.; Sinha, R.P.; Moh, S.H.; Lee, T.K.; Kottuparambil, S.; Kim, Y.J.; Rhee, J.S.; Choi, E.M.; Brown, M.T.; Häder, D.P.; et al. Ultraviolet Radiation and Cyanobacteria. *J. Photochem. Photobiol. B Biol.* **2014**, *141*, 154–169. [CrossRef]
6. Rastogi, R.P. UV-Induced Oxidative Stress in Cyanobacteria: How Life Is Able to Survive? *Biochem. Anal. Biochem.* **2015**, *4*, 2–5. [CrossRef]
7. Marangoni, R.; Paris, D.; Melck, D.; Fulgentini, L.; Colombetti, G.; Motta, A. In Vivo NMR Metabolic Profiling of Fabrea Salina Reveals Sequential Defense Mechanisms against Ultraviolet Radiation. *Biophys. J.* **2011**, *100*. [CrossRef]
8. Sinha, R.P.; Häder, D.P. UV-Protectants in Cyanobacteria. *Plant Sci.* **2008**, *174*, 278–289. [CrossRef]
9. Wada, N.; Sakamoto, T.; Matsugo, S. Multiple Roles of Photosynthetic and Sunscreen Pigments in Cyanobacteria Focusing on the Oxidative Stress. *Metabolites* **2013**, *3*, 463–483. [CrossRef]
10. Rastogi, R.P.; Incharoensakdi, A. Characterization of UV-Screening Compounds, Mycosporine-like Amino Acids, and Scytonemin in the Cyanobacterium Lyngbya Sp. CU2555. *FEMS Microbiol. Ecol.* **2014**, *87*, 244–256. [CrossRef]
11. Ehling-Schulz, M.; Bilger, W.; Scherer, S. UV-B-Induced Synthesis of Photoprotective Pigments and Extracellular Polysaccharides in the Terrestrial Cyanobacterium Nostoc Commune. *J. Bacteriol.* **1997**, *179*, 1940–1945. [CrossRef] [PubMed]
12. Singh, S.P.; Klisch, M.; Sinha, R.P.; Hader, D.-P. Effects of Abiotic Stressors on Synthesis of the Mycosporine-like Amino Acid Shinorine in the Cyanobacterium Anabaena Variabilis PCC 7937. *Photochem. Photobiol.* **2008**, *84*, 1500–1505. [CrossRef] [PubMed]
13. Hartmann, A.; Albert, A.; Ganzera, M. Effects of Elevated Ultraviolet Radiation on Primary Metabolites in Selected Alpine Algae and Cyanobacteria. *J. Photochem. Photobiol. B Biol.* **2015**, *149*, 149–155. [CrossRef] [PubMed]
14. Portwich, A.; Garcia-Pichel, F. A Novel Prokaryotic UVB Photoreceptor in the Cyanobacterium Chlorogloeopsis PCC 6912. *Photochem. Photobiol.* **2000**, *71*, 493–498. [CrossRef]

15. Ehling-Schulz, M.; Schulz, S.; Wait, R.; Görg, A.; Scherer, S. The UV-B Stimulon of the Terrestrial Cyanobacterium Nostoc Commune Comprises Early Shock Proteins and Late Acclimation Proteins. *Mol. Microbiol.* **2002**, *46*, 827–843. [CrossRef] [PubMed]
16. Shrivastava, A.K.; Chatterjee, A.; Yadav, S.; Singh, P.K.; Singh, S.; Rai, L.C. UV-B Stress Induced Metabolic Rearrangements Explored with Comparative Proteomics in Three Anabaena Species. *J. Proteomics* **2015**, *127*, 122–133. [CrossRef] [PubMed]
17. Shen, S.G.; Jia, S.R.; Yan, R.R.; Wu, Y.K.; Wang, H.Y.; Lin, Y.H.; Zhao, D.X.; Tan, Z.L.; Lv, H.X.; Han, P.P. The Physiological Responses of Terrestrial Cyanobacterium: Nostoc Flagelliforme to Different Intensities of Ultraviolet-B Radiation. *RSC Adv.* **2018**, *8*, 21065–21074. [CrossRef]
18. Wase, N.; Pham, T.K.; Ow, S.Y.; Wright, P.C. Quantitative Analysis of UV-A Shock and Short Term Stress Using ITRAQ, Pseudo Selective Reaction Monitoring (PSRM) and GC-MS Based Metabolite Analysis of the Cyanobacterium Nostoc Punctiforme ATCC 29133. *J. Proteomics* **2014**, *109*, 332–355. [CrossRef] [PubMed]
19. Fiore, C.L.; Longnecker, K.; Kido Soule, M.C.; Kujawinski, E.B. Release of Ecologically Relevant Metabolites by the Cyanobacterium S Ynechococcus Elongatus CCMP 1631. *Environ. Microbiol.* **2015**, *17*, 3949–3963. [CrossRef]
20. Stuart, R.K.; Mayali, X.; Lee, J.Z.; Craig Everroad, R.; Hwang, M.; Bebout, B.M.; Weber, P.K.; Pett-Ridge, J.; Thelen, M.P. Cyanobacterial Reuse of Extracellular Organic Carbon in Microbial Mats. *ISME J.* **2016**, *10*, 1240–1251. [CrossRef] [PubMed]
21. Kujawinski, E.B. The Impact of Microbial Metabolism on Marine Dissolved Organic Matter. *Ann. Rev. Mar. Sci.* **2011**, *3*, 567–599. [CrossRef]
22. Yadav, S.; Sinha, R.P.; Tyagi, M.B.; Kumar, A. Cyanobacterial Secondary Metabolites. *Int. J. Pharma Bio Sci.* **2011**, *2*, 144–167.
23. Mota, R.; Guimarães, R.; Büttel, Z.; Rossi, F.; Colica, G.; Silva, C.J.; Santos, C.; Gales, L.; Zille, A.; De Philippis, R.; et al. Production and Characterization of Extracellular Carbohydrate Polymer from Cyanothece Sp. CCY 0110. *Carbohydr. Polym.* **2013**, *92*, 1408–1415. [CrossRef] [PubMed]
24. Sue, T.; Obolonkin, V.; Griffiths, H.; Villas-Bôas, S.G. An Exometabolomics Approach to Monitoring Microbial Contamination in Microalgal Fermentation Processes by Using Metabolic Footprint Analysis. *Appl. Environ. Microbiol.* **2011**, *77*, 7605–7610. [CrossRef] [PubMed]
25. Pinu, F.; Villas-Boas, S. Extracellular Microbial Metabolomics: The State of the Art. *Metabolites* **2017**, *7*, 43. [CrossRef] [PubMed]
26. Granucci, N.; Pinu, F.R.; Han, T.-L.; Villas-Boas, S.G. Can We Predict the Intracellular Metabolic State of a Cell Based on Extracellular Metabolite Data? *Mol. Biosyst.* **2015**, *11*, 3297–3304. [CrossRef] [PubMed]
27. Balasundaram, B.; Skill, S.C.; Llewellyn, C. a. A Low Energy Process for the Recovery of Bioproducts from Cyanobacteria Using a Ball Mill. *Biochem. Eng. J.* **2012**, *69*, 48–56. [CrossRef]
28. Evans, E.H.; Foulds, I.; Carr, N.G. Environmental Conditions and Morphological Variation in the Blue-Green Alga Chlorogloea Fritschii. *J. Gen. Microbiol.* **1976**, *92*, 147–155. [CrossRef]
29. Portwich, A.; Garcia-Pichel, F. Ultraviolet and Osmotic Stresses Induce and Regulate the Synthesis of Mycosporines in the Cyanobacterium Chlorogloeopsis PCC 6912. *Arch. Microbiol.* **1999**, *172*, 187–192. [CrossRef] [PubMed]
30. Airs, R.L.; Temperton, B.; Sambles, C.; Farnham, G.; Skill, S.C.; Llewellyn, C.A. Chlorophyll f and Chlorophyll d Are Produced in the Cyanobacterium Chlorogloeopsis Fritschii When Cultured under Natural Light and Near-Infrared Radiation. *FEBS Lett.* **2014**, *588*, 3770–3777. [CrossRef] [PubMed]
31. Stanier, R.Y.; Deruelles, J.; Rippka, R.; Herdman, M.; Waterbury, J.B. Generic Assignments, Strain Histories and Properties of Pure Cultures of Cyanobacteria. *Microbiology* **1979**, *111*, 1–61.
32. Singh, S.P.; Häder, D.P.; Sinha, R.P. Cyanobacteria and Ultraviolet Radiation (UVR) Stress: Mitigation Strategies. *Ageing Res. Rev.* **2010**, *9*, 79–90. [CrossRef]
33. Latifi, A.; Ruiz, M.; Zhang, C.C. Oxidative Stress in Cyanobacteria. *FEMS Microbiol. Rev.* **2009**, *33*, 258–278. [CrossRef]
34. Kumar, A.; Sinha, R.P.; Häder, D.-P. Effect of UV-B on Enzymes of Nitrogen Metabolism in the Cyanobacterium Nostoc Calcicola. *J. Plant Physiol.* **1996**, *148*, 86–91. [CrossRef]
35. Kultschar, B.; Llewellyn, C. Secondary Metabolites in Cyanobacteria. In *Secondary Metabolites—Sources and Applications*; InTech: London, UK, 2018; Volume 2, p. 64.

36. Cameron, J.C.; Pakrasi, H.B. Essential Role of Glutathione in Acclimation to Environmental and Redox Perturbations in the Cyanobacterium Synechocystis Sp. PCC 6803. *Plant Physiol.* **2010**, *154*, 1672–1685. [CrossRef] [PubMed]

37. Liang, X.; Zhang, L.; Natarajan, S.K.; Becker, D.F. Proline Mechanisms of Stress Survival. *Antioxid. Redox Signal.* **2013**, *19*, 998–1011. [CrossRef] [PubMed]

38. Verbruggen, N.; Hermans, C. Proline Accumulation in Plants: A Review. *Amino Acids* **2008**, *35*, 753–759. [CrossRef] [PubMed]

39. Chris, A.; Zeeshan, M.; Abraham, G.; Prasad, S.M. Proline Accumulation in *Cylindrospermum* sp. *Environ. Exp. Bot.* **2006**, *57*, 154–159. [CrossRef]

40. Kusano, M.; Tohge, T.; Fukushima, A.; Kobayashi, M.; Hayashi, N.; Otsuki, H.; Kondou, Y.; Goto, H.; Kawashima, M.; Matsuda, F.; et al. Metabolomics Reveals Comprehensive Reprogramming Involving Two Independent Metabolic Responses of Arabidopsis to UV-B Light. *Plant J.* **2011**, *67*, 354–369. [CrossRef] [PubMed]

41. Pinu, F.R.; Granucci, N.; Daniell, J.; Han, T.-L.; Carneiro, S.; Rocha, I.; Nielsen, J.; Villas-Boas, S.G. Metabolite Secretion in Microorganisms: The Theory of Metabolic Overflow Put to the Test. *Metabolomics* **2018**, *14*, 43. [CrossRef]

42. Shastri, A.A.; Morgan, J.A. A Transient Isotopic Labeling Methodology for 13C Metabolic Flux Analysis of Photoautotrophic Microorganisms. *Phytochemistry* **2007**, *68*, 2302–2312. [CrossRef] [PubMed]

43. Styczynski, M.P.; Moxley, J.F.; Tong, L.V.; Walther, J.L.; Jensen, K.L.; Stephanopoulos, G.N. Systematic Identification of Conserved Metabolites in GC/MS Data for Metabolomics and Biomarker Discovery. *Anal. Chem.* **2007**, *79*, 966–973. [CrossRef] [PubMed]

44. Kopka, J.; Schauer, N.; Krueger, S.; Birkemeyer, C.; Usadel, B.; Bergmüller, E.; Dörmann, P.; Weckwerth, W.; Gibon, Y.; Stitt, M.; et al. GMD@CSB.DB: The Golm Metabolome Database. *Bioinformatics* **2005**, *21*, 1635–1638. [CrossRef] [PubMed]

45. Chong, J.; Soufan, O.; Li, C.; Caraus, I.; Li, S.; Bourque, G.; Wishart, D.S.; Xia, J. MetaboAnalyst 4.0: Towards More Transparent and Integrative Metabolomics Analysis. *Nucleic Acids Res.* **2018**, *46*, W486–W494. [CrossRef]

46. Xia, J.; Sinelnikov, I.V.; Wishart, D.S. MetATT: A Web-Based Metabolomics Tool for Analyzing Time-Series and Two-Factor Datasets. *Bioinformatics* **2011**, *27*, 2455–2456. [CrossRef]

47. Fiehn, O. Metabolomics by Gas Chromatography-Mass Spectrometry: Combined Targeted and Untargeted Profiling. In *Current Protocols in Molecular Biology*; John Wiley & Sons, Inc.: Hoboken, NJ, USA, 2016; Volume 131 A, pp. 30.4.1–30.4.32.

48. Sumner, L.W.; Amberg, A.; Barrett, D.; Beale, M.H.; Beger, R.; Daykin, C.A.; Fan, T.W.M.; Fiehn, O.; Goodacre, R.; Griffin, J.L.; et al. Proposed Minimum Reporting Standards for Chemical Analysis: Chemical Analysis Working Group (CAWG) Metabolomics Standards Initiative (MSI). *Metabolomics* **2007**, *3*, 211–221. [CrossRef]

49. Rastogi, R.P.; Madamwar, D.; Incharoensakdi, A. Sun-Screening Bioactive Compounds Mycosporine-like Amino Acids in Naturally Occurring Cyanobacterial Biofilms: Role in Photoprotection. *J. Appl. Microbiol.* **2015**, *119*, 753–762. [CrossRef] [PubMed]

50. Ritchie, R.J. Consistent Sets of Spectrophotometric Chlorophyll Equations for Acetone, Methanol and Ethanol Solvents. *Photosynth. Res.* **2006**, *89*, 27–41. [CrossRef] [PubMed]

51. Henriques, M.; Silva, A.; Rocha, J. Extraction and Quantification of Pigments from a Marine Microalga: A Simple and Reproducible Method. *Commun. Curr. Res. Educ. Top. Trends Appl. Microbiol.* **2007**, *2*, 586–593.

 *metabolites*

Article

Effects of Copper and pH on the Growth and Physiology of *Desmodesmus* sp. AARLG074

Nattaphorn Buayam [1,2], Matthew P. Davey [3,*], Alison G. Smith [3] and Chayakorn Pumas [2,*]

[1] Master's Degree Program in Applied Microbiology, Department of Biology, Faculty of Science, Chiang Mai University, Chiang Mai 50200, Thailand; nattaphorn.buayam@gmail.com
[2] Center of Excellence in Bioresources for Agriculture, Industry and Medicine, Department of Biology, Faculty of Science, Chiang Mai University, Chiang Mai 50200, Thailand
[3] Plant Metabolism Group, Department of Plant Sciences, University of Cambridge, Cambridge CB2 3EA, UK; as25@hermes.cam.ac.uk
* Correspondence: mpd39@cam.ac.uk (M.P.D.) (metabolite analysis only); chayakorn.pumas@gmail.com (C.P.); Tel.: +66-53941948 (ext. 116) (C.P.)

Received: 3 April 2019; Accepted: 27 April 2019; Published: 30 April 2019

Abstract: Copper (Cu) is a heavy metal that is widely used in industry and as such wastewater from mining or industrial operations can contain high levels of Cu. Some aquatic algal species can tolerate and bioaccumulate Cu and so could play a key role in bioremediating and recovering Cu from polluted waterways. One such species is the green alga *Desmodesmus* sp. AARLG074. The aim of this study was to determine how *Desmodesmus* is able to tolerate large alterations in its external Cu and pH environment. Specifically, we set out to measure the variations in the Cu removal efficiency, growth, ultrastructure, and cellular metabolite content in the algal cells that are associated with Cu exposure and acidity. The results showed that *Desmodesmus* could remove up to 80% of the copper presented in Jaworski's medium after 30 min exposure. There was a decrease in the ability of Cu removal at pH 4 compared to pH 6 indicating both pH and Cu concentration affected the efficiency of Cu removal. Furthermore, Cu had an adverse effect on algal growth and caused ultrastructural changes. Metabolite fingerprinting (FT-IR and GC-MS) revealed that the polysaccharide and amino acid content were the main metabolites affected under acid and Cu exposure. Fructose, lactose and sorbose contents significantly decreased under both acidic and Cu conditions, whilst glycerol and melezitose contents significantly increased at pH 4. The pathway analysis showed that pH had the highest impact score on alanine, aspartate and glutamate metabolism whereas Cu had the highest impact on arginine and proline metabolism. Notably both Cu and pH had impact on glutathione and galactose metabolism.

Keywords: algae; copper; FT-IR; metabolite fingerprinting; pathway analysis; TEM

1. Introduction

Copper (Cu) is widely used in industry [1] and as a result there is a very high demand of raw Cu extracted from mines. This demand is projected to increase 2.8 to 3.5 fold by 2050 [2]. Consequently, Cu containing waste (in particular wastewater) is generated daily from mining and industrial operations. The concentration of Cu in such mine drainage and water systems can vary substantially [3]. Additionally, the pH of the water systems is lowered (pH 3–5) resulting in acid mine drainage (AMD) [4,5]. A range of studies [3,6,7] have determined the level of Cu directly from industrial and acid mine run-off is around 10 to 80 times higher (up to 160 mg/L) than control level, indicating an urgent need for remediation. For example, the run-off from an abandoned Cu mine in Norway has a Cu concentration of 13.9 mg/L (pH 2.9), whereas the nearby receiving lake has just 0.04 mg Cu/L (pH 6.6) [3].

High levels of Cu exposure can be toxic to most living organisms [8–11] as heavy metals are non-biodegradable and environmentally persistent, which may be deposited on surfaces and then absorbed into the tissues of organisms [12]. Although the toxicity of Cu is recognized, the permissible amount of Cu in effluents slightly differs around the world. For example, the Ministry of Industry of Thailand has announced that the maximum permissible limit of Cu in industrial effluent is 2.0 mg Cu/L, which is the same as the permissible limit of Cu ions in drinking water permitted by the World Health Organization (WHO). The United States Environmental Protection Agency (USEPA) regulations state that Cu in industrial effluents must not exceed 1.3 mg Cu/L [1].

Despite the toxicity, some micro-algal species are able to grow in waterways that contain high concentrations of Cu, alongside the associated altered pH. This is in part due to Cu being an essential trace element that is required for normal algal growth [13,14] and also due to some algae (both macro- and micro-algae) being highly efficient in bioaccumulating heavy metals [15]. As such, bioaccumulation of the Cu from these polluted waterways is a potential approach for heavy metal bioremediation. Although the efficiency of heavy metal removal by algae and its effect on algal growth has been well characterized for some species [8,10,16–19] the biochemical mechanisms that are associated with algae tolerating high concentrations of Cu tolerance are not clearly defined or understood. In addition, *Desmodesmus* (Chlorophyta, Chlorophyceae) is the genera that frequently found distributed in the freshwater resources over the north and north-eastern of Thailand [20].

The aim of this study was to determine how microalgae are able to tolerate large alterations in their external Cu and pH environment. To this end we studied a green microalga *Desmodesmus* sp. (Chlorophyta, Chlorophyceae) AARLG074, a species isolated from and commonly found in a natural water reservoir in the north of Thailand. Specifically, we set out to measure the variations in the Cu removal efficiency, growth, ultrastructure, and cellular metabolite content in algal cells that are associated with Cu exposure and acidity. The Cu concentrations used in this study were set to the permittable Cu concentration in wastewater of Thailand (2 mg Cu/L) and Cu concentrations in the acid mine drainage system and industrial effluents (14–164 mg Cu/L) [3,6]. The pH values were also set according to the pH of AMD [4,5].

2. Results

2.1. Copper Removal Efficiency

The Cu removal efficiency of the cultures was assessed by growing *Desmodesmus* sp. AARLG074 in JM medium supplemented with 0, 2, 20, and 50 mg Cu/L at pH4 and pH6 for 168 h (7 days). The Cu removal efficiency was highest at pH6 where up to 83% of Cu in the media was absorbed within 30 min after inoculation (Table 1). The highest Cu absorption (93.5%) was measured 168 h after exposure to Cu in medium supplemented with 2 mg Cu/L, pH 6. Even in medium supplemented with 20 mg Cu/L the absorption after 24 h exposure at pH 6 was 92.8%.

In contrast, when algae were grown at pH 4 only 16.5% of Cu in the medium was removed at a Cu concentration of 20 mg Cu/L. The Cu absorption fluctuated over seven days with approximately 4–40% absorption at pH 4 and 64–94% at pH 6. Nonetheless, statistically significant lower Cu removal efficiency values were observed in all time points at pH 4, compared to pH 6 (Table 1).

Table 1. Percent removal of copper from growth media containing *Desmodesmus* sp. AARLG074 over 168 h (7 days) growth at pH4 and pH6. Each value represents the mean ± SD (n = 3) with different superscript letter in the same time point indicating statistically significant differences (Two-way ANOVA, Tukey HSD, $p < 0.05$) using R version 3.4.3.

	Mean Percentage of Copper Removal from the Media ± SD (n = 3)					
pH	4			6		
Cu (mg/L)	2	20	50	2	20	50
0 h	0	0	0	0	0	0
0.5 h	20.3 ± 26.6 [a]	13.1 ± 1.8 [a]	17.7 ± 3.8 [a]	57.6 ± 2.8 [b]	82.6 ± 5.2 [b]	82.5 ± 1.6 [b]
8 h	27.6 ± 24.7 [a]	15.5 ± 2.2 [a]	13.0 ± 1.8 [a]	67.2 ± 24.3 [b]	82.6 ± 5.2 [b]	81.0 ± 1.9 [b]
16 h	4.1 ± 5.7 [a]	14.0 ± 4.6 [a]	14.6 ± 1.7 [a]	64.6 ± 1.5 [b]	88.9 ± 2.2 [d]	77.2 ± 6.2 [c]
24 h	10.7 ± 6.1 [a]	16.5 ± 4.0 [ab]	27.7 ± 9.1 [b]	64.2 ± 1.4 [c]	92.9 ± 4.0 [d]	81.6 ± 2.6 [d]
72 h	34.1 ± 19.4 [a]	15.2 ± 4.9 [a]	16.4 ± 2.9 [a]	65.5 ± 0.5 [b]	91.3 ± 2.1 [c]	77.0 ± 5.0 [bc]
120 h	40.6 ± 16.5 [b]	18.0 ± 5.6 [a]	15.8 ± 1.7 [a]	64.6 ± 1.5 [c]	83.7 ± 2.7 [cd]	85.7 ± 1.2 [d]
168 h	39.2 ± 18.9 [a]	34.6 ± 0.6 [a]	20.3 ± 0.6 [a]	93.5 ± 1.2 [b]	73.9 ± 1.1 [b]	85.9 ± 0.1 [b]

2.2. Copper and Acid Exposure Affected the Cell Density, Pigment Content and Ultrastructure in Desmodesmus sp. AARLG074

Cell and colony density—The cell density was calculated based on the number of colonies within four categories, based on number of cells in each colony (single, duplet, triplet, quadruplet) (Figure 1). The cell density of algae grown under control conditions (no Cu, pH 6) gradually increased and reached the highest cell density after 120 h cultivation (Figure 1A). However, the cell density of Cu-treated algae cultures at pH 6 was indicated a longer stationary phase (up to 72 h) with the highest cell densities occurring after 120 h cultivation. The cell density of algae grown in non-Cu supplemented JM at pH 4, which reflects that effects of acidic stress (pH 4) on algal growth, was significantly less than the cell densities of algae grown pH 6 (Figure 1A) (Supplementary Materials Table S1). As measured in cultures grown at pH 6, the cell density of cultures grown in combined Cu and pH 4 media also had a statistically-significantly negative effect on the cell density of *Desmodesmus* sp. AARLG074, when compared to no Cu media controls (Figure 1A).

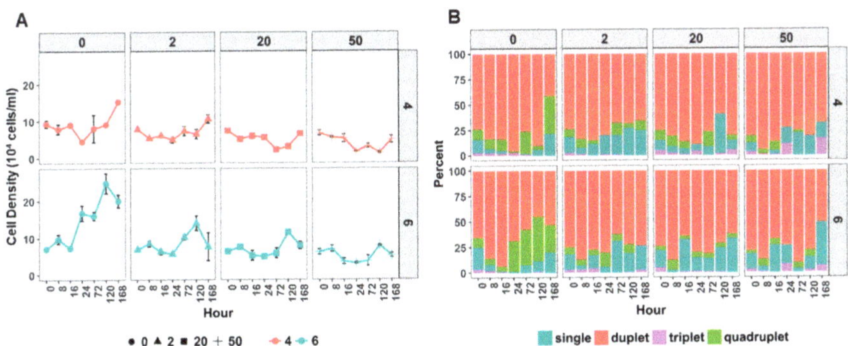

Figure 1. The effects of copper and pH on cell density and the percentage of number of cells per colony of *Desmodesmus* sp. AARLG074. (**A**) The mean cell density of *Desmodesmus* sp. AARLG074 grown at pH 4 (red line) and pH 6 (blue line) and copper concentrations (0, 2, 20, 50 mg/L) over 168 h. (**B**) The mean percentage of single (blue), duplet (pink), triplet (purple) and quadruplet (green) cells per colony of *Desmodesmus* sp. AARLG074 growing at pH 4 and pH 6 and copper concentration over 168 h. Values are mean ± SD (n = 3).

The majority of colonies over the testing period were duplet (Figure 1B). The percentage of duplet colonies increased over 16 h cultivation under both control (0 mg Cu/L) and low Cu (2 mg Cu/L) conditions at pH 4 or pH 6. Additionally, the percentage of triplet colonies (three-cells-colony) significantly increased when exposed to high copper (50 mg Cu/L) for 168 h. The percentage of quadruplet colonies (four-cells-colony) also increased after 24 h cultivation, but this effect was negated when Cu was added to the media (Figure 1B and Supplementary Materials Tables S2–S5).

Pigments—The pigment content (chlorophyll *a*, chlorophyll *b* and total carotenoids) of the cultures significantly decreased in the algal cultures containing Cu (Figure 2). This effect was more severe at pH 4 compared to pH 6 (Figure 2A and Supplementary Materials Tables S7–S9). The amount of pigments in the control (0 mg Cu/L) and low Cu (2 mg Cu/L) supplemented cultures gradually increased over time. This was not observed under high Cu supplemented conditions (20 and 50 mg Cu/L), where pigment concentrations were statistically lower than cultures grown at 0 and 2 mg Cu/L.

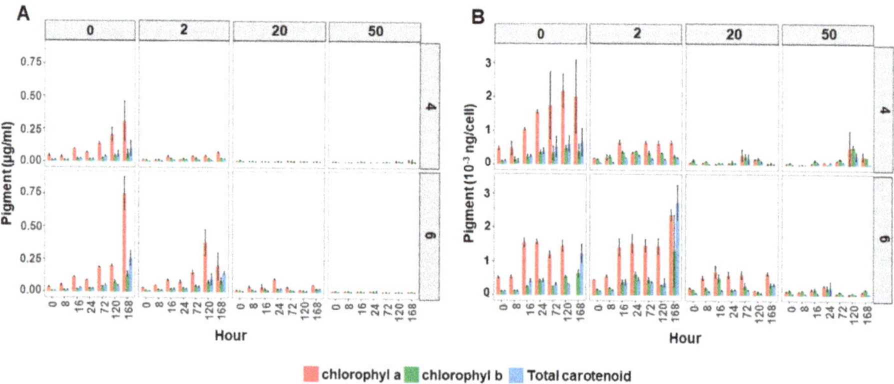

Figure 2. Pigment content (chlorophyll *a* (pink), chlorophyll *b* (green) and total carotenoids (blue)) of *Desmodesmus* sp. AARLG074 grown at pH 4 and pH 6 with varied copper concentrations (0, 2, 20, 50 mg/L) over 168 h. Data expressed as (**A**) pigments per volume of culture (ug per ml) and (**B**) per algal cell (ng per cell). Values are mean ± SD ($n = 3$).

This lower pigment content in the culture would partly be due to the lower cell count per ml of culture, but also a lower pigment concentration per cell as there was a similar response to Cu and pH when the pigment values were expressed as amount of pigments per cell (Figure 2B and Supplementary Materials Tables S7–S9). The amount of pigments in un-supplemented Cu cultures grown in either pH 4 or 6 gradually increased over time (Figure 2B). Also, under low Cu supplemented conditions (2 mg Cu/L) the pigment content per cell increased until 16 h and remained stable until the end of experiment, in both pH 4 and 6. However, the amount of chlorophyll *a* in cultures grown at 50 mg Cu/L was significantly lower (Supplementary Materials Table S9). The amount of chlorophyll *a* in pH 4 grown cultures was lower than pH 6 but not statistically different.

Ultrastructure—The ultrastructure of *Desmodesmus* sp. AARLG074 was altered after 24 h of Cu exposure as observed using TEM (Figure 3). Control algal cells that were grown in non-copper supplemented JM media at pH 4 had few starch granules with little or no periplasmic space (PM) (Figure 3A). However, control algal cells grown in non-copper supplemented JM media at pH 6 had many starch granules (S), again with little or no periplasmic space (PM) (Figure 3B).

In Cu supplemented media, the largest change in ultrastructure was observed in *Desmodesmus* sp. AARLG074 exposed to 50 mg Cu/L in both pH 4 (Figure 3C) and 6 (Figure 3D). The body of the cell had retracted from the cell wall resulting in an increased periplasmic space (PM) indicating that plasmolysis had occurred (Figure 3C,D). In addition, fewer starch grains (S) and membrane whorls (MW) were observed in the 50 mg Cu/L-treated cells at pH 6 (Figure 3D). The thylakoidal space was also more apparent under the high Cu and pH 4 (Figure 3C) conditions, compared to those grown under the high Cu and pH 6 conditions (Figure 3D).

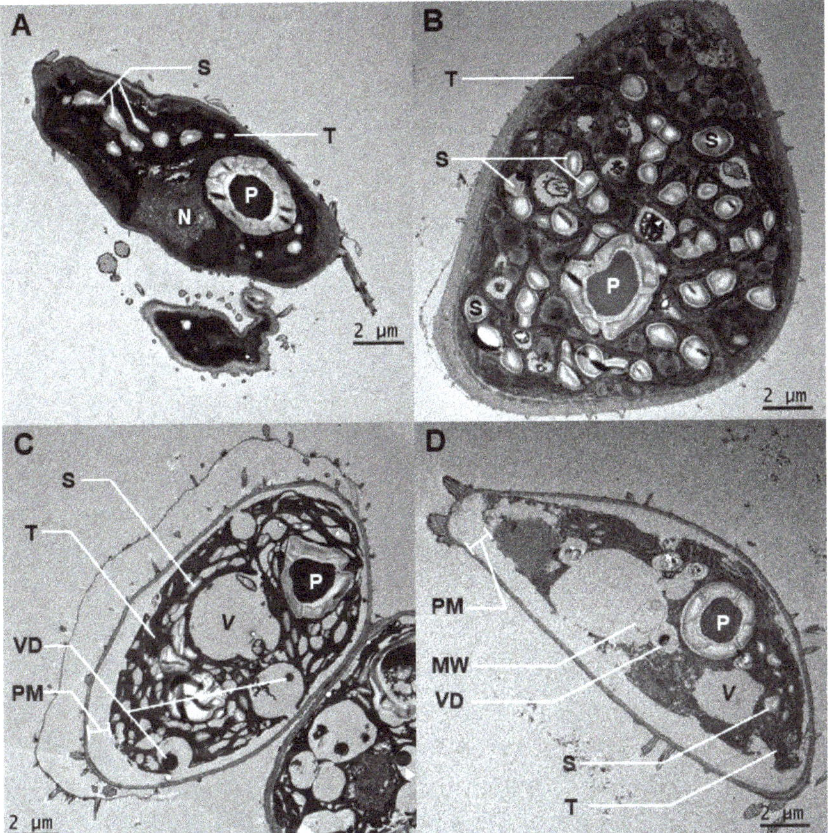

Figure 3. Electron micrographs of *Desmodesmus* sp. AARLG074 after 24 h growth in JM media at pH 4 and pH 6 at various Cu concentrations. The ultracellular structure of *Desmodesmus* sp. AARLG074 cultivated in the 0 mg Cu/L media at pH 4 (**A**) and pH 6 (**B**). Numerous starch granules (S) were observed. The ultrastructure of *Desmodesmus* sp. AARLG074 that grew in 50 mg Cu/L supplemented media at pH 4 (**C**) and pH 6 (**D**). Abbreviations: P, pyrenoids; PM, periplasmic space; S, starch granule; TH, thylakoids; V, vacuole; VD, vacuole deposit.

2.3. The Effects of Copper and pH on Metabolite Composition of Desmodesmus sp. AARLG074

Fourier transform-infrared spectrometry (FT-IR) was used to study the effects of different Cu concentrations and pH conditions on the metabolite composition of *Desmodesmus* sp. AARLG074 over 168 h of cultivation. The results, based on the principal component analysis (PCA) (Figure 4), showed that the metabolite fingerprints of *Desmodesmus* sp. AARLG074 cultivated in high Cu (20 and 50 mg Cu/L) at pH 6 were different from the metabolite fingerprints of algae that were cultivated in any other conditions. The specific wavenumber regions that were strongly associated with high copper concentration treated *Desmodesmus* sp. AARLG074 at pH 6 only was in the range of 835–1090 cm^{-1} (Figure 5B), which was associated with wavenumbers relating to polysaccharides.

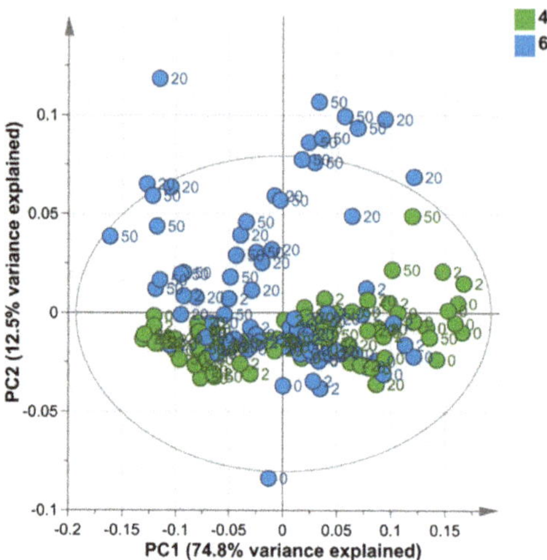

Figure 4. Score scatter plot from Principal Component Analysis (PCA) plot of FT-IR scans of *Desmodesmus* sp. AARLG074 cultured in JM under various copper concentrations (0, 2, 20 and 50 mg Cu/L) at pH 4 and pH 6. Each dot represents one biological replicate. The dot color represents the cultivation pH condition-pH 4 (green) and pH 6 (blue) and the labelled number beside dots represent the copper concentration as mg/L in the JM. The results shown here were analyzed using SIMCA-P (Version 15.0.2).

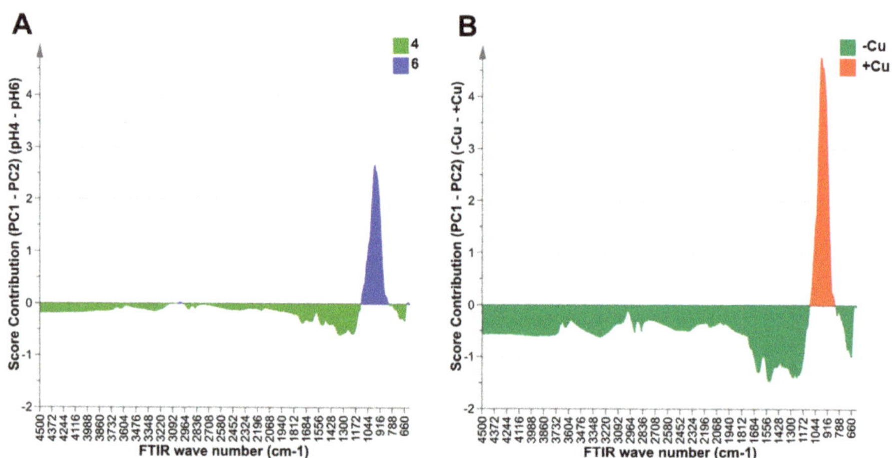

Figure 5. Score contribution plot values from PCA (PC1 and PC2 loading) from Fourier transform-infrared spectrometry (FT-IR) scans of *Desmodesmus* sp. AARLG074. FT-IR wavenumber score values are negative if they contribute towards PCA loadings associated with (**A**) growing in media at pH 4 (green) and positive for pH 6 (blue) and (**B**) negative if they contribute towards PCA loadings associated with no copper (green) and positive for copper supplemented media (red).

The detailed metabolite profiling by GC-MS did not reveal any major clustering of samples based on the pH and Cu concentration growth conditions in any of the principal components (Figure 6). Even though the PCA of the GC-MS data did not cluster by treatment, the score contribution plots of 40 key metabolites between absent Cu and Cu-supplemented treatment and between pH 4 and pH6 (Figure 7) were calculated to indicate the main metabolite differences between culture conditions. The score contribution of the full list of metabolites associated with the different Cu or pH are shown in Supplementary Materials Figures S1 and S2, respectively. The key metabolites based on the score contribution plots of both pH and Cu were polysaccharides, which corresponded to FT-IR results above (Figure 7). The other metabolites listed were lipids, amino and organic acids and vitamins. In pH 4 grown cells, the dominant metabolites were melezitose, galactose, hexadecane and mannose whereas in cells grown at pH 6 the dominant metabolites were phosphoric acid and cystathionine (Figure 7). The dominant metabolites in algae grown under high Cu supplemented media were sorbose, octacosane, hexadecanol, cystathionine and xylofuranose. The key metabolites under control conditions (zero Cu) were threose, lactose, galactose, maltose, ribose mannitol and lipids (e.g., decane and digalactosylglycerol). The statistical analysis of the detected metabolites showed that there were five statistically different metabolites among the treatments which were fructose, glycerol, lactose, melezitose and sorbose (Supplementary Materials Figure S3). Four of five statistically different metabolites among the treatments belonged to polysaccharides which supported the FT-IR results.

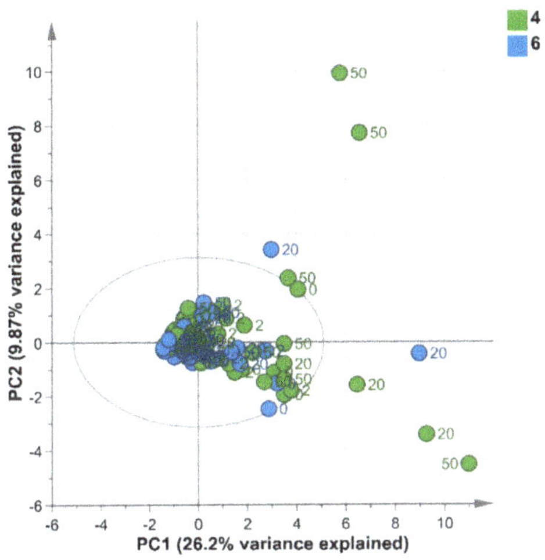

Figure 6. Metabolic profiling of the *Desmodesmus* sp. AARLG074 cultivated under different copper and pH conditions. The score scatter plot from PCA plot of GC-MS results of *Desmodesmus* sp. AARLG074 that were cultured in JM under various copper concentration (0, 2, 20 and 50 mg/L) at pH 4 and pH 6. Each dot represents one biological replicate. The colour of dot represents the cultivation pH condition-pH 4 (green) and pH 6 (blue) and the labelled number beside dots represent the copper concentration as mg/L in the JM.

The impact of Cu and pH on the metabolic pathways (based on the number of metabolites per pathway being identified and changing in intensity using the GC-MS data) were analysed using the MetaboAnalyst Pathway-Analysis function [21]. The results which had P values less than 0.05 are shown in Table 2. The pathway analysis of different pH condition revealed that pH had impacted on multiple amino acid, sugar and lipid metabolic pathways. The greatest impact of pH was shown on alanine, aspartate and glutamate metabolism. Additionally, Cu exposure had also affected various protein metabolic pathways with the greatest impact on arginine and proline metabolism. Moreover, the result indicated that both pH and Cu had an effect on glutathione metabolism and galactose metabolism. Although the impact levels were low there were altered metabolic pathways related to photosynthesis- pigment synthesis and carbon fixation in photosynthesis. The results showed that seven amino acid or protein synthesis pathways had been affected by pH. However, only four amino acid synthesis pathways (arginine and proline metabolism, lysine biosynthesis, beta-alanine metabolism and butanoate metabolism) had been affected by Cu (Table 2).

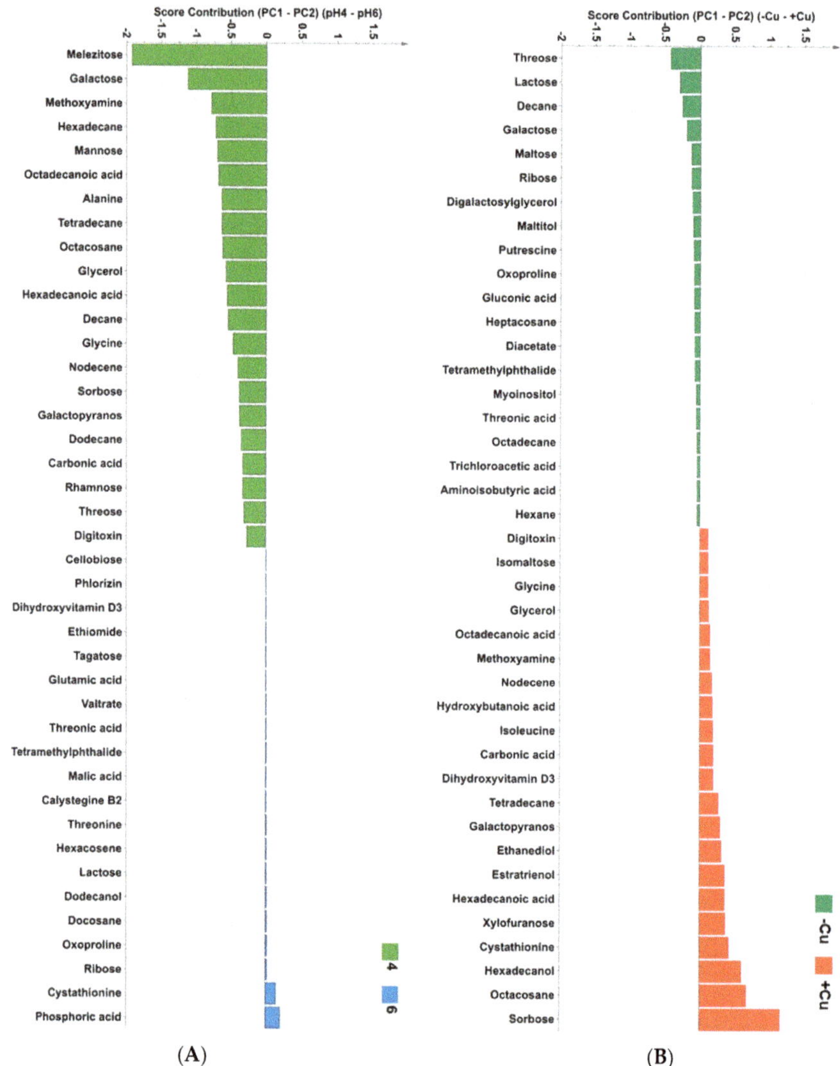

Figure 7. The score contribution plot values (top 40) that contribute towards GC-MS PCA loading plots of *Desmodesmus* sp. AARLG074 under different treatments (**A**) pH and (**B**) copper. The score contribution plot values (top 40) were ranked in order of importance and are negative if they contribute towards PCA loading plot for the pH 4 (green) and positive if they contribute towards the pH 6 (blue) and positive for copper stress (red) and negative if they contribute towards control (without copper) (green). The full list of the metabolites is presented in supplementary (Figures S1 and S2).

Table 2. Pathway analysis of the putatively identified metabolites by GC-MS in MetaboAnalyst open source 237 software (v4, pathway analysis tool) using the *Arabidopsis thaliana* metabolic 238 pathway library. The pathway analysis shows the effect of copper and pH on *Desmodesmus* sp. AARLG074 based on the number of metabolites matched in a metabolic pathway (Match), the statistical significance of the number of metabolites and their treatment response in that pathway (Raw P) and the importance of that pathway change in the network (Impact) [21].

Group of Metabolites	Pathway	pH			Copper		
		Match	Raw P	Impact	Match	Raw P	Impact
Lipid	Glycerolipid metabolism	3/13	2.86×10^{-5}	0.08		-	
Amino acid	Glycine, serine and threonine metabolism	4/30	8.78×10^{-3}	0.53		-	
	Cyanoamino acid metabolism	3/11	9.88×10^{-3}	0.00		-	
	Selenoamino acid metabolism	2/19	3.98×10^{-2}	0.00		-	
	Valine, leucine and isoleucine degradation	4/34	4.02×10^{-2}	0.00		-	
	Alanine, aspartate and glutamate metabolism	4/22	4.04×10^{-2}	0.54		-	
	Aminoacyl-tRNA biosynthesis	10/67	1.17×10^{-2}	0.09		-	
	Nitrogen metabolism	3/15	2.06×10^{-2}	0.00		-	
	Arginine and proline metabolism		-		5/38	2.15×10^{-6}	0.27
	Lysine biosynthesis		-		1/10	1.40×10^{-2}	0.00
	beta-Alanine metabolism		-		1/12	1.40×10^{-2}	0.00
	Butanoate metabolism		-		2/18	4.01×10^{-2}	0.00
Antioxidant or Homeostasis	Glutathione metabolism	4/26	1.20×10^{-2}	0.09	4/26	2.84×10^{-3}	0.09
	Ascorbate and aldarate metabolism		-		1/15	8.24×10^{-3}	0.00
Carbohydrate (Sugar)	Galactose metabolism	5/26	1.65×10^{-2}	0.09	5/26	3.03×10^{-3}	0.09
	Inositol phosphate metabolism		-		1/24	8.24×10^{-3}	0.25
Photosynthesis	Carbon fixation	2/21	3.94×10^{-2}	0.00		-	
	Porphyrin and chlorophyll metabolism		-		1/29	4.32×10^{-3}	0.00
Energy	Pentose phosphate pathway		-		2/18	7.60×10^{-5}	0.00
	Glyoxylate and dicarboxylate metabolism		-		1/17	2.78×10^{-3}	0.00
	Nicotinate and nicotinamide metabolism		-		1/12	1.40×10^{-2}	0.00
	Methane metabolism	2/11	9.98×10^{-3}	0.17		-	

3. Discussion

3.1. Tolerance of High Cu Exposure and Acidity in Desmodesmus sp. AARLG074—Growth and Structural Alterations Associated with Cu Removal

The aim of this study was to determine how *Desmodesmus* are able to physiologically tolerate large alterations in their external Cu and pH environment. *Desmodesmus* was indeed able to grow, although less so, in media supplemented with Cu. The Cu in the growth media was removed in the algae cultures with up to 83% of the copper present in the medium within 30 min after exposure, after which the absorption stabilizing over 168 h. We did not measure the Cu content in the cells but previous studies have shown that this biphasic absorption process may be due to a rapid non-metabolic dependent adsorption followed by a slow metabolic dependent uptake process [9,15]. The early absorption (<30 min) could be due to adsorption of Cu to the outer cell components (e.g., polysaccharides, mucilage and cell walls). The slower fluctuation in Cu removal efficiency (Table 1) would then be uptake into the living algal cell. In addition, the reduction in Cu removal efficiency at 20 mg Cu/L, pH 6 after 168 h exposure might be due to cell death and lysis, releasing the Cu back to the medium, or the living algae by export Cu out to the environment [15]. Moreover, the results also indicate that the pH is a critical factor that influences copper removal. The maximum removal, 93%, was when the algae were grown in JM media containing 2 mg Cu/L at pH 6. The efficiency of Cu absorption at pH 4 was significantly lower than pH 6, which was similar to several previous studies showing that heavy metal biosorption is a pH dependent process [22–26].

Despite the ability to tolerate Cu in the growth media, the exposure to Cu had an adverse effect on growth and cellular pigment content. Such effects have been previously observed and reported in other marine and freshwater algal species [8,10,16–19]. Although overall cell growth was reduced in cultures containing Cu we also found that the percentage of cell type was changed. There was a higher proportion of triplet cells in cultures that were cultivated in high Cu media at both pH 4 and pH 6 (Figure 1B). We can speculate that this might be one of the mechanisms that *Desmodesmus* sp. AARLG074 use for acclimating to live in a copper containing environment as multiple cell colonies have reduced cell surface area being exposed to the surrounding copper. Furthermore, the pigment analysis showed that chlorophyll *a*, *b* and total carotenoids content in algal cultures (Figure 2A) or per cell (Figure 2B), was lowered when cultivated in Cu supplemented media, this effect was much more severe in pH 4, a phenomena similar to previous studies [18,27–29]. Several studies have shown that the pigment level was reduced when algae were exposed to heavy metals [18,27–29], but none of them explain the reason of this phenomena. According to our TEM, ultrastructural changes of the algal cells occurred when exposed to Cu and acidity (Figure 3) notably at the high Cu concentration (50 mg Cu/L). One of the most drastic changes was the increase of intra-thylakoid space and disorganization of membrane which was similar to previous studies [19,24,29–32]. The alteration of the chloroplast ultrastructure was correlated with a lower chlorophyll content value (Figure 2). In addition, increasing the thylakoid space might be due to an excess of Cu accumulated inside due to Cu transporters locate on the pyrenoid and thylakoid membranes [33–35].

Furthermore, the Cu treated algae had fewer starch granules compared to the control, which might be due to an imbalance of the energy generated and used under Cu exposure as a consequence of chloroplast disorganization and fewer pigments [19,36]. Studies on another species of algae exposed to Cu, Desmidium swartzii, also had electron-dense inclusions in chloroplast and starch grains, of which Cu were detected in those compartments [37]. The presence of membrane whorls (MW) has previously been observed in heavy metal treated algae, where metals were detected inside the membrane whorl and as such has been proposed as a possible metal detoxification mechanism [38]

3.2. Metabolic Alterations Associated with Exposure to Cu and Acidity in Desmodesmus sp. AARLG074

The FT-IR results showed that different Cu and pH conditions affected a similar class of metabolites within the cells (Figures 4 and 5). The wavenumbers that differentiated treatments were 835–1090 cm^{-1}

(Figure 5A,B) which coincide with polysaccharide metabolites [39,40]. However, the GC-MS metabolite profiling showed that most of the key metabolites differed under the different pH or Cu conditions were sugars and amino acids (Figure 7). The PCA plots of both FT-IR metabolites fingerprinting and GC-MS metabolites profiling did indicate that there were relatively small differences in the metabolome of *Desmodesmus* sp. AARLG074 under various pH and copper treatments. This corresponds to a previous report on the effect of Cu on the transcriptome and metabolome of *Ectocarpus siliculosus* (Ochrophyta, Phaeophyceae) [41].

Five metabolites (glycerol, melezitose, lactose, fructose and sorbose) were highlighted as statistically different between treatments, which four of five are sugars (Figure S3) indicating some agreement between the FT-IR and GC-MS findings. From this list, the pH might be one of the crucial factors that affected the level of glycerol as there was considerably more glycerol in cells grown under pH 4 whereas there was no change under pH 6. The high level of glycerol at pH 4 might be explained for several reasons, largely osmoregulatory and membrane composition. Accumulation of intracellular glycerol in algae function as an osmoregulator which lowers internal osmotic potential and so prevent excessive water loss during high Cu exposure [42]. Additionally, released glycerol may also serve as a sink for products of photosynthesis [42]. Glycerol is also part of plasma membrane component and high glycerol induce membrane stiffening [43,44] and changes in membrane composition will affect the membrane permeability. The plasma membrane of algae living in acidic environments is fairly impermeable for protons, requiring relatively little energy for active pumping against a passive proton influx [45]. Thus, the higher level of glycerol detected in pH 4 might be one of the mechanisms that this green microalgae *Desmodesmus* sp. AARLG074 used to cope with the acidic and/or Cu stress. Moreover, melezitose (syn: melicitose), is an allelopathic chemical in marine planktonic microalgae [46]. This chemical is also found to be upregulated under osmotic stress in higher plants [47,48]. However, to our knowledge there is no previous report on an association between melezitose and Cu or pH in algae.

Beyond the top five metabolites mentioned above, one of metabolites that was highlighted in the score contribution plots as being important in cells grown in both pH 6 and Cu conditions was cystathionine (Figure 7). The cystathionine is an important intermediate compound in the biosynthesis of cysteine which is linked to glutathione metabolism [49]. The addition of glutamate to cysteine leads to form glutamyl cysteine, which the addition of glycine to this glutamyl cysteine leads to form glutathione [49]. This corresponded to our pathway analysis that showed Cu had impacted on glutathione metabolism (Table 2). Importantly, glutathione have been reported as an important intracellular ligands involved in metal sequestration and detoxification in algae [50].

Although the top five metabolites that were statistically different between the treatments were sugars the pathways that had the highest impact score for different pH and Cu were alanine, aspartate and glutamate metabolism (impact = 0.54) and arginine and proline metabolism (impact = 0.27) (Table 2). The pH treatment had the biggest impact on alanine, aspartate and glutamate metabolism, whereas Cu had the biggest impact on arginine and proline metabolism. The impact scores are a combination of the number of metabolites detected in the pathway and the fold change between treatments of those metabolites. This implies that although the metabolites may not statistically different between treatments, the impact score may still be on when taking all the metabolites in that pathway into consideration. Care should be taken when interpreting the in silico pathway analysis scores as the pathways were not measured directly, only the individual metabolites that could be in one or more metabolic pathway. However, such pathway analyses do provide us with an initial snapshot of which pathways were the most likely to be changing, which provides useful information for further targeted studies. There are similarities with the metabolites and pathways highlighted in the analysis in other organisms, especially bacteria [51–55]. Glutamate has been reported to be involved in acid tolerance in *Streptococcus mutans* [52], in both acid and Cu tolerance of *Escherichia coli* [53] and the accumulation of potassium glutamate in *E. coli* has been shown to occur immediately after osmotic shock to provide temporary protection to the cells [53,55]. In addition, aspartic acid and glutamic acid are also reported to be involved in acid tolerance in *Acetobacter pasteurianus* [51]. Proline is well known as playing an important role in plants and microorganism osmoprotection, metal chelation and general antioxidative defense molecule and signaling molecule [52,56,57]. Enhancement and accumulation of proline in heavy metals and acid exposed algae and bacteria has also been previously reported [48,57–59].

To conclude, we have shown that cultures of are able to tolerate and grow on media in both high Cu and acidic conditions (pH 4) and indeed removes over 80% Cu from the media within 30 min. The Cu concentration did affect Cu removal percentages. We showed that at the highest Cu removal efficiency and cell density, approximately 0.002 ng Cu could be removed per cell. This information shows that *Desmodesmus* sp. AARLG074 has the potential to industrially recover Cu from the algae that would otherwise be lost. However, despite its tolerance, the higher concentrations of Cu did negatively affect its growth rate. Additionally, we have shown that exposure and tolerance to Cu and acid conditions was associated with changes in its morphology, ultrastructure and overall metabolite composition of the cells, with polysaccharide and amino acid biochemistry changing the most.

4. Materials and Methods

4.1. Organism and Culture Conditions

The unicellular green microalga *Desmosdesmus* sp. AARLG074 was obtained from the Applied Algal Research Laboratory (AARL), Department of Biology, Faculty of Science, Chiang Mai University. This strain was isolated from the northern Thailand natural water reservoir [60]. The algal feed stock was maintained in Jaworski's medium (JM), pH 7.0 under continuous white light emitting diode (LED) illumination (75 µmol photon/m^2/s) at 25 °C and shaken by hand twice daily. The optical density 650 nm and cell density was measured until it reached the log phase after which the cultures were used as stock for further experiments. To study the effects of pH and Cu on Cu removal efficiency, algal growth, ultrastructure and metabolites the algae were inoculated into 1l flasks containing 800 mL Jaworski's medium (JM) supplemented with 0, 2, 20 and 50 mg Cu/L by using CuSO$_4$·5H$_2$O as a Cu source. The pH of medium was adjusted to pH 4 and pH 6 using hydrochloric acid (HCl). Each treatment was conducted in triplicate biological samples ($n = 3$), in total 24 flasks were used. Atomic absorption spectroscopy (AAS) (see below method) confirmed the initial copper concentration in the treatment of 0, 2, 20, and 50 mg Cu/L was 0, 1.97 ± 0.11, 20.59 ± 0.55, and 50.66 ± 1.03 mg/L respectively.

4.2. Copper Absorption Efficiency

To determine the copper absorption efficiency, 10 mL of culture from three biological replicate samples ($n = 3$) was collected using an autopipette at various time points from 0 to 168 h (0, 0.5, 8, 16, 24, 72, 120 and 168 h) and then centrifuged at 3000 rpm for 5 min. The Cu concentration in the supernatant was measured using flame atomic absorption spectroscopy (AAS) as described in Mota et al. (2015) [61] and the removal efficiency (%) was calculated using the following equation:

$$\text{Removal efficiency (\%)} = (100/\text{Ci}) \times (\text{Ci} - \text{Cf}) \tag{1}$$

where: Ci is the initial copper concentration (mg/L), Cf is the final (residual) copper concentration (mg/L).

4.3. The Effects of Copper and pH on Cell Density, Pigments and Ultrastructure

The three biological replicates of algae that were cultured in different conditions were collected at the same time point as described above. *Desmodesmus* sp. are colony forming. As 1, 2 and 4 cell colonies were observed (3 cell colonies were rarely found) the cell density (cells per ml of culture) was counted into four categories based on the number of cells per colony—single, duplet, triplet and quadruplet using a hematocytometer. The percentage of each type of colony found in the cultures was calculated to investigate the effect of pH and Cu on the change of percentage of each colony presented in algal community over time. Chlorophylls and carotenoids were extracted using 90% methanol as modified method from Saijo (1975) [62] and Lichtenthaler and Buschmann (2001) [63]. In addition, the amount of pigment per cell was calculated based on the amount of pigment in culture (μg/mL) and cell density (ng/cell). Transmission electron microscopy (TEM) was used to investigate the ultrastructural change of *Desmodesmus* sp. AARLG074 after 24 h exposed to Cu and acidic stress [38].

4.4. Metabolite Analysis

Ten ml of samples were collected at 0, 0.5, 8, 16, 24, 72, 120 and 168 h into the treatment phase and centrifuged at 3000 rpm at 4 °C for 5 min. The supernatant was discarded and the pellets were freeze-dried (freezone labconco, USA) and transferred to the Department of Plant Sciences, University of Cambridge, UK. Metabolite fingerprints of freeze-dried samples were obtained using a Perkin-Elmer Spectrum Two FT-IR, within the wavenumbers of 600–4500 cm^{-1}. The spectra were normalized against air.

The metabolite profiling of freeze-dried samples was further investigated in more detail using GC-MS by extracting soluble polar and non-polar metabolites in methanol-chloroform-water method as described in Davey et al. (2005) [64]. The compounds within the polar methanol-water phase were derivatized by N-Methyl-N-trimethylsilyl-trifluoroacetamide (MSTFA) and Trimethylsilyl (TMS) as described by Dunn et al. (2011) [65]. Then, 1 μL (splitless) of the derivatized extracts were separated and profiled by GC-MS (GC-MS (Thermo Scientific Trace 1310 GC with 211 ISQ LT MS, Xcaliber v2.2) with a ZB-5MSi column (30 m, 0.25 mm ID, 0.25 μm film 212 thickness, Phenomenex, UK). GC-MS spectra were aligned to an internal standard (phenyl-β-D-glucopyranoside hydrate 98%) and processed using Thermo TraceFinder (v3.1) using genesis peak search method to aid identification based on molecular mass as previously described [64,66]. Pathway analysis of the identified metabolites were done both on SIMCA-P program (v15.0.2 Umetrics, Sweden) and MetaboAnalyst open source software (version 4.0, pathway analysis tool) using the *Arabidopsis thaliana* metabolic pathway library [21], www.metaboanalyst.ca). The results of pathway analysis that have *p* value ≤ 0.05 only were shown in the Table 2. R-script for the metaboanalyst software can be downloaded at https://github.com/xia-lab/MetaboAnalystR [21].

4.5. Statistical Analyses

To indicate the effect of initial Cu and pH on copper absorption efficiency, cell density and pigments of *Desmodesmus* sp. AARLG074 a two-way ANOVA (after a Lavene's normality distribution) with Tukey HSD test (*p* ≤ 0.05) was performed using car, multcompView and lsmeans library on R version 3.4.3. as previously described [67]. Multivariate analyses to test whether the effect of initial Cu and pH on the algae could be discriminated based on their identified and unidentified metabolites (from FT-IR fingerprints or GC-MS profiling datasets) were performed using Principal Component Analysis (PCA) [68] on Pareto Scaled data (FT-IR absorbance values or GC-MS identified peak area units) within the SIMCA-P v14.1 PCA analysis pipeline (Umetrics, Sweden) to produce standard score scatter plots and ranked score contribution plots of how each variable (FT-IR wavenumber or GC-MS metabolite) contributed to clustering within the PCA score scatter plot.

Supplementary Materials: The following are available online at http://www.mdpi.com/2218-1989/9/5/84/s1, Table S1: The effects of copper and pH over 7 days on cell density (cells/mL) of *Desmodesmus* sp. AARLG074, Table S2: The effects of copper and pH over 7 days on percentage of single colony (1-cell-colony) of *Desmodesmus* sp. AARLG074, Table S3: The effects of copper and pH over 7 days on percentage of duplet colony (two-cells-colony) of *Desmodesmus* sp. AARLG074, Table S4: The effects of copper and pH over 7 days on percentage of triplet colony (three-cells-colony) of *Desmodesmus* sp. AARLG074, Table S5: The effects of copper and pH over 7 days on percentage of quadruplet colony (four-cells-colony) of *Desmodesmus* sp. AARLG074, Table S6: The effects of copper and pH over 7 days on chlorophyll *a* (μg/mL) of *Desmodesmus* sp. AARLG074, Table S7: The effects of copper and pH over 7 days on chlorophyll *b* (μg/mL) of *Desmodesmus* sp. AARLG074, Table S8: The effects of copper and pH over 7 days on total carotenoids (μg/m) of *Desmodesmus* sp. AARLG074, Table S9: The effects of copper and pH over 7 days on chlorophyll *a* (10^{-3} ng/cell) of *Desmodesmus* sp. AARLG074, Table S10: The effects of copper and pH over 7 days on chlorophyll *b* (10^{-3} ng/cell) of *Desmodesmus* sp. AARLG074, Table S11: The effects of copper and pH over 7 days on total carotenoids (10^{-3} ng/cell) of *Desmodesmus* sp. AARLG074, Figure S1: Score contribution plot values showing the full list which contribute towards GC-MS PCA loading plots of

Desmodesmus sp. AARLG074 under different pH condition, Figure S2: Score contribution plot values showing the full list which contribute towards GC-MS PCA loading plots of *Desmodesmus* sp. AARLG074 under different copper condition, Figure S3: Statistical analysis of GC-MS metabolite profiling of Desmodesmus sp. AARLG074 under copper and acidic stress.

Author Contributions: All authors have contributed to the conceptualization, investigation, data analysis and writing, reviewing and editing of this work.

Funding: We would like to thank the Development and Promotion of Science and Technology Talents Project (DPST Scholar), which is jointly supported by the Ministry of Science and Technology, the Ministry of Education, and the Institute for the Promotion of Teaching Science and Technology (IPST) who provided financial support to conduct this research in Thailand and aboard.

Acknowledgments: We truly appreciated the kindly help, support and advice of Jeeraporn Pekkoh. Also thank you all supportive and cheerful lab members of the Applied Algal Research Laboratory (AARL), Department of Biology, Faculty of Science, Chiang Mai University, Thailand and lab members of 220 Plant Metabolism Group, Department of Plants Sciences, University of Cambridge, United Kingdom. This research work was partially supported by Chiang Mai University.

Conflicts of Interest: The authors declare no conflict of interest.

References

1. Al-saydeh, S.A.; El-naas, M.H.; Zaidi, S.J. Copper removal from industrial wastewater: A comprehensive review. *J. Ind. Eng. Chem.* **2017**, *56*, 35–44. [CrossRef]

2. Elshkaki, A.; Graedel, T.E.; Ciacci, L.; Reck, B.K. Copper demand, supply, and associated energy use to 2050. *Glob. Environ. Chang.* **2016**, *39*, 305–315. [CrossRef]

3. Breedveld, G.D.; Klimpel, F.; Kvennås, M.; Okkenhaug, G. Stimulation of natural attenuation of metals in acid mine drainage through water and sediment management at abandoned copper mines. In Proceedings of the International Mine Water Association 2017 Congress (IMWA 2017)-Mine Water and Circular Economy, Lappeenrante, Finland, 25–30 June 2017; pp. 1133–1137.

4. Liang-qi, L.; Ci-an, S. Acid mine drainage and heavy metal contamination in groundwater of metal sulfide mine at arid territory (BS mine, Western Australia). *Trans. Nonferrous Met. Soc. China* **2010**, *20*, 1488–1493.

5. Ehrlich, H.L. Microorganisms in Acid Drainage from a Copper Mine. *J. Bacteriol.* **1963**, *86*, 350–352.

6. Basha, C.A.; Bhadrinarayana, N.S.; Anantharaman, N.; Sheriffa, K.M.M. Heavy metal removal from copper smelting effluent using electrochemical cylindrical flow reactor. *J. Hazard Mater.* **2008**, *152*, 71–78. [CrossRef] [PubMed]

7. Larsson, M.; Nosrati, A.; Kaur, S.; Wagner, J.; Baus, U.; Nydén, M. Copper removal from acid mine drainage-polluted water using glutaraldehyde-polyethyleneimine modified diatomaceous earth particles. *Heliyon* **2018**, *10*. [CrossRef] [PubMed]

8. Zou, H.; Pang, Q.; Zhang, A.; Lin, L.; Li, N.; Yan, X. Excess copper induced proteomic changes in the marine brown algae *Sargassum fusiforme. Ecotoxicol. Environ. Saf.* **2015**, *111*, 271–280. [CrossRef] [PubMed]

9. Anu, P.R.; Bijoy Nandan, S.; Jayachandran, P.R.; Don Xavier, N.D. Toxicity effects of copper on the marine diatom, *Chaetoceros calcitrans. Reg. Stud. Mar. Sci.* **2016**, *8*, 498–504. [CrossRef]

10. Debelius, B.; Forja, J.M.; DelValls, Á.; Lubián, L.M. Toxicity and bioaccumulation of copper and lead in five marine microalgae. *Ecotoxicol. Environ. Saf.* **2009**, *72*, 1503–1513. [CrossRef] [PubMed]

11. Johnson, H.L.; Stauber, J.L.; Adams, M.S.; Jolley, D.F. Copper and zinc tolerance of two tropical microalgae after copper acclimation. *Environ. Toxicol.* **2007**, *22*, 234–244. [CrossRef]

12. Singh, R.; Gautam, N.; Mishra, A.; Gupta, R. Heavy metals and living systems: An overview. *Indian J. Pharmacol.* **2011**, *43*, 246–253. [CrossRef]

13. Usak, V.I.H. Copper and copper-containing pesticides: Metabolisms, toxicity and oxidative stress. *J. Vasyl. Stefanyk Precarpathian Natl. Univ.* **2015**, *2*, 38–50.

14. Hill, K.; Hassett, R.; Kosman, D.; Merchant, S. Regulated copper uptake in Chlamydomonas reinhardtii in response to copper availability. *Plant Physiol.* **1996**, *112*, 697–704. [CrossRef] [PubMed]

15. Kaplan, D. Absorption and Adsorption of Heavy Metals by Microalgae. In *Handbook of Microalgal Culture: Applied Phycology and Biotechnology*, 2nd ed.; Richmond, A., Hu, Q., Eds.; John Wiley & Sons Ltd.: Oxford, UK, 2013; pp. 602–611. [CrossRef]

16. Levy, J.L.; Angel, B.M.; Stauber, J.L.; Poon, W.L.; Simpson, S.L.; Cheng, S.H.; Jolley, D.F. Uptake and internalisation of copper by three marine microalgae: Comparison of copper-sensitive and copper-tolerant species. *Aquat. Biosyst.* **2008**, *89*, 82–93. [CrossRef]

17. Cao, D.J.; Xie, P.P.; Deng, J.W.; Zhang, H.M.; Ma, R.X.; Liu, C.; Liu, R.; Liang, Y.; Li, H.; Shi, X. Effects of Cu^{2+} and Zn^{2+} on growth and physiological characteristics of green algae, *Cladophora*. *Environ. Sci. Pollut. Res.* **2015**, *22*, 16535–16541. [CrossRef]

18. Sabatini, S.E.; Juárez, Á.B.; Eppis, M.R.; Bianchi, L.; Luquet, C.M.; Ríos de Molina M del, C. Oxidative stress and antioxidant defenses in two green microalgae exposed to copper. *Ecotoxicol. Environ. Saf.* **2009**, *72*, 1200–1206. [CrossRef]

19. Volland, S.; Bayer, E.; Baumgartner, V.; Andosch, A.; Lütz, C.; Sima, E.; Lütz-Meindl, U. Rescue of heavy metal effects on cell physiology of the algal model system *Micrasterias* by divalent ions. *J. Plant Physiol.* **2014**, *171*, 154–163. [CrossRef] [PubMed]

20. Phinyo, K.; Pekkoh, J.; Peerapornpisal, Y. Distribution and ecological habitat of *Scenedesmus* and related genera in some freshwater resources of Northern and North-Eastern Thailand. *Biodiversitas* **2017**, *18*, 1092–1099. [CrossRef]

21. Chong, J.; Soufan, O.; Li, C.; Caraus, I.; Li, S.; Bourque, G.; Wishart, D.S.; Xia, J. MetaboAnalyst 4.0: Towards more transparent and integrative metabolomics analysis. *Nucleic Acids Res.* **2018**, *46*, 486–494. [CrossRef]

22. Kleinübing, S.J.; Vieira, R.S.; Beppu, M.M.; Guibal, E.; da Silva, M.G.C. Characterization and evaluation of copper and nickel biosorption on acidic algae *Sargassum Filipendula*. *Mater. Res.* **2010**, *13*, 541–550. [CrossRef]

23. Al-Homaidan, A.A.; Al-Houri, H.J.; Al-Hazzani, A.A.; Elgaaly, G.; Moubayed, N.M.S. Biosorption of copper ions from aqueous solutions by *Spirulina* a biomass. *Arab. J. Chem.* **2014**, *7*, 57–62. [CrossRef]

24. Rangsayatorn, N.; Upatham, E.S.; Kruatrachue, M.; Pokethitiyook, P.; Lanza, G.R. Phytoremediation potential of *Spirulina* (*Arthrospira*) *platensis*: Biosorption and toxicity studies of cadmium. *Environ. Pollut.* **2002**, *119*, 45–53. [CrossRef]

25. Wilde, K.L.; Stauber, J.L.; Markich, S.J.; Franklin, N.M.; Brown, P.L. The effect of pH on the uptake and toxicity of copper and zinc in a tropical freshwater alga (*Chlorella* sp.). *Arch. Environ. Contam. Toxicol.* **2006**, *51*, 174–185. [CrossRef]

26. Zeraatkar, A.K.; Ahmadzadeh, H.; Talebi, A.F.; Moheimani, N.R. Potential use of algae for heavy metal bioremediation, a critical review. *J. Environ. Manag.* **2016**, *181*, 817–831. [CrossRef] [PubMed]

27. Kováčik, J.; Klejdus, B.; Hedbavny, J.; Bačkor, M. Effect of copper and salicylic acid on phenolic metabolites and free amino acids in *Scenedesmus quadricauda* (*Chlorophyceae*). *Plant Sci.* **2010**, *178*, 307–311. [CrossRef]

28. Kondzior, P.; Butarewicz, A. Effect of heavy metals (Cu and Zn) on the content of photosynthetic pigments in the cells of algae *Chlorella vulgaris*. *J. Ecol. Eng.* **2018**, *19*, 18–28. [CrossRef]

29. Rodrigo, W.; Schmidt, É.C.; Felix, M.R.D.L.; Polo, L.K.; Kreusch, M.; Pereira, D.T.; Costa, G.B.; Simioni, C.; Chow, F.; Ramlov, F.; et al. Bioabsorption of cadmium, copper and lead by the red macroalga *Gelidium floridanum*: Physiological responses and ultrastructure features. *Ecotoxicol. Environ. Saf.* **2014**, *105*, 80–89. [CrossRef]

30. Surosz, W.; Palinska, K.A. Ultrastructural changes induced by selected cadmium and copper concentrations in the cyanobacterium *Phormidium*: Interaction with salinity. *J. Plant Physiol.* **2000**, *157*, 643–650. [CrossRef]

31. Rachlin, J.W.; Jensen, T.E.; Baxter, M.; Jani, V. Utilization of Morphometric Analysis in Evaluating Response of *Plectonema boryanum* (*Cyanophyceae*) to Exposure to Eight Heavy Metals. *Arch. Environ. Contam. Toxicol.* **1982**, *11*, 323–333.

32. Nie, M.; Wang, Y.; Yu, J.; Xiao, M.; Jiang, L.; Yang, J.; Fang, C.; Chen, J.; Li, B. Understanding Plant-Microbe Interactions for Phytoremediation of Petroleum-Polluted Soil. *PLoS ONE* **2011**, *6*, e17961. [CrossRef]

33. Marchand, J.; Heydarizadeh, P.; Schoefs, B.; Spetea, C. Ion and metabolite transport in the chloroplast of algae: Lessons from land plants. *Cell. Mol. Life Sci.* **2018**, *75*, 2153–2176. [CrossRef] [PubMed]

34. Kropat, J.; Gallaher, S.D.; Urzica, E.I.; Nakamoto, S.S.; Strenkert, D.; Tottey, S.; Andrew ZMason, A.Z.; Merchant, S.S. Copper economy in *Chlamydomonas*: Prioritized allocation and reallocation of copper to respiration vs. photosynthesis. *Proc. Natl. Acad. Sci. USA* **2015**, *112*, 2644–2651. [CrossRef]

35. Pfeil, B.E.; Schoefs, B.; Spetea, C. Function and evolution of channels and transporters in photosynthetic membranes. *Cell. Mol. Life Sci.* **2014**, *71*, 979–998. [CrossRef] [PubMed]

36. Cid, A.; Herrero, C.; Torres, E.; Abalde, J. Copper toxicity on the marine microalga *Phaeodactylum tricornutum*: Effects on photosynthesis related parameters. *Aquat. Toxicol.* **1995**, *31*, 165–174. [CrossRef]

37. Andosch, A.; Höftberger, M.; Lütz, C.; Lütz-meindl, U. Subcellular Sequestration and Impact of Heavy Metals on the Ultrastructure and Physiology of the Multicellular Freshwater Alga *Desmidium swartzii*. *Int. J. Mol. Sci.* **2015**, *16*, 10389–10410. [CrossRef] [PubMed]
38. Nishikawa, K.; Yamakoshi, Y.; Uemura, I.; Tominaga, N. Ultrastructural changes in *Chlamydomonas acidophila* (*Chlorophyta*) induced by heavy metals and polyphosphate metabolism. *FEMS Microbiol. Ecol.* **2003**, *44*, 253–259. [CrossRef]
39. Benning, L.G.; Phoenix, V.R.; Yee, N.; Tobin, M.J. Molecular characterization of cyanobacterial silicification using synchrotron infrared micro-spectroscopy. *Geochim. Cosmochim. Acta* **2004**, *68*, 729–741. [CrossRef]
40. Mayers, J.J.; Flynn, K.J.; Shields, R.J. Rapid determination of bulk microalgal biochemical composition by Fourier-Transform Infrared spectroscopy. *Bioresour. Technol.* **2013**, *148*, 215–220. [CrossRef] [PubMed]
41. Ritter, A.; Dittami, S.M.; Goulitquer, S.; Correa, J.A.; Boyen, C.; Potin, P. Transcriptomic and metabolomic analysis of copper stress acclimation in *Ectocarpus siliculosus* highlights signaling and tolerance mechanisms in brown algae. *BMC Plant Biol.* **2014**, *14*, 1–17. [CrossRef] [PubMed]
42. Demmig-adams, B.; Burch, T.A.; Stewart, J.J.; Savage, E.L.; Iii, W.W.A. Algal glycerol accumulation and release as a sink for photosynthetic electron transport. *Algal Res.* **2017**, *21*, 161–168. [CrossRef]
43. Thompson, G.A.J. Lipids and membrane function in green algae. *Biochim. Biophys. Acta* **1996**, *1302*, 17–45. [CrossRef]
44. Pocivavsek, L.; Gavrilov, K.; Cao, K.D.; Chi, E.Y.; Li, D.; Lin, B.; Meron, M.; Majewski, J.; Lee, K.Y. Glycerol-Induced Membrane Stiffening: The Role of Viscous Fluid Adlayers. *Biophys. J.* **2011**, *101*, 118–127. [CrossRef]
45. Gross, W. Ecophysiology of algae living in highly acidic environments. *Hydrobiologia* **2000**, *433*, 31–37. [CrossRef]
46. Eick, K.C. Metabolomic Analysis of the Allelopathic Interactions between Marine Planktonic Microalgae. Ph.D. Dissertation, Friedrich-Schiller-Universität Jena, Jena, Germany, 1987.
47. Zhao, X.; Wang, W.; Zhang, F.; Deng, J.; Li, Z.; Fu, B. Comparative Metabolite Profiling of Two Rice Genotypes with Contrasting Salt Stress Tolerance at the Seedling Stage. *PLoS ONE* **2014**, *9*, 1–7. [CrossRef]
48. Vílchez, J.I.; Niehaus, K.; Dowling, D.N.; González-lópez, J. Protection of Pepper Plants from Drought by *Microbacterium* sp. 3J1 by Modulation of the Plant's Glutamine and α -ketoglutarate Content: A Comparative Metabolomics Approach. *Front. Microbiol.* **2018**, *9*, 1–17. [CrossRef]
49. Kumaresan, V.; Nizam, F.; Ravichandran, G.; Viswanathan, K.; Palanisamy, R.; Bhatt, P.; Arasu, M.V.; Al-Dhabi, N.A.; Mala, K.; Arockiaraj, J. Transcriptome changes of blue-green algae, *Arthrospira* sp. in response to sulfate stress. *Algal Res.* **2017**, *23*, 96–103. [CrossRef]
50. Smith, C.L.; Steele, J.E.; Stauber, J.L.; Jolley, D.F. Copper-induced changes in intracellular thiols in two marine diatoms: *Phaeodactylum tricornutum* and *Ceratoneis closterium*. *Aquat. Toxicol.* **2014**, *156*, 211–220. [CrossRef] [PubMed]
51. Yin, H.; Zhang, R.; Xia, M.; Bai, X.; Mou, J.; Zheng, Y.; Wang, M. Effect of aspartic acid and glutamate on metabolism and acid stress resistance of *Acetobacter pasteurianus*. *Microb. Cell Fact.* **2017**, *16*, 1–14. [CrossRef]
52. Krastel, K.; Senadheera, D.B.; Mair, R.; Downey, J.S.; Goodman, S.D.; Cvitkovitch, D.G. Characterization of a Glutamate Transporter Operon, glnQHMP, in *Streptococcus mutans* and Its Role in Acid Tolerance. *J. Bacteriol.* **2010**, *192*, 984–993. [CrossRef]
53. Djoko, K.Y.; Phan, M.; Peters, K.M.; Walker, M.J.; Schembri, M.A.; Mcewan, A.G. Interplay between tolerance mechanisms to copper and acid stress in *Escherichia coli*. *Proc. Natl. Acad. Sci. USA* **2017**, *114*, 6818–6823. [CrossRef]
54. Hamada, A.; Hibino, T.; Nakamura, T.; Takabe, T. Na+/H+ Antiporter from *Synechocystis* species PCC6803, Homologous to SOS1, Contains an Aspartic Residue and Long C-Terminal Tail Important for the Carrier Activity. *Plant Physiol.* **2001**, *15*, 437–446. [CrossRef]
55. Gralla, J.D.; Vargas, D.R. Potassium glutamate as a transcriptional inhibitor during bacterial osmoregulation. *Eur. Mol. Biol. Organ.* **2006**, *25*, 1515–1521. [CrossRef]
56. Hayat, S.; Hayat, Q.; Alyemeni, M.N.; Wani, A.S.; Pichtel, J.; Ahmad, A. Role of proline under changing environments A review. *Plant Signal Behav.* **2012**, *7*, 1456–1466. [CrossRef]
57. Siripornadulsil, S.; Traina, S.; Verma, P.S.; Sayre, R.T. Molecular Mechanisms of Proline-Mediated Tolerance to Toxic Heavy Metals in Transgenic Microalgae. *Plant Cell* **2002**, *14*, 2837–2847. [CrossRef]

58. Zhang, L.P.; Mehta, S.K.; Liu, Z.P.; Yang, Z.M. Copper-Induced Proline Synthesis is Associated with Nitric Oxide Generation in Chlamydomonas reinhardtii. *Plant Cell Physiol.* **2008**, *49*, 411–419. [CrossRef]

59. Wu, J.-T.; Chang, S.-J.; Chou, T.-L. Intracellular proline accumulation in some algae exposed to copper and caddmium. *Bot. Bull. Acad. Sin. Taipei* **1995**, *36*, 89–93.

60. Phinyo, K. Diversity of Microalgae Genus *Scenedesmus* and Related Genera from Some Freshwater Resources in Thailand and Some Biochemical Properties of Selected Strains. Ph.D. Thesis, Graduate School, Chiang Mai University, Chiang Mai, Thailand, 2018.

61. Mota, R.; Pereira, S.B.; Meazzini, M.; Fernandes, R.; Santos, A.; Evans, C.A.; Philippisd, R.D.; Wright, P.C.; Tamagnini, P. Effects of heavy metals on *Cyanothece* sp. CCY 0110 growth, extracellular polymeric substances (EPS) production, ultrastructure and protein profiles. *J. Proteom.* **2015**, *120*, 75–94. [CrossRef]

62. Saijo, Y. Methods for quantitative determination of chlorophyll. *Jpn. J. Limonol.* **1975**, *36*, 103–109. [CrossRef]

63. Lichtenthaler, H.K.; Buschmann, C. Chlorophylls and Carotenoids: Measurement and Characterization by UV-VIS Spectroscopy. *Curr. Protoc. Food Anal. Chem.* **2001**, *1*, F4.3.1–F4.3.8. [CrossRef]

64. Davey, M.P.; Burrell, M.M.; Woodward, F.I.; Quick, W.P. Population-specific metabolic phenotypes of *Arabidopsis lyrata* ssp. *petraea. New Phytol.* **2005**, *177*, 380–388. [CrossRef]

65. Dunn, W.B.; Broadhurst, D.; Begley, P.; Zelena, E.; Francis-mcintyre, S.; Anderson, N.; Brown, M.; Knowles, J.D.; Halsall, A.; Haselden, J.N.; et al. The Human Serum Metabolome (HUSERMET) Consortium. Procedures for large-scale metabolic profiling of serum and plasma using gas chromatography and liquid chromatography coupled to mass spectrometry. *Nat. Protoc.* **2011**, *6*, 1060–1083. [CrossRef] [PubMed]

66. Davey, M.P.; Woodward, F.I.; Quick, W.P. Intraspecfic variation in cold-temperature metabolic phenotypes of *Arabidopsis lyrata* ssp. *petraea. Metabolomics* **2009**, *5*, 138–149. [CrossRef]

67. Mangiafico, S.S. An R Companion for the Handbook of Biological Statistics, version 1.3.2. 2015. Available online: Rcompanion.org/documents/RCompanionBioStatistics.pdf (accessed on 24 November 2018).

68. Paliy, O.; Shankar, V. Application of multivariate statistical techniques in microbial ecology. *Mol. Ecol.* **2016**, *25*, 1032–1057. [CrossRef] [PubMed]

 *metabolites*

Article

Improved Algal Toxicity Test System for Robust *Omics*-Driven Mode-of-Action Discovery in *Chlamydomonas reinhardtii*

Stefan Schade [1], Emma Butler [2], Steve Gutsell [2], Geoff Hodges [2], John K. Colbourne [1] and Mark R. Viant [1,*]

[1] School of Biosciences, University of Birmingham, Edgbaston, Birmingham B15 2TT, UK; sxs1318@bham.ac.uk (S.S.); J.K.Colbourne@bham.ac.uk (J.K.C.)

[2] Safety and Environmental Assurance Centre, Unilever, Colworth Science Park, Sharnbrook MK44 1LQ, UK; Emma.Butler@unilever.com (E.B.); Steve.Gutsell@unilever.com (S.G.); Geoff.Hodges@unilever.com (G.H.)

* Correspondence: m.viant@bham.ac.uk; Tel.: +44-121-414-2219

Received: 12 April 2019; Accepted: 7 May 2019; Published: 10 May 2019

Abstract: Algae are key components of aquatic food chains. Consequently, they are internationally recognised test species for the environmental safety assessment of chemicals. However, existing algal toxicity test guidelines are not yet optimized to discover molecular modes of action, which require highly-replicated and carefully controlled experiments. Here, we set out to develop a robust, miniaturised and scalable *Chlamydomonas reinhardtii* toxicity testing approach tailored to meet these demands. We primarily investigated the benefits of synchronised cultures for molecular studies, and of exposure designs that restrict chemical volatilisation yet yield sufficient algal biomass for omics analyses. Flow cytometry and direct-infusion mass spectrometry metabolomics revealed significant and time-resolved changes in sample composition of synchronised cultures. Synchronised cultures in sealed glass vials achieved adequate growth rates at previously unachievably-high inoculation cell densities, with minimal pH drift and negligible chemical loss over 24-h exposures. Algal exposures to a volatile test compound (chlorobenzene) yielded relatively high reproducibility of metabolic phenotypes over experimental repeats. This experimental test system extends existing toxicity testing formats to allow highly-replicated, *omics*-driven, mode-of-action discovery.

Keywords: synchronisation; algae; bioassay; biomarker; key event; adverse outcome pathway

---

1. Introduction

Algae are internationally established test organisms in chemical risk assessment [1–3]. Toxicity test guidelines incorporating algae are largely focused on traditional apical endpoints such as growth rates following a 72-h chemical exposure. By consequence, no knowledge is obtained on the causes of chemical toxicity, nor on the generality of the chemical effects on other species. Increasingly, chemical risk assessment includes New Approach Methodologies (NAMs) to improve precision in the prioritisation and chemical categorisation of chemicals [4]. One application of NAMs is to increase confidence in cause–effect analyses by providing mechanistic evidence on the harmful effects of chemicals, e.g., in the form of rapid and large-scale discovery of molecular key events (mKE) in an adverse outcome pathway (AOP) framework [5]. This transition, from traditional whole organism endpoint-based to molecular pathway-based (eco)toxicology, is motivated by the application of high-content technologies such as transcriptomics and metabolomics, which have demonstrated a capability to discover biological processes [6]. A robust, scalable test guideline for *omics*-driven discovery of toxicological key events in algae first requires a rigorously characterised algae culturing and exposure protocol with biological processes synchronised across all cells within a test population, so to minimise molecular variation for the pathway-based cause-effect assessments. Ideally, such an exposure system features minimal test duration to facilitate rapid measurements, and is matched to an algal lifecycle to allow accurate linking of the molecular perturbations to growth inhibition to address regulatory requirements of phenotypic anchoring [6–10]. Additionally, the exposure system should be capable of testing both non-volatile and volatile toxicants, and providing highly-replicated generation of sufficient algal biomass for multi-*omics* measurements. Design of such a robust test guideline should facilitate the generation of highly reproducible, high-quality multi-*omics* data.

Current algal growth inhibition test guidelines [2,3] recommend culturing in constant lighting, a condition that induces rapid drift and loss of cell-cycle alignment over a population of algae that are responsive to light-regulation of the cell cycle [11]. In such a culturing regimen, samples for molecular analyses taken at any given time-point over the test duration will contain a cross-section of individual cells performing simultaneously all of the biological functions in the algal cell cycle, leading to averaged molecular signatures that are non-specific to biological functions or condition dependent perturbations (Figure 1A). Equally, apical endpoints (e.g., inhibition of cell division in algae) will be observable throughout this time course, preventing the possibility of biological and temporal anchoring of hypothesised mKEs to the adverse outcome (AO). To emphasize the challenging nature of interpreting molecular results from such experiments, an analogy is drawn to a hypothetical rodent bioassay with inclusion of ante-natal, neonatal, adolescent, middle aged, old aged and pregnant animal subjects, and pooling tissue samples from all of these life-stages to infer a chemical mode of action (MoA). There is a logical argument that just as life stage is controlled in vertebrate and invertebrate animal testing, similar considerations should be given to algal toxicity testing, which regrettably have so far been limited in scope [12,13].

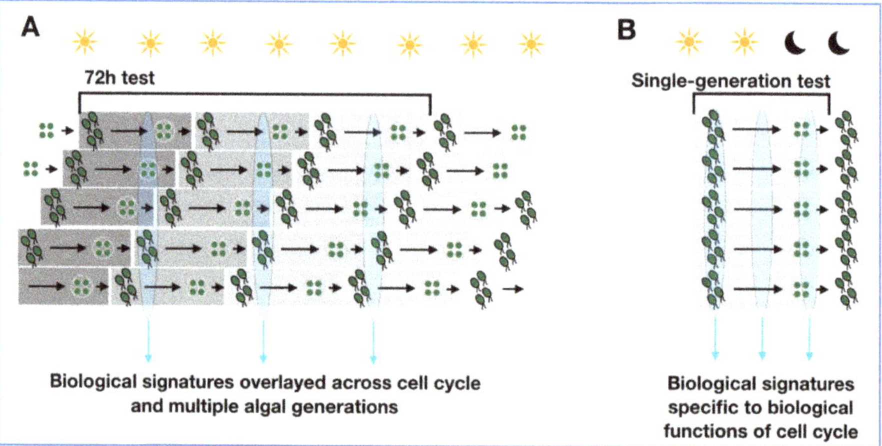

Figure 1. Abstract representation of the differences in sample composition between algal cell cultures grown in (**A**) continuous lighting over 72 h—as routinely used in OECD toxicity testing, and (**B**) alternating light:dark cycles for a single algal generation—as we propose and investigate in this study.

Conventional algal growth inhibition tests commonly apply a 72 h exposure duration. Following introduction of cell cycle synchronisation, we aim to reduce the duration of exposure, and adjust patterns of sampling time-test duration for high-throughput screening, to ultimately enable discovery of mKEs occurring before (and that are predictive of) acute growth inhibition. Here, the prime consideration is the biological anchoring of the test duration to the life-cycle of a single algal generation (Figure 1B).

While *omics* technologies hold considerable promise for the discovery of MoA(s) [14–17], to date the majority of *omics* studies have been too small-scale to deliver on this promise in algae research. Seminal studies in microalgal toxico-*omics* include those by Nestler et al.,

Jamers et al. [18,19] and Pillai et al. [20]. However, the majority of past study designs either tended to (i) scrutinise only single snapshots of molecular perturbation [16,18,21–25], (ii) apply test durations which lack a biologically-justified anchoring of sampling time-points [19,26–31], and/or (iii) utilise non-synchronised cultures over time-course experiments [20,32–35]. Overall, new robust, miniaturised and scalable designs in routine algal toxicity testing are required [36]. However, large scale studies factoring time and toxicant exposure are currently not supported by conventional microalgal toxicity test practices. For instance, recent multi-*omics* studies on *C. reinhardtii* feature experimental designs that are restricted in sample size, analytical methodology, and time-points due to lack of biomass [37]. This challenge of generating sufficient biological material for *omics* analysis is particularly aggravated during the testing of volatile toxicants. Due to the CO_2-dependence of natural photoautotrophic growth, existing volatile test systems are incompatible with above-mentioned experimental designs.

The overarching aim of the current study was to overcome the existing limitations in the algal growth inhibition test in the context of extending its applicability to *omics*-driven toxicological biomarker research. Following some initial modifications to the culture media, the first objective was to demonstrate the benefits of synchronised algae cultures for molecular studies using a combination of untargeted metabolomics, cell counting and flow cytometry. The second objective was to implement a highly replicable, closed-vial chemical exposure system with optimised *C. reinhardtii* growth at high inoculation cell densities, while minimising chemical volatilisation and test duration. The third objective was to demonstrate the application of untargeted metabolomics to this closed-vial test system by characterising the repeatability of the metabolic perturbations to a model narcotic, chlorobenzene, across three independent exposure experiments.

2. Results

2.1. Synchronised Versus Non-Synchronised Algae Cultures for Molecular Studies

Cells grown in constant lighting or alternating 12 h:12 h light:dark conditions were compared regarding time-point specificity of zoospore release (i.e., increase in cell number and density) and mitotic activity, over a time-course of 24 h. A clear logarithmic increase of cell density over the whole measurement period (thus a constant presence of zoospore hatching events) were observed in the cell populations grown in constant light (Figure 2A). By contrast, for cell cultures grown in alternating 12 h:12 h light:dark conditions, an increase in cell number occurred exclusively after the light:dark transition in a narrow time window of approximately 2.5–3 h starting around the 16 h time-point, indicating a high population-wide coordination of zoospore release from zoosporangia within the experimental time-frame.

A similar, although phase-shifted, dynamic process was observed when mitotic activity between the two culturing regimens was compared with flow cytometric measurements (Figure 2). Cells were analysed for fluorescence of DNA-intercalating fluorophore propidium iodide after a nucleotide staining procedure and enzymatic RNA degradation. Cell cultures grown in constant lighting conditions comprised either single cell zoospores, zoosporangia enclosing two or four zoospores, or cell particles undergoing S-phase, to narrowly-defined percentages over the whole course of the experiment (single genome copy zoospores 92.07 ± 0.77% of population, two-genome copy zoosporangia 3.25 ± 0.26%, four copy 3.21 ± 0.77%, cells in S-phase 1.48 ± 0.26%). Cell cultures from alternating light:dark conditions were composed of highly variable and time-dependent cell population, dominated by cells containing a single genome copy (SGC zoospores) at all time-points before the light:dark transition at 12 h (97.06 ± 1.53%). Yet at the 13 h time-point, these SGC zoospores comprised only 58% of the population, with a concordant increase in cell particles in S-phase, or zoosporangia containing two or four genome copies. The number of cells containing ever greater numbers of genome copies grew exponentially at later time-points, concordant with the occurrence of multiple-fission cycles of the cells. Replication activity started as early as 11 h into the light-phase (cells in S-phase at 11 h significantly increased from 9 h, Welch's t-test $p = 2.7 \times 10^{-3}$). Nuclear division events were observed at merely 1 h after the light:dark transition (13 h time-point). By comparison, zoospore hatching and respective increase in cell density was only observed after 15.5 h.

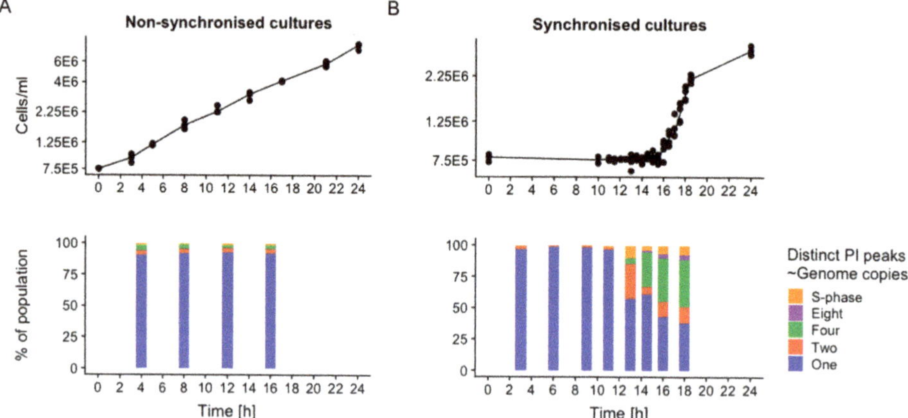

Figure 2. Cell density (**Top**) and number of distinct peaks in PI fluorescence intensity (**Bottom**; ~ amount of genomic material found in each zoosporangium or zoospore) indicating genomic division events during multiple-fission cycles for non-synchronised (**A**) or synchronised (**B**) cultures. For synchronised cultures, light:dark transition occurred at 12 h.

To analyse the impact of light-induced cell cycle synchronisation on the metabolic phenotypes of algal cultures, direct infusion mass spectrometry (DIMS) based metabolomics was applied to cells grown in alternating light:dark conditions (synchronised cell cycles) versus grown in constant lighting (non-synchronised). Initially, two small optimisation studies were conducted; the first to determine the number of washes required during the extraction of metabolites from the algal cells; the second to optimise the dilution of the metabolite extract that was infused into the mass spectrometer (see Supp. Info SI-1 and SI-T1). Basal (unexposed) algal cultures were sampled from each culturing regimen at 4 h 15 min (±15 min), 8 h 15 min (±15 min) and 12 h 15 min (±15 min) post-seeding (n = 8 for each light regime and time-point). Intra-study quality control samples (QC) were derived from a single pool of aliquots from each biological sample. Using the established measure of median relative standard deviation (mRSD) of measured metabolite features as a descriptor of variance [38], technical variability over the dataset was estimated from QC samples [39], and total (dominated by biological) metabolic variability was assessed from groups of biological samples. mRSD of all m/z features for the QC samples in the analysis of the synchronised cell cultures was 9.1%, while QC mRSD for analysis of non-synchronised cell cultures was 12.1%. These results indicate low technical variability and hence high quality of the DIMS metabolomics data, as previously described by Parsons et al. [38]. Total mRSD for synchronised cultures was as follows: 25.1% (4 h 15 min ± 15 min), 27.4% (8 h 15 min ± 15 min) and 30.6% (12 h 15 min ± 15 min); while for non-synchronised cultures: 36.2% (4 h 15 min ±15 min), 27.6% (8 h 15 min ±15 min) and 33.8% (12 h 15 min ± 15 min). These values indicate a slightly higher biological variability in non-synchronised cultures *versus* synchronised cultures. Unsupervised multivariate analysis (principal components analysis, PCA) compared the metabolic phenotypes of the algae as a function of lighting regime and time-course (Figure 3). Outliers were removed if their measurements exceeded the 95% confidence interval derived for all the biological samples, within each dataset (4 samples in non-synchronised study, 2 samples in synchronised). No metabolic separation of time-points was observed for non-synchronised cell cultures grown in constant lighting conditions. However, for cells grown in alternating light:dark conditions, a progression of their metabolic phenotypes was discovered over time (along PC1). Statistical analysis of the PC scores for each sample indicated significant differences in PC1 (ANOVA, $p = 1.8 \times 10^{-11}$) and PC2

($p = 6.39 \times 10^{-3}$) scores between metabolic phenotypes of the time-points for cells with synchronised cell-cycles, but not for non-synchronised cell cultures (ANOVA, PC1 $p = 0.84$, PC2 $p = 0.85$). This result indicates homogeneity of the metabolic composition of samples from non-synchronised cultures, and discrete compositions over time of those from synchronised cultures. Given the data, all subsequent studies were conducted using synchronised cell cultures.

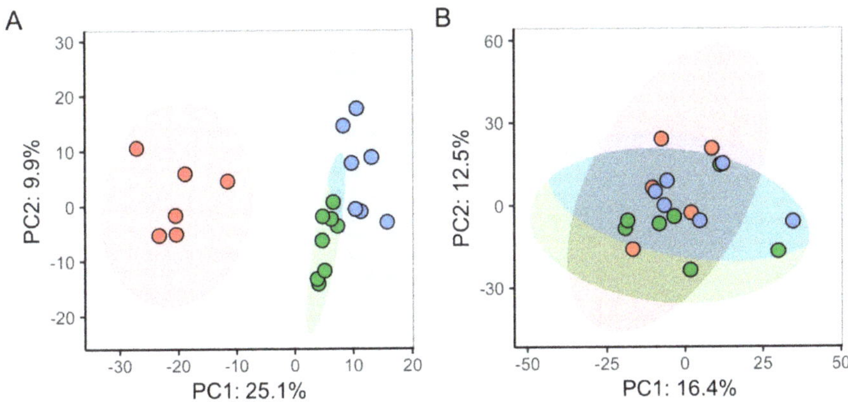

Figure 3. PCA scores plots comparing the DIMS metabolic phenotypes of (**A**) synchronised algal cell cultures grown in alternating light:dark conditions, and of (**B**) non-synchronised algal cell cultures grown in constant light. Plotted are PCA scores for samples taken at 4 h 15 min ±15 min (Red), 8 h 15 min ±15 min (Green)) and 12 h 15 min ±15 min (Blue).

2.2. Validation of C. reinhardtii Test System for Volatile Substances

To establish the validity of the medium and vial culturing system, growth rates of synchronised *C. reinhardtii* cultures and medium pH in capped-vials (7.5×10^5 cells/mL inoculation density, 10% vial air space) were checked against OECD 201 Test Guideline criteria [3] prior to initiating dose-response experiments in the vials. Adequate growth rates of >0.92^{-d} (mean 1.14^{-d}; $n = 4$) and a pH drift < 1.5 (mean 0.25; $n = 4$) over a 24 h incubation period were achieved, meeting the required specification (Supplementary Figure S2).

Subsequently, the modified capped-vial volatile test system was checked for robust and repeatable generation of concentration-response data, specifically for the traditional apical endpoint of algal growth, for both an OECD-recommended reference test substance 3,5-dichlorophenol (Supplementary Figure S3; [3]) and the volatile model toxicant chlorobenzene. The exposure period used was 24 h, aligning with both the cell culture studies presented above. An EC$_{50}$ = 0.93 ± 0.09 mg/L was derived for 3,5-dichlorophenol, while growth data from three independent chlorobenzene exposure experiments yielded an effective concentration estimate of EC$_{50}$ = 32.5 ± 3.6 mg/L (Figure 4). Of particular note is the high repeatability (low variation) in the growth data derived from the synchronised culture test system.

To assess the suitability of the modified capped-vial exposure system for maintaining stable concentrations of a volatile toxicant in the medium, over the test duration, extra vials were added to the experimental design and GC-MS analysis conducted on chlorobenzene levels in the medium of these sacrificial vials sampled at the beginning (0 h) and after 12 h 15 min incubation. Exposures were conducted and samples taken at each time-point for both low (14 mg/L) and high (24 mg/L) nominal chlorobenzene concentrations ($n = 4$ for each time-point and concentration; Figure 5). Concentrations of chlorobenzene from medium sampled at 0 h were slightly lower than nominal (low group: 12.32 ± 0.31 mg/L; high group 20.87 ± 1.78 mg/L). However, they remained stable until 12 h 15 min (Low: 12.51 ± 0.13 mg/L, High: 22.43 ± 0.78 mg/L). No significant differences were observed between the 0 h and 12 h 15 min medium chlorobenzene concentrations for both levels (Welch's t-test, Low $p = 0.39$, High $p = 0.24$).

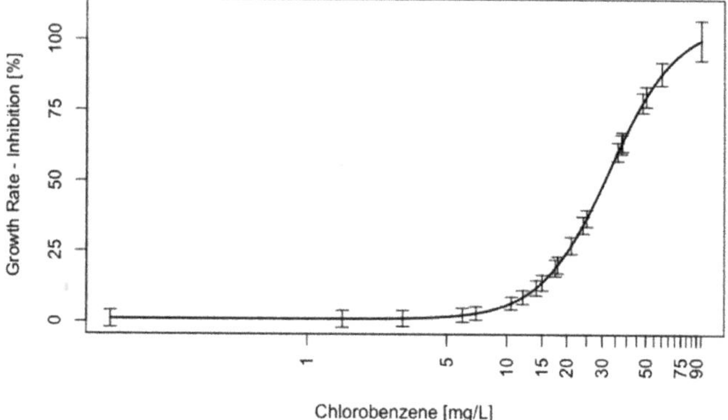

Figure 4. Dose-response curve (4 parameter log-logistic model) of 24 h algal growth inhibition experiments of chlorobenzene. The graph represents data from three independent experiments ($n = 4$ per concentration), error bars correspond to 95%-confidence intervals of fitted curve.

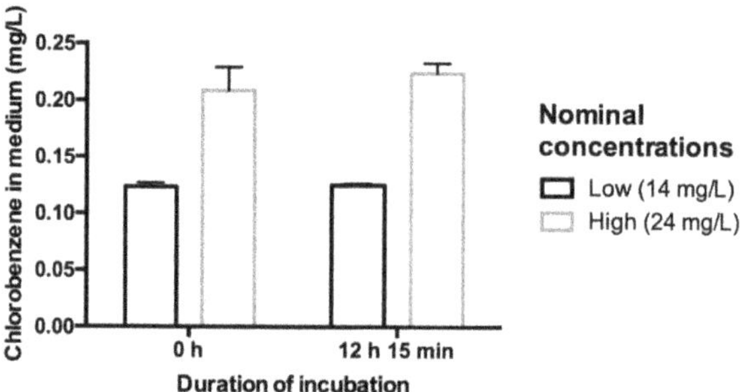

Figure 5. Barplots comparing chlorobenzene concentrations in test system medium, measured using GC-MS, to assess whether any loss occurs over the test duration. At 0 h and after a 12 h 15 min exposure period, two concentration groups 'Low' (nominal 14 mg/L) and 'High' (nominal 24 mg/L) test groups were measured. Bars represent mean ±1sd ($n = 4$ per group).

2.3. Repeatability of C. reinhardtii Metabolic Phenotypes in Test System

To characterise the reproducibility of the metabolite phenotypes of synchronised *C. reinhardtii* cultures grown in the capped-vial exposure system, DIMS metabolomics was conducted on samples generated over three independent experiments. Each biological batch comprised of highly replicated algal cultures in pure growth medium (control, $n = 10$) and cultures exposed to 25 mg/L chlorobenzene (CB, $n = 10$), with all samples harvested after a 3 h incubation. This exposure concentration was selected based on the previous dose-response study (Figure 4) with the 3 h time-point assumed to capture early metabolic perturbations induced by the narcotic substance; exposures were conducted early in the light phase of the 12 h:12 h light:dark cycle. Algal cell samples of each biological batch were stored at −80 °C, then all samples were extracted and DIMS metabolomics analysis (positive ion mode, polar metabolite fraction) performed as a single analytical batch. This design allowed biological variation between the exposure experiments to be isolated and identified.

First, RSD and median RSD values were calculated for all batches and groups to characterise technical and total (dominated by biological) metabolic variability. mRSD of all *m/z* features within intra-study QC samples was 9.3%, confirming low technical variability and high quality of the instrumental analysis (Supplementary Figure S4). RSD distributions and mRSD values suggested observable, although small, differences between biological batches, with batches 2 and 3 displaying lower mRSD than batch 1. mRSD values in biological batches 1, 2 and 3 were—for Control groups—26.0%, 23.1% and 22.4%, and in CB groups 26.7%, 24.1% and 24.1%.

PCA was applied to study the metabolic variability inherent to the complete dataset. Two outliers were removed (batch 1, CB samples) as they exceeded the 95% confidence interval of the complete dataset, and PCA conducted on the 56 remaining samples. The PCA scores plot (Figure 6A) revealed that the three batches were not identical, with batch 1 CB samples showing the highest intra-group variance as well as differing from the batch 2 and 3 CB samples in multivariate space. Additionally, while the metabolic effects of CB relative to the control samples were apparent for batches 1 and 3, a minimal difference between controls and CB-treated samples was observed for batch 2.

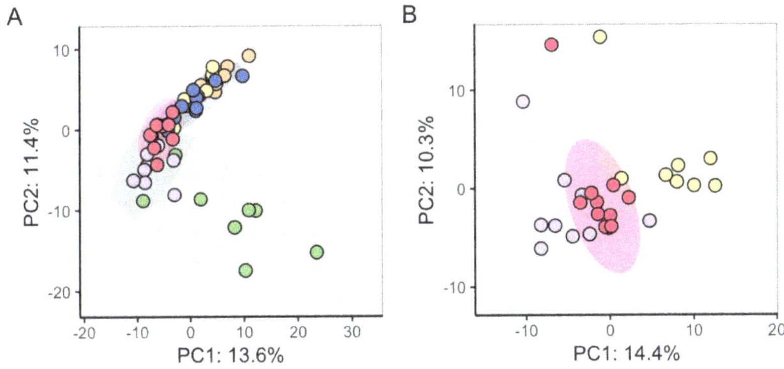

Figure 6. PCA scores plots (PC1 vs. PC2) visualising (**A**) the metabolic differences across three exposure studies (termed biological batches), with each batch comprising of control ($n = 10$) and CB-exposed ($n = 10$) samples; (**B**) only the control samples from each batch. Sample groups are biological batch 1 Control (Purple) and CB (Green), batch 2 Control (Yellow) and CB (Orange), and batch 3 Control (Red) and CB (Blue).

To further investigate, PCA was conducted on the control samples alone, from all three biological batches (Figure 6B and Supplementary Figure S5), to determine the extent to which the baseline algal metabolome varied across studies. ANOVA of the PC scores for the three control groups indicated significant differences along PC1 and PC3 ($p = 7.15 \times 10^{-6}$ and 3.84×10^{-4}, respectively; Table 1), indicating certain inter-batch metabolic variation between the three biological batches. Similarly, PCA and subsequent ANOVA of PC scores were applied to just the CB samples, which also revealed significance of metabolic differences between the three biological batches (Supplementary Figure S5 and Table 1). Potential sources of this batch effect included metabolic variation in the cultures themselves (over the duration of the three studies), the sampling of the algae, and/or other operator effects.

To achieve the high level of repeatability required for toxicity testing, we explored the effect of normalising each batch of exposure data to the corresponding control metabolic phenotypes. First, we calculated the $\log_2$ fold-change (LFC) for every m/z feature (peak), specifically by determining $\log_2$ of the ratio of the median feature intensity in the CB group over the median intensity in the respective control group ($\log_2$ (median[Int_{CB}]/median[$\text{Int}_{Control}$]), for each m/z feature). This calculation was repeated for each of the three batches. The resulting density distributions of LFC values, one for each biological batch, were visually similar (Figure 7A). Yet upon statistical testing, these LFC batch values were significantly different (Kruskal-Wallis test, $p = 1.38 \times 10^{-12}$).

Table 1. ANOVA results comparing PC scores of control samples only, CB samples only, and CB samples normalised to the batch-specific controls, in each case over the three independent biological batches.

Samples Analysed by PCA (All Three Batches)	PC1	PC2	PC3
Controls only	$p = 7.15 \times 10^{-6}$	$p = 0.11$	$p = 3.84 \times 10^{-4}$
CB only	$p = 1.72 \times 10^{-7}$	$p = 0.20$	$p = 0.048$
CB normalized to batch-specific controls	$p = 0.20$	$p = 0.015$	$p = 0.11$

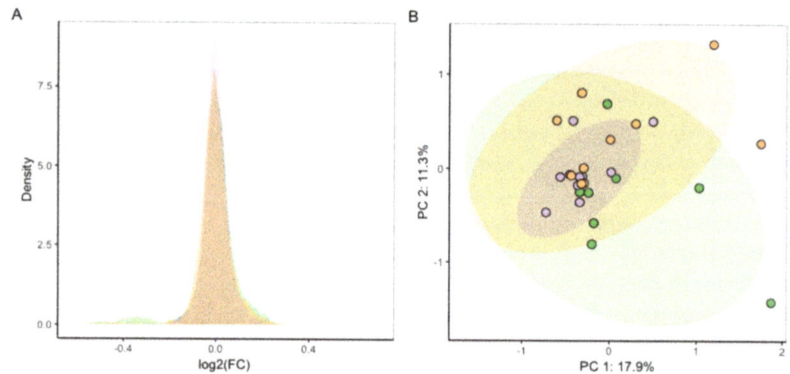

Figure 7. To increase comparability of exposure effects over the three repeated experiments, *m/z* feature intensities in CB groups were normalised to their batch-respective control groups by calculating the $\log_2$ fold-change of *m/z* features: (**A**) density distribution of LFC values for each batch ($\log_2$ (median[Int_{CB}]/median[$Int_{Control}$]), per *m/z* feature); (**B**) PCA scores plot from analysis of LFC values of all three batches ($\log_2$ (Int_{CB}/median[$Int_{Control}$]), for each *m/z* feature and for each of 10 CB samples).

To further explore the effect of normalising each batch of exposure data to the corresponding control metabolic phenotypes, specifically using PCA, $\log_2$ fold-changes were recalculated individually for the metabolite features of each of the $n = 10$ samples in the CB group relative to the median feature intensity of their respective control group ($\log_2$ (Int_{CB}/median[$Int_{Control}$])). This allowed PCA of these LFC values to reveal any metabolic differences between the effects of CB-treatment normalised to each batch-control, for all three batches (Figure 7B), with the scores plot now highlighting the significant overlap and therefore consistency of the batches. Indeed, ANOVA indicated no significant differences between PC1 and PC3 scores (PC1 $p = 0.20$, PC3 $p = 0.11$). However, slight batch differences were still detectable along PC2 ($p = 0.015$). Importantly, initial batch effects in PCA (Figure 6A,B) were greatly decreased by applying this normalisation strategy to the relevant batch-controls (Figure 7B).

We next evaluated the repeatability of the CB-induced metabolic perturbations in the three exposure experiments from the context of molecular biomarker discovery. This was achieved by applying set enrichment analysis [40] to the metabolomics data. This statistical process of comparing the rank order of genes, based on the magnitude or significance of their responses to multiple experimental conditions, was here applied on the rank order of mass over charge (*m/z*) metabolic features. Three comparisons were made (Table 2).

First, the top 100 LFC *m/z* features from batch 1 were designated as metabolite set (termed set batch 1, or S_{B1}), and used to interrogate the rank order of *m/z* metabolic features obtained from the other two batches. The significance of this set enriching the leading edges (LE) of each of the two batches was calculated using the normalized enrichment score (NES) and by estimating a false discovery rate (FDR). This enrichment test was twice repeated, for S_{B2} and S_{B3}. The results indicate that top 100 LFC *m/z* features for from each batch are strongly and significantly enriching all biological batch-metabolite set combinations (Table 2), concluding that a common and consistent subset of metabolic markers as a toxicological signal was discovered across all the biological batches, despite small batch effects in PCA.

Table 2. Enrichment analysis applied to three metabolomics datasets following exposure of *C. reinhardtii* to chlorobenzene in independent experiments (batches 1, 2, 3). Ranked lists of *m/z* feature LFC were calculated (S_{B1}–S_{B3} for batches 1–3) and every list was compared against every other list to determine whether S_{B1}-S_{B3} were enriched with elements of the other sets. Significance of enrichment was assessed using 1000 random permutations of *m/z* values in LFC-ranked lists, and false-discovery rates are reported. NES = normalised enrichment score, degree of metabolite enrichment of S within dataset. LE = leading edge, subset of *m/z* features in S that are most impactful towards a high NES.

Metabolite Set	Metric	Batch 1	Batch 2	Batch 3
S_{B1}	NES	-	1.39	1.33
	LE	-	11 features	72 features
	FDR	-	0.004	0.006
S_{B2}	NES	1.41	-	1.96
	LE	53 features	-	43 features
	FDR	<0.001	-	<0.001
S_{B3}	NES	1.38	2.02	-
	LE	61 features	50 features	-
	FDR	<0.001	<0.001	-

Finally, we sought to evaluate the metabolic perturbations in the three exposure experiments using a supervised multivariate analysis, which focuses on the most consistent changes induced in the CB-treated samples. Specifically, partial least squares-discriminant analysis (PLS-DA) was conducted on *m/z* features of all of the samples from the three biological batches. The algorithm selected an optimum of 5 latent variables for the classification model. While the control and CB samples were clearly separated in the PLS-DA scores plot of latent variable 1 vs. 2 (Figure 8), a robust interpretation of these results required any over-fitting of the PLS model to be assessed. Following 10-fold internal cross-validation, the PLS model yielded an R^2 = 0.99 (amount of variation in data explained by model) and Q^2 = 0.72 (goodness of fit), which indicated good generalisability.

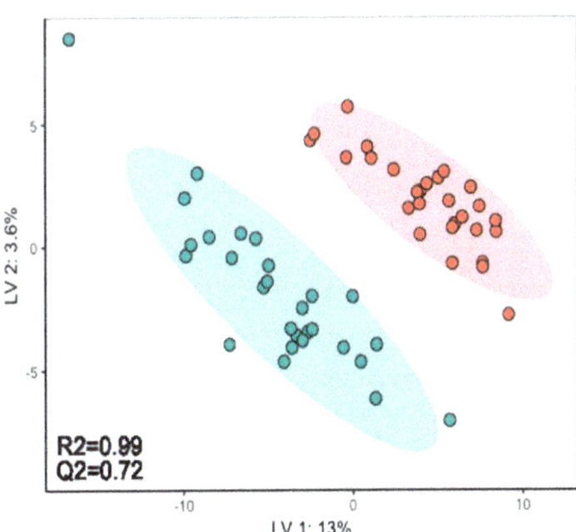

Figure 8. PLS-DA scores plot visualising the metabolic effects in *C. reinhardtii* following exposure to 25 mg/L chlorobenzene (red circles) relative to unexposed controls (green circles). All samples from three independent exposure experiments (biological batches) were pooled in this analysis.

3. Discussion

Chemical risk assessment by new approach methodologies requires test protocols with experimental designs that are tailored to applying *omics*, so to gain insights into toxicity mechanisms on a global molecular level and for discovering mKEs [6,15,36,41]. We here developed an extended variant of the regulatory algal growth inhibition test towards routine large-scale investigation of molecular toxicological processes. We achieved this by combining population-wide synchronisation of molecular functions in algal cell cultures with an easily scalable and abbreviated single-generation test system, suitable for testing of both soluble and volatile toxicants, and characterised the quality of generated molecular data using metabolomics.

3.1. Synchronised Cultures in Algal Toxicity Testing

For various unicellular green algae commonly applied in growth inhibition tests (e.g., genera *Chlamydomonas*, *Raphidocelis*, *Desmodesmus*, *Scenedesmus*, *Chlorella*), the phenomenon of light-driven cell cycle synchronisation is a known manifestation of evolutionary adaptation to scarcity of light in natural environments that can be experimentally induced by introduction of alternating light:dark cycles during incubation [42–44]. The degree of induced coordination between cells is close to absolute [45], nonetheless the benefits of this algae culture synchronisation to the context of environmental toxicology have previously only been given brief consideration [12]. Exposure and effect analysis of multiple life-stages within the same experiment are commonly avoided in toxicity assessments due to introduction of various confounding factors [46]. Initial matching of test organisms by life-stage is prescribed in the standardised OECD test criteria for most environmental model organisms, such as water flea, birds, bee, annelids, springtail, molluscs, amphibia and fish [47]. As one of the few cell-based in vivo systems, microalgal bioassays do not feature the requirement of life-stage/cell-cycle matching [2,3]. Commonly prescribed test conditions, including constant lighting, may further be considered non-representative of natural environments, and drives suboptimal metabolic efficiency and rapid cell cycle misalignment [11,48,49].

In the presented data, coordination of genomic division and sporulation events over the complete algal population in cultures grown in light:dark cycles, and lack thereof in cultures grown in constant light, was confirmed via electronic cell counting and flow-cytometry. This difference was indicative of cell cycle synchronisation in algae grown under alternating light cycles [45]. From the perspective of mechanistic (eco-)toxicology, one significant benefit of the observed population-wide coordination of nuclear division and sporulation is the ensuing occurrence and measurability of the algal adverse phenotypic endpoint within a narrow time-window across the cell population. This in turn enables temporal phenotypic anchoring of the AO to prior molecular perturbations, thereby defining the AOPs [9]. In contrast, described coordination and opportunity of mechanistic anchoring is lost in algal cell culturing using constant light regimes [11], due to consistent occurrences of nuclear division and sporulation over the entire experimental time-frame.

In mammalian cell models, synchronisation is routinely applied to enable detailed investigations into specific phases of the cell cycle [50]. In microalgae such as *C. reinhardtii*, synchronisation of cell cycle was applied to study various biological processes [51–55], however its explicit importance for toxicological investigations remains undervalued [12,13]. Our experiments comparing the metabolite phenotypes from *C. reinhardtii* grown under the two culturing regimens emphasised substantial and significant changes in the molecular composition in the synchronised *C. reinhardtii* cultures through time. Cells are tightly aligned in their progress through the entire cell cycle up to nuclear and cellular division [56] and characteristically evolving metabolic and transcriptomic patterns have previously been linked to this progression in *C. reinhardtii* [51,57]. In both pro- and eukaryotic cells, evidence is mounting that metabolic activity is not merely affected by position in the cell cycle, but a major driver of it [58,59]. Accordingly, distinct changes of metabolic patterns along the cell cycle were discernable in the scope of metabolomics analysis of synchronised cell populations. However, a complete lack of such an effect was apparent in cultures grown in constant light as prescribed by regulatory guidelines. Under non-nutrient-limited exponential growth conditions, non-synchronised cultures grown in constant light will comprise algal cells from every possible stage of the cell cycle, irrespective of experimental time-point. Although individual cells still progress through cell cycle, extracted samples for molecular studies comprise cross-sections over the whole complement of cell cycle-dependent biological functions, as evidenced by the high similarity of metabolic profiles between successive time-points over the experimental time-frame.

Metabolic phenotypes from samples of synchronised algae cultures taken at specific time-points during the exposure period were representative of narrow windows of biological functions anchored to cell cycle. The equivalent is true for perturbations of cellular functions (mKEs), which can be precisely measured over respective time-course analyses. Light:dark cycle-induced algal cell cycle synchronisation should thus be able improve *omics*-driven time-course analysis of mKEs by avoiding phase-shifts of molecular signals in the test system [60,61]. In non-synchronised cultures, discrete sampling intervals cannot be precisely rooted to cell-cycle-dependent biological processes. Interpretation of toxicological mechanisms remains limited to snapshots of overlaying molecular cross-sections of algal populations, and molecular phenotypes are averaged over the heterogeneous cell population within a sampling time-point. It is concluded that synchronised cell cultures in this context represent an approach to study molecular biological processes and their perturbations with strongly reduced variability, enabling increased analytical resolution of biological processes and their perturbances.

3.2. Design and Validation of an Alternative Volatile Testing System

Reducing the duration of a conventional algal test protocol would increase sample throughput and thus facilitate applicability for high throughput *omics* technologies, e.g., *for* applications in read-across and prioritisation. For the purpose of discovering mKEs, the commonly prescribed 72 h test duration represents an arbitrarily-selected timeframe to capture low-concentration delayed-onset mechanisms of toxicity. A warranted question in this context is why the standard test duration should not be increased to even longer durations to optimise sensitivity of the system to lower concentration effects (bioaccumulation or damage). Given the selection of inhibition of cellular division as the AO for microalgal toxicity, it can be assumed that AO-inducing key events must occur and be measurable within perturbations of preceding biological functions [62,63] and thus within previous cell cycle stages. Justification of the test duration in microalgal exposure studies aiming to discover mKE and infer chemical MoA, therefore, should not hinge upon absolute length, but on biological anchoring, i.e., the minimal time required to link molecular stress responses to successive phenotypic effects. The reduction of test duration encompassing a single generation of synchronised cells (Figure 1B) is proposed in this context. It may be argued that an optimal sampling strategy would extend exposure and sampling to the respective dark phase of the alternating light:dark cycle after sporulation, allowing the effects of a chemical on dark cycle processes to be examined. The practical implementation of such a protocol is not readily feasible, risking activation of biological functions in algae exposed even to minimal light indicative of a commencing cell growth cycle [51,64]. Furthermore, algal biological activity in complete darkness is indicated to be largely limited to mitochondrial energy metabolism, which is maintained throughout the light phase [51,65–67]. Consequently, the proposed time-frame from commencement of light phase to sporulation covers the single largest set of molecular and cellular functions within a cell cycle that chemicals may perturb.

Algal toxicity testing of volatile chemicals has historically proven problematic when combined with requirements for test miniaturisation and increased harvestable biomass, due to natural autotrophic requirements. Conventional agitation- and aeration-based exposure vessel systems become highly cumbersome to maintain with increasing experimental scale, and the stability of mid-concentration exposures is difficult in open flask culturing. A variety of solutions has been proposed to address the issue, including tailored complex flasks [68–70] and variants of bottle or tube testing [71–79]. The lower limit of *C. reinhardtii* biomass to achieve reproducible data in metabolomics and lipidomics studies has been reported at ca. 2–2.5 × 10^6 cells per sample [80,81], and similar or higher for transcriptomics [51,82–84] or proteomics [22,85]. This precludes the application of existing methods to large-scale (multi-)*omics* studies due to low cell densities, problematic physical scalability, disturbance of phase equilibria during repeat-sampling procedures, strong pH drift, or a combination thereof.

Here we developed and validated a hybrid sacrificial capped-vial exposure system to address existing shortcomings. The achieved biomass of $>7.5 \times 10^6$ harvestable cells per test vial has been substantially increased from comparable vial systems [75–77] and is sufficient to generate biological material for the demands of multiple *omics* technologies in parallel. In the exposure system, synchronous *C. reinhardtii* cultures performed well with mean growth rates at 1.14^{-d}, well above the validity requirements of 0.92^{-d} [3]. Low pH drift of 0.25 over the test duration was achieved (required <1.5; [3], precluding risk of changing bioavailability due to photosynthesis-driven pH changes, an issue more likely observed in other volatile test systems [68,71,86].

Past studies to hypothesise on chemical MoA in algae were frequently limited in exploring the temporal dynamics of toxicity. Many existing multi-*omics* studies attempting to deduce a MoA of chemicals as a dynamic processes in *C. reinhardtii* have employed experimental designs with relatively low numbers of sampling time-points [18,19,21–24,37,87]. Again, the required experimental designs are not sufficiently supported by conventional microalgal toxicity test practices, and the scarcity of existing publications with time-series analyses of algae is partly due to a relative lack of physical scalability. The volatile testing system reported here represents a miniaturised and abbreviated variant of the conventional algal growth inhibition test, reorienting its focus on molecular events predictive of a measurable acute adverse outcome after a single clonal generation of *C. reinhardtii*. By utilising small glass vials, parallelisability of test vessels and thus factorisation of experimental variables can in principle be drastically increased in individual experiments. Capacity to test hundreds of vial replicates in parallel, per individual toxicity test, has been achieved in multiple other studies in the author's laboratory (in prep.), enabling experimental designs to accommodate for high replication, temporal resolution and multiple exposure concentrations.

To demonstrate the generation of reproducible results by the designed testing approach for soluble and volatile toxicants, dose-response experiments of 3,5-DCP and chlorobenzene were performed to compare generated data to existing knowledge. EC_{50} value for the reference substance 3,5-DCP was estimated at 0.93 ± 0.09 mg/L, well within the range of 0.5 to 6.1 mg/L (48 to 72 h-EC_{50}) observed for *C. reinhardtii* and other green algae [88]. The determined 24 h-EC_{50} of the highly volatile chemical chlorobenzene for *C. reinhardtii* was estimated at 32.52 ± 3.64 mg/L (data of three repeats), again within range of published closed-test system data (*P. subcapitata* 48 h-EC_{50} = 14.4 mg/L [79]; *Selenastrum capricornutum* 96 h-EC_{50} = 12.5 mg/L, [69]), and drastically reduced compared to results derived from open-flask testing (*S. capricornutum* 96 h-EC_{50} = 224 mg/L [89]). When accounting for inter-species variability in sensitivity and differences in test exposure durations and physical setups [69], the estimated EC_{50} suggests adequate performance of the vial exposure system within expected ranges of variation.

Another requirement of the exposure system was to enable stable exposure concentrations of volatile toxicants over the experimental timeframe. Measurements of chlorobenzene levels at the start of the experiment were lower than nominal (14 mg/L nominal measured at 12.32 ± 0.31 mg/L; 24 mg/L nominal measured at 20.87 ± 1.78 mg/L), however the levels did not change significantly over the 12 h 15 min incubation period. This indicates chemical loss during initial stock preparations, however high stability of concentration levels over the incubation period. According to Henry's law constant, fraction of chlorobenzene in the gas phase is calculated as $\%CB_g = V_g/(V_g + (V_L \cdot R \cdot T/K_H))$ with V_g = 10% vial gas phase, V_L = 90% vial liquid phase, T = 298.15 K, R = 8.205 ± 10^{-5} m^3·atm/(K·mol), $K_H = 3.7 \times 10^{-3}$ atm·m^3/mol [90]. At equilibrium, only 1.65% of chlorobenzene theoretically partitions into the gas phase. Similar calculation for the expected partitioning of 204 Volatile Organic Compounds [91] suggests that 85% would partition less than 6% into the gas phase of the proposed exposure system. Concluding, stable exposure over the time-course of the experiment was well achievable using the designed capped-vial testing system, and minimal partitioning into the gas phase occurs during the experimental timeframe. Anticipation of and correction for initial volatilisation loss during stock preparation of highly volatile chemicals at inoculation is advised for further studies.

3.3. Characterising Variability and Repeatability of Metabolomics Data

Robust and reproducible determination of biomarkers and toxicological key events predictive of chemical insult from the algal metabolome is paramount for downstream utilisation of the generated data to support environmental hazard assessments. To characterise the reproducibility of data gained from our exposure approach, differences in the metabolite phenotypes between three repeated exposure studies were compared. Ranges of recorded median RSD values (20.86% to 24.99%) lay within the expected ranges for test organisms and cells, e.g., median RSD values have been reported for human cell lines between 20–22% [38].

Small batch effects could be observed between the baseline metabolic phenotypes across the three repeats of the exposure experiment when compared in PCA. A number of possible factors might have contributed to these differences, including subtle changes in laboratory maintenance that unknowingly affected the growth and biological activity of the sensitive algal cultures, as reported previously for microalgal bioassays [92–95]. We demonstrated that normalisation of the observed differences in algal stress response, by calculating log_2 fold-change of the measured m/z feature intensities to their batch-specific biological baseline (controls), proved to mitigate existing batch differences.

Despite these small batch effects in PCA, the algal metabolic stress response appeared to be regulated by a consistent mechanism across studies, as indicated by PLS-DA on pooled data. Additionally, we applied set enrichment analysis to metabolomics data, an approach originally developed for analysing gene set enrichment [40] which lately has become increasingly applied in metabolomics [96–99]. Set enrichment analysis was conducted to characterise the overall degree of similarity of whole sets of metabolic markers associated with chlorobenzene exposure between the repeat experiments. Metabolite sets containing the most important class markers within each of the biological batches were found to be significantly and strongly enriched within the top ranks of m/z features within the other batches, adding further support to our conclusion that a consistent toxicological effect could be measured.

4. Materials and Methods

4.1. Algae Culturing

Axenic *Chlamydomonas reinhardtii* Dangeard (1888) CC-125 wild type mt+ was purchased from the Culture Collection of Algae and Protozoa (CCAP, Dunbeg, UK). The strain was maintained heterotrophically on agar plates (1.5% *w/v* agar in tris-acetate-phosphate media) on laboratory shelves at 23 °C until liquid inoculation. A modified growth medium (termed *Chlamydomonas* growth medium; CGM) was developed here based on Sueoka's high salt medium supplemented with Kropat's trace elements [100,101] with the following modifications. The final concentrations of the divalent inorganic salts $CaCl_2*2H_2O$ and $MgSO_4*7H_2O$ were increased to 50 mg/L and 100 mg/L, and the concentration of total phosphates was reduced by 92.5% to 54 mg/L KH_2PO_4 and 108 mg/L K_2HPO_4 to decrease both palmelloid formation and mass spectrometric ion suppression due to phosphate accumulation in *C. reinhardtii* cells [102–104]. To increase inorganic carbon levels, the growth medium was supplemented with 500 mg/L $Na-HCO_3$. As *C. reinhardtii* growth is optimal between pH 5.5 to 9 [105,106], pH was adjusted to 6 to shift carbon equilibrium towards photosynthetically-available carbonic acid species [107]. To retain buffer capacity for a photosynthesis-driven change in medium pH, CGM was supplemented with 20 mM MOPS buffer, similar to Renberg et al. [102].

The growth medium was sterilised by 0.22 μm-filtration to avoid degradation of MOPS, salt precipitation and shift of the carbon equilibrium caused by autoclaving [108]. For initial liquid stock cultures, agar cultures were inoculated into 35 mL CGM within foam-bunged 250 mL wide-neck glass flasks. Cultures were incubated in a Multitron Pro rotary shaking incubator (Infors HT, Bottmingen-Basel, Switzerland) at 25 °C and 200 rpm orbital shaking. Lighting conditions were either set as constant lighting (effecting non-synchronised cell cycles), or as a 12 h:12 h light:dark cycle (inducing population-wide synchronization of cell cycle). Light colour temperature was 8500 K.

Algae were adapted to this lighting and autotrophic growth conditions in liquid cultures for at least 120 h prior to experiments.

The first study compared the metabolic phenotypes and mitotic activity of cell cultures grown under continuous lighting versus those grown under alternating 12 h:12 h light:dark cycles. Algae grown in either condition were seeded in vials as described below. Samples were taken at 4 h 15 min (±15 min), 8 h 15 min (±15 min) and 12 h 15 min (±15 min) post-seeding. 5.25×10^6 (7 mL) were taken from synchronised samples at every time-point, and sampling volume of the non-synchronised samples were adjusted to match this cell number by prior electronic cell counting. The latter was performed using CASY TT cell counting technology (Roche Innovatis AG, Basel, Switzerland) which operates via electric pulse area analysis to count suspended particle. As described below, cells were harvested and stored at −80 °C until combined metabolite extractions.

Over the course of this study, further samples were taken for flow cytometric measurements of genomic material in cell particles at 4 h, 8 h, 12 h and 16 h post-seeding for non-synchronised cultures, and 3 h, 6 h, 9 h, 11 h, 13 h, 14.5 h, 16 h, 18 h post-seeding for synchronised cultures. 1×10^6 cells were sampled ($n = 3$), centrifuged at $7000 \times g$ for 3 min (23 °C), pellets washed once in 1 mL phosphate-buffered saline (PBS), and cells fixed in 70/30 ethanol/water (both HPLC-grade, Fisher Scientific, Loughborough, UK) for 24 h at 4 °C. This last step served to permeabilise cells for propidium iodide (PI) and to extract chlorophylls which might interfere with the PI emission spectrum [109] The cells were washed again in 1 mL PBS and incubated in the dark for 30 min in 500 µL PBS containing 0.1 mg/mL RNAse-A (Qiagen, Hilden, Germany) and 0.01 mg/mL PI in water (94%, Sigma-Aldrich, Gillingham, UK). For 1×10^4 particles per sample, PI fluorescence was measured using a Becton Dickinson FACS-Calibur (488 nm excitation, bandpass filter 610/10), and analysed using open-source Flowing v2.5.1 (flowingsoftware.btk.fi). Absence of chlorophyll interference was confirmed with non-stained samples. Due to haploidy of *C. reinhardtii* vegetative cells, discrete peaks in per-particle fluorescence intensity histograms from cells sampled at a given time were equated either to the number of genome copies present within each cell particle (zoosporangia or zoospores) post nuclear division. Fluorescence areas in between discrete peaks were assigned as cells currently undergoing DNA replication and assigned as 'S-phase'. Growth curves of both culture regimen were determined using CASY cell-counting, at 2–4 h intervals for non-synchronised cultures and at higher resolution (30 min intervals) at the light:dark transition for synchronised cultures.

For each sampling time-point, vials were of sacrificial nature to maintain carbon equilibrium. Synchronised cell culturing was determined as preferable to non-synchronised cultures, as described in results, and used in all further studies.

4.2. Chlorobenzene Exposures

To expose cells, quadruplicate independent culture batches of synchronised cultures were grown in open flask culture, and at commencement of the light phase the cell cultures were concentrated individually by centrifugal separation in 50 mL tubes at $1200 \times g$ for 3 min (Eppendorf 5920 r, Eppendorf, Stevenage, UK). The supernatant was decanted, cells were diluted in toxicant-spiked CGM to yield a final cell concentration of 7.5×10^5 cells/mL, and seeded in clear 10.5 mL glass vials with aluminium cap inlay (N° T101/V4, Scientific Glass Laboratories, Stoke-On-Trent, UK) to establish 10% total gas phase volume within each vial (Figure 9). Vials were capped and incubated for 24 h, rolling free on a lined tray in the rotary incubator to maintain cell suspension. Preparation steps were restricted to a maximum duration of 30 min within the onset of light phase. Culture cell density for inoculation was scrupulously controlled and growth rates measured in technical duplicate using CASY cell counting.

A study conducting toxicity tests on synchronised cell cultures was performed using chlorobenzene (CB; 99.9%, Sigma-Aldrich, Gillingham, UK) as a volatile substance and 3,5-dichlorophenol (3,5-DCP; 97%, Sigma-Aldrich, Gillingham, UK) as the reference substance. All tests were performed with biological quadruplicates. *C. reinhardtii* was exposed to CB in three independent experiments at individual concentration ranges: 0, 15, 24, 38, 60, 95 mg/L (repeat 1, for range-finding); 0, 3.5, 7, 10.5, 14,

17.5, 21, 25, 37.5, 50 mg/L (repeat 2, adjusted range to encompass curve slope); 0, 1.5, 3, 6, 12, 18, 25, 36, 48 mg/L (repeat 3, for focus on environmentally-relevant lower end of dose-response curve). Cells were exposed to 3,5-DCP at 0, 0.6, 1.2, 2.4, 4.8 and 9.6 mg/L. A 4-parameter log-logistic dose-response model and effective concentration (EC) estimation was performed using [110] and drc package.

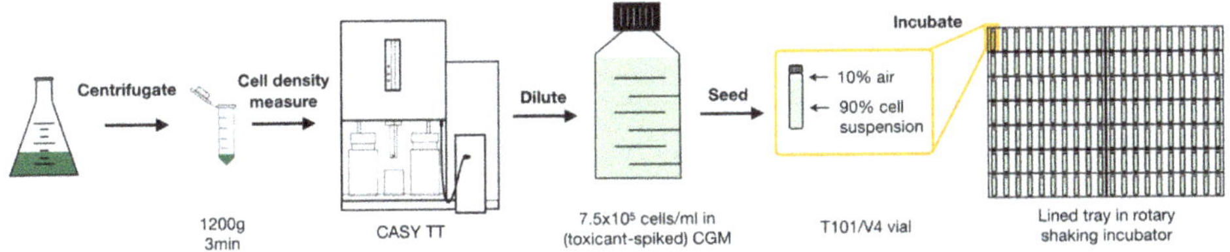

Figure 9. Individual steps for centrifugal concentration, cell density measurement, dilution, seeding of *C. reinhardtii* cell cultures into T101/V4 vials, and incubation of vials on a lined tray. Biological replicates in vials (n = 4 per experimental group) were generated from independent flask cultures.

A further exposure study was conducted to compare the metabolite phenotypes of synchronised cell cultures in pure growth medium (control) versus cells exposed to 25 mg/L CB, repeated three times, to characterise the metabolic variability and repeatability of data generated in the exposure vial system. Ten biological replicates for each of the control and CB groups were seeded in capped vials, incubated for 3 h, harvested as described below, and cell pellets stored at −80 °C until metabolite extraction. This experiment was repeated three times in identical set-up with a duration of 5d between experiments. Metabolite extraction was performed for all samples of the three repeats combined.

4.3. Sampling and Metabolite Extraction for DIMS Metabolomics

4.3.1. Sampling Procedure

Modifying the procedure by Lee and Fien [80], algae cell suspensions were quenched by injection into an equivalent volume of −78 °C 70:30 methanol:water (HPLC-grade, Fisher Scientific, Loughborough, UK) solution in 15 mL tubes over dry ice. Quenched cell suspensions were centrifuged at 3800× g and −11 °C for 3 min, and the supernatant aspirated. The number of washes was first optimised in a pilot study by comparing 10 samples from each of three groups: no wash, single wash of cell pellet, or two washes. Each wash comprised of resuspension of the cell pellet in 10 mL of −20 °C 35% methanol:water, then repeating the centrifugation and supernatant aspiration (described above). Between centrifugation and aspiration steps, samples were stored in a custom-manufactured mobile freezer unit between −20 °C pre-cooled aluminum beads to minimise metabolic activity and avoid freezing of suspension and thus risk of cell lysis at lower temperatures. After aspiration, pellets were flash-frozen in liquid nitrogen, and all samples stored at −80 °C until metabolite extraction. A single pellet wash was determined as optimal (see results in Supplementary Figure S1) and used for all further experiments in this study.

4.3.2. Metabolite Extractions

Biphasic metabolite extractions were performed with a final solvent ratio of methanol:chloroform:water (chloroform HPLC-grade, Fisher Scientific, Loughborough, UK) of 2:2:1.8 [111]. Cell pellets were thawed in randomised subsets of 24 tubes on ice. Cell pellets were resuspended in 970 μL ice-cold 70:30 methanol:water, transferred into homogenisation tubes with 0.5 mm glass beads (VK05, Stretton Scientific, Stretton, UK) on ice, and immediately homogenised at 6800 rpm for 2 × 15 sec bursts using a Precellys-24 tissue homogeniser (Bertin Instruments). Samples were placed back on ice for 2 min, then the supernatant (742 μL) was transferred to a 1.8 mL glass vial (aluminum-lined caps) containing 795 μL of ice-cold 1:2 water:chloroform, to yield the final solvent ratio of 2:2:1.8. Vials were vortexed for 30 sec and centrifuged for 15 min at 2585× g at 4 °C to achieve phase separation. Half the polar phase was aliquoted to a 2 mL Eppendorf tube using a glass syringe for untargeted metabolomics of the polar metabolite fraction (the other half retained for a second metabolomics assay, i.e., a different ion mode mass spectrometry). Each aliquot was dried in a SPD11V SpeedVac sample concentrator (Thermo Scientific, Rugby, UK) at room temperature. Extracts were stored at −80 °C until reconstitution in injection solvent (below). The same procedure was performed to generate an extraction blank sample.

4.4. Direct Infusion Mass-Spectrometry (DIMS) Metabolomics

Sample preparation, direct infusion mass spectrometric analyses and data processing were performed after the procedure of Southam et al. [39] To maximise the number of reproducibly detected m/z features, the dilution of the reconstituted algal extracts was first optimised (see results in Supplementary Table S1). A reconstitution volume of 50 μL injection solvent (80:20 methanol:water containing 0.25% v/v formic acid (98%, VWR)) was found to result in the maximal number of unique m/z features post-processing, therefore this option was used for preparation of all further samples. Ten μL of each reconstituted sample was pooled into an intrastudy quality control (QC) sample. All samples (biological, extraction blank, intrastudy QC) were analysed as technical quadruplicates (10 μL) on an Orbitrap Elite mass spectrometer (Thermo Scientific) with direct infusion nanoelectrospray ionisation (nESI) source (Triversa, Advion Bisociences, Ithaca, NY, USA) in positive ionisation mode. Individual SIM-windows of samples were stitched into single mass spectra, ≥75% replicate filtered, global peak aligned with a 3 ppm error tolerance, extraction blank filtered, QC-based locally-estimated scatterplot smoothing (LOESS) signal corrected [112], and ≥80% QC sample-based peak filtered. Features with ≥20% missing-values were filtered out, and probabilistic quotient normalisation applied [113]. RSD values were calculated for each m/z feature, and features in QC samples with RSD values >30% omitted from the data. Missing values imputation (k-NN) was performed and the generalised log transformation applied for multivariate data analysis.

Statistical tests (t-test, Kruskal-Wallis test, ANOVA) were performed using R and the R-Stats base package, PCA and PLS-DA conducted using MetaboAnalyst v4.0 [114]. Set enrichment analysis conducted with a java-based tool [40]. The method calculates an enrichment score of a supplied set S of variables against test data, to determine whether variables of S occur tendentially within the m/z features that are high-ranking in importance for separation of phenotype in the test data. S commonly comprises elements of a biological pathway (pathway enrichment analysis), however may contain individually defined variables. Here, S comprised m/z features with high LFC of one biological batch, which were checked for enrichment against LFC-ranked lists of m/z-values in the other biological batches. Thus, it was determined whether those m/z-values that were responsible for LFC-driven phenotype-separation in one biological batch were equally important in the other biological batches. Strength of the enrichment was calculated using NES metric (normalised enrichment score), significance determined by estimating false-discovery rate corrected p-values using 1000-fold permutation of the m/z features. m/z features within S deemed most impactful towards a high enrichment metric NES are referred to as the leading edge subset (LE, [40]).

4.5. Water Chemistry of Chlorobenzene

To measure concentrations of chlorobenzene in test medium over time, algae were incubated in vials with medium containing 14 mg/L or 24 mg/L chlorobenzene (nominal) for 12 h 15 min. At 0 h and 12 h 15 min, 100 μL (n = 4) was injected into 20 mL headspace vials with PTFE-lined septa (Scientific Glass Laboratories, UK) containing 9.7 mL CGM and 100 μL 0.8 M H_2SO_4. Samples were stored in the dark at 4 °C. One hundred μL of 10 mg/L d_5-chlorobenzene (99 atom-%, Sigma-Aldrich) in CGM was added immediately prior to analysis (final concentration 0.1 mg/L). Headspace GC-MS was performed on an Agilent 5975B inert XL MSD using DB-624 column (30 m × 0.25 mm i.d. × 1.4 μm) and Gerstel Autosampler MPS2. Analytes were identified using target m/z = 112 and qualifier m/z = 77 (chlorobenzene); target m/z = 117 and qualifier m/z = 82 (d_5-chlorobenzene). Sample sequence was randomised and quantification of chlorobenzene was performed relative to d_5-chlorobenzene.

5. Conclusions

Firstly, we designed an alternative closed-vial chemical exposure system that allows for sufficient growth of *C. reinhardtii* (at high inoculation cell densities) for a metabolomics assay, while minimising test duration and chemical volatilisation. The latter was proven by stable exposure concentrations to chlorobenzene over the test duration. We also examined the repeatability of this *C. reinhardtii* exposure system, identifying batch variation when applying unsupervised multivariate analyses to the metabolomics data, yet reducing these unwanted effects by normalising each biological batch to its respective control data. Furthermore, we used set enrichment analysis as well as supervised multivariate analysis to demonstrate the consistency of the metabolic markers discovered in three repeat exposures to chlorobenzene. Secondly, we have demonstrated that the application of an *omics* technology to synchronised algal cell cultures grown in alternating light:dark cycles can detect characteristic metabolic phenotypes that evolve through the cell-cycle. This should benefit the interpretability of molecular data in terms of discovering early mKEs that are anchored to, and predictive of, the algal adverse phenotype of reduced growth.

Supplementary Materials: The following are available online at http://www.mdpi.com/2218-1989/9/5/94/s1, Figure S1: Metabolic fingerprints for pellet wash optimisation, Table S1: Unique m/z features in direct infusion mass spectra in dilution study, Figure S2: Growth rates and pH drift over 24 h growth of synchronised *C. reinhardtii* in CGM medium, Figure S3: Dose-response curve of 3,5-dichlorophenol, Figure S4: Relative standard deviation of *m/z* features per group between three independent repeated exposure studies, Figure S5: PCA scores plots (PC1 vs. PC3) visualising the similarities and differences between three independent repeated exposure studies.

Author Contributions: Conceptualization, experimental design, data analysis, review and editing, S.S., E.B., J.K.C., G.H., M.R.V.; Conceptualization, experimental design, S.G.; Experimentation, S.S.

Funding: The authors thank Unilever for funding and BBSRC for a PhD studentship (BB/N503587/1).

Acknowledgments: We thank Alessio Perroti for help with conducting flow cytometric measurements. We also thank Chris Sparham, Mike Pleasants and Alexandre Teixeira for their major support in conducting gas-chromatography mass-spectrometry.

Conflicts of Interest: The authors declare no conflict of interest.

References

1. ECHA. *Guidance on Information Requirements and Chemical Safety Assessment. Chapter R.10: Characterisation of Dose [Concentration]-Response for Environment*; ECHA: Helsinki, Finland, 2008.

2. ISO (International Organisation for Standardisation). *ISO Water quality—Algal Growth Inhibition Test*; ISO/DIS 8692; ISO: Geneva, Switzerland, 2012.

3. OECD. Test No. 201: Alga, Growth Inhibition Test. In *OECD Guidelines for the Testing of Chemicals, Section 2: Effects on Biotic Systems*; OECD Publishing: Paris, France, 2011.

4. ECHA. *Usage of (Eco)Toxicological Data for Bridging Data Gaps between and Grouping of Nanoforms of the Same Substance*; Elements to Consider ED-02-16-228-EN-N; ECHA: Helsinki, Finland, 2016.

5. Ankley, G.T.; Bennett, R.S.; Erickson, R.J.; Hoff, D.J.; Hornung, M.W.; Johnson, R.D.; Mount, D.R.; Nichols, J.W.; Russom, C.L.; Schmieder, P.K.; et al. Adverse outcome pathways: A conceptual framework to support ecotoxicology research and risk assessment. *Environ. Toxicol. Chem.* **2010**, *29*, 730–741. [CrossRef]

6. Sauer, U.G.; Deferme, L.; Gribaldo, L.; Hackermüller, J.; Tralau, T.; van Ravenzwaay, B.; Yauk, C.; Poole, A.; Tong, W.; Gant, T.W. The challenge of the application of omics technologies in chemicals risk assessment: Background and outlook. *Regul. Toxicol. Pharmacol.* **2017**, *91*, 14–26. [CrossRef]

7. Keller, D.A.; Juberg, D.R.; Catlin, N.; Farland, W.H.; Hess, F.G.; Wolf, D.C.; Doerrer, N.G. Identification and Characterization of Adverse Effects in 21st Century Toxicology. *Toxicol. Sci.* **2012**, *126*, 291–297. [CrossRef] [PubMed]

8. Shostak, S. *Exposed Science: Genes, the Environment and the Politics of Population Health*; University of California Press: Berkeley, CA, USA, 2013.

9. ECETOC. *Workshop Report No. 25: Omics and Risk Assessment Science*; ECETOC: Málaga, Spain, 2013.

10. ICCVAM. Validation and Regulatory Acceptance of Toxicological Test Methods. In *A Report of the ad hoc Interagency Coordinating Committee on the Validation of Alternative Methods*; NIEHS: Durham, NC, USA, 1997;

p. 123. Available online: https://ntp.niehs.nih.gov/iccvam/docs/about_docs/validate.pdf (accessed on 19 February 2019).

11. Kanesaki, Y.; Imamura, S.; Minoda, A.; Tanaka, K. External Light Conditions and Internal Cell Cycle Phases Coordinate Accumulation of Chloroplast and Mitochondrial Transcripts in the Red Alga *Cyanidioschyzon merolae*. *DNA Res.* **2012**, *19*, 289–303. [CrossRef] [PubMed]

12. Kluender, C.; Sans-Piché, F.; Riedl, J.; Altenburger, R.; Härtig, C.; Laue, G.; Schmitt-Jansen, M. A metabolomics approach to assessing phytotoxic effects on the green alga *Scenedesmus vacuolatus*. *Metabolomics* **2009**, *5*, 59–71. [CrossRef]

13. Vogs, C.; Bandow, N.; Altenburger, R. Effect propagation in a toxicokinetic/toxicodynamic model explains delayed effects on the growth of unicellular green algae *Scenedesmus vacuolatus*. *Environ. Toxicol. Chem.* **2013**, *32*, 1161–1172. [CrossRef]

14. Antczak, P.; White, T.A.; Giri, A.; Michelangeli, F.; Viant, M.R.; Cronin, M.T.D.; Vulpe, C.; Falciani, F. Systems Biology Approach Reveals a Calcium-Dependent Mechanism for Basal Toxicity in Daphnia magna. *Environ. Sci. Technol.* **2015**, *49*, 11132–11140. [CrossRef]

15. Bridges, J.; Sauer, U.G.; Buesen, R.; Deferme, L.; Tollefsen, K.E.; Tralau, T.; van Ravenzwaay, B.; Poole, A.; Pemberton, M. Framework for the quantitative weight-of-evidence analysis of 'omics data for regulatory purposes. *Regul. Toxicol. Pharmacol.* **2017**, *91*, 46–60. [CrossRef] [PubMed]

16. Du, C.; Zhang, B.; He, Y.; Hu, C.; Ng, Q.X.; Zhang, H.; Ong, C.N. ZhifenLin Biological effect of aqueous C60 aggregates on Scenedesmus obliquus revealed by transcriptomics and non-targeted metabolomics. *J. Hazard. Mater.* **2017**, *324*, 221–229. [CrossRef]

17. Norris, J.L.; Farrow, M.A.; Gutierrez, D.B.; Palmer, L.D.; Muszynski, N.; Sherrod, S.D.; Pino, J.C.; Allen, J.L.; Spraggins, J.M.; Lubbock, A.L.R.; et al. Integrated, High-Throughput, Multiomics Platform Enables Data-Driven Construction of Cellular Responses and Reveals Global Drug Mechanisms of Action. *J. Proteome Res.* **2017**, *16*, 1364–1375. [CrossRef]

18. Nestler, H.; Groh, K.J.; Schönenberger, R.; Eggen, R.I.L.; Suter, M.J.F. Linking proteome responses with physiological and biochemical effects in herbicide-exposed *Chlamydomonas reinhardtii*. *J. Proteom.* **2012**, *75*, 5370–5385. [CrossRef]

19. Jamers, A.; Van der Ven, K.; Moens, L.; Robbens, J.; Potters, G.; Guisez, Y.; Blust, R.; De Coen, W. Effect of copper exposure on gene expression profiles in *Chlamydomonas reinhardtii* based on microarray analysis. *Aquat. Toxicol.* **2006**, *80*, 249–260. [CrossRef]

20. Pillai, S.; Behra, R.; Nestler, H.; Suter, M.J.F.; Sigg, L.; Schirmer, K. Linking toxicity and adaptive responses across the transcriptome, proteome, and phenotype of *Chlamydomonas reinhardtii* exposed to silver. *Proc. Natl. Acad. Sci. USA* **2014**, *111*, 3490–3495. [CrossRef]

21. Beauvais-Flück, R.; Slaveykova, V.I.; Cosio, C. Cellular toxicity pathways of inorganic and methyl mercury in the green microalga *Chlamydomonas reinhardtii*. *Sci. Rep.* **2017**, *7*, 8034. [CrossRef] [PubMed]

22. Esperanza, M.; Seoane, M.; Rioboo, C.; Herrero, C.; Cid, Á. Early alterations on photosynthesis-related parameters in *Chlamydomonas reinhardtii* cells exposed to atrazine: A multiple approach study. *Sci. Total Environ.* **2016**, *554*, 237–245. [CrossRef] [PubMed]

23. Simon, D.F.; Domingos, R.F.; Hauser, C.; Hutchins, C.M.; Zerges, W.; Wilkinson, K.J. Transcriptome sequencing (RNA-seq) analysis of the effects of metal nanoparticle exposure on the transcriptome of *Chlamydomonas reinhardtii*. *Appl. Environ. Microbiol.* **2013**, *79*, 4774–4785. [CrossRef] [PubMed]

24. Simon, D.F.; Descombes, P.; Zerges, W.; Wilkinson, K.J. Global expression profiling of *Chlamydomonas reinhardtii* exposed to trace levels of free cadmium. *Environ. Toxicol. Chem.* **2008**, *27*, 1668–1675. [CrossRef] [PubMed]

25. Kim, Y.K.; Yoo, W.I.; Lee, S.H.; Lee, M.Y. Proteomic analysis of cadmium-induced protein profile alterations from marine alga *Nannochloropsis oculata*. *Ecotoxicol. Lond. Engl.* **2005**, *14*, 589–596. [CrossRef]

26. Pan, C.G.; Peng, F.J.; Shi, W.J.; Hu, L.X.; Wei, X.D.; Ying, G.G. Triclosan-induced transcriptional and biochemical alterations in the freshwater green algae *Chlamydomonas reinhardtii*. *Ecotoxicol. Environ. Saf.* **2018**, *148*, 393–401. [CrossRef] [PubMed]

27. González-Pleiter, M.; Rioboo, C.; Reguera, M.; Abreu, I.; Leganés, F.; Cid, Á.; Fernández-Piñas, F. Calcium mediates the cellular response of *Chlamydomonas reinhardtii* to the emerging aquatic pollutant Triclosan. *Aquat. Toxicol.* **2017**, *186*, 50–66. [CrossRef]

28. Jamers, A.; Blust, R.; De Coen, W.; Griffin, J.L.; Jones, O.A. Copper toxicity in the microalga *Chlamydomonas reinhardtii*: An integrated approach. *BioMetals* **2013**, *26*, 731–740. [CrossRef] [PubMed]

29. Walliwalagedara, C.; Keulen, H.; Willard, B.; Wei, R. Differential Proteome Analysis of *Chlamydomonas reinhardtii* Response to Arsenic Exposure. *Am. J. Plant Sci.* **2012**, *3*, 764–772. [CrossRef]

30. Patel, N.; Cardoza, V.; Christensen, E.; Rekapalli, B.; Ayalew, M.; Stewart, C.N.N. Differential gene expression of *Chlamydomonas reinhardtii* in response to 2,4,6-trinitrotoluene (TNT) using microarray analysis. *Plant Sci.* **2004**, *167*, 1109–1122. [CrossRef]

31. Wang, S.B.; Chen, F.; Sommerfeld, M.; Hu, Q. Proteomic analysis of molecular response to oxidative stress by the green alga *Haematococcus pluvialis* (Chlorophyceae). *Planta* **2004**, *220*, 17–29. [CrossRef] [PubMed]

32. Taylor, N.S.; Merrifield, R.; Williams, T.D.; Chipman, J.K.; Lead, J.R.; Viant, M.R. Molecular toxicity of cerium oxide nanoparticles to the freshwater alga *Chlamydomonas reinhardtii* is associated with supra-environmental exposure concentrations. *Nanotoxicology* **2015**, *5390*, 32–41.

33. Jiang, Y.; Xiao, P.; Shao, Q.; Qin, H.; Hu, Z.; Lei, A.; Wang, J. Metabolic responses to ethanol and butanol in *Chlamydomonas reinhardtii*. *Biotechnol. Biofuels* **2017**, *10*, 239. [CrossRef] [PubMed]

34. Rubinelli, P.; Siripornadulsil, S.; Gao-Rubinelli, F.; Sayre, R.T. Cadmium- and iron-stress-inducible gene expression in the green alga *Chlamydomonas reinhardtii*: Evidence for H43 protein function in iron assimilation. *Planta* **2002**, *215*, 1–13. [CrossRef] [PubMed]

35. Gillet, S.; Decottignies, P.; Chardonnet, S.; Le Maréchal, P. Cadmium response and redoxin targets in *Chlamydomonas reinhardtii*: A proteomic approach. *Photosynth. Res.* **2006**, *89*, 201–211. [CrossRef] [PubMed]

36. Brockmeier, E.K.; Hodges, G.; Hutchinson, T.H.; Butler, E.; Hecker, M.; Tollefsen, K.E.; Garcia-Reyero, N.; Kille, P.; Becker, D.; Chipman, K.; et al. The Role of Omics in the Application of Adverse Outcome Pathways for Chemical Risk Assessment. *Toxicol. Sci.* **2017**, *158*, 252–262. [CrossRef] [PubMed]

37. Jamers, A.; Blust, R.; De Coen, W.; Griffin, J.L.; Jones, O.A. An omics-based assessment of cadmium toxicity in the green alga *Chlamydomonas reinhardtii*. *Aquat. Toxicol.* **2013**, *126*, 355–364. [CrossRef]

38. Parsons, H.M.; Ekman, D.R.; Collette, T.W.; Viant, M.R. Spectral relative standard deviation: A practical benchmark in metabolomics. *Analyst* **2009**, *134*, 478–485. [CrossRef] [PubMed]

39. Southam, A.D.; Weber, R.J.M.; Engel, J.; Jones, M.R.; Viant, M.R. A complete workflow for high-resolution spectral-stitching nanoelectrospray direct-infusion mass-spectrometry-based metabolomics and lipidomics. *Nat. Protoc.* **2017**, *12*, 255–273. [CrossRef]

40. Subramanian, A.; Tamayo, P.; Mootha, V.K.; Mukherjee, S.; Ebert, B.L.; Gillette, M.A.; Paulovich, A.; Pomeroy, S.L.; Golub, T.R.; Lander, E.S.; et al. Gene set enrichment analysis: A knowledge-based approach for interpreting genome-wide expression profiles. *Proc. Natl. Acad. Sci. USA* **2005**, *102*, 15545–15550. [CrossRef] [PubMed]

41. Corvi, R.; Ahr, H.J.; Albertini, S.; Blakey, D.H.; Clerici, L.; Coecke, S.; Douglas, G.R.; Gribaldo, L.; Groten, J.P.; Haase, B.; et al. Meeting report: Validation of toxicogenomics-based test systems: ECVAM-ICCVAM/NICEATM considerations for regulatory use. *Environ. Health Perspect.* **2006**, *114*, 420–429. [CrossRef] [PubMed]

42. Faetsch, S.; Matzke, M.; Stolte, S. Application Note Omni Life Sciences: Monitoring the Cell Cycle of the Unicellular Green Algae *Raphidocelis subcapitata*. Analysis of Cell Growth and Proliferation Using CASY. 2017. Available online: https://www.ols-bio.de/media/pdf/Algae-CellCycle_CASY-AppNote_OLS.pdf (accessed on 10 May 2019).

43. Grabski, K.; Aksmann, A.; Mucha, P.; Tukaj, Z. Conditioned medium factor produced and released by *Desmosdemus subspicatus* and its effect on the cell cycle of the producer. *J. Appl. Phycol.* **2010**, *22*, 517–524. [CrossRef]

44. Vítová, M.; Bišová, K.; Umysová, D.; Hlavová, M.; Kawano, S.; Zachleder, V.; Čížková, M. *Chlamydomonas reinhardtii*: Duration of its cell cycle and phases at growth rates affected by light intensity. *Planta* **2011**, *233*, 75–86. [CrossRef] [PubMed]

45. Bisova, K.; Zachleder, V. Cell-cycle regulation in green algae dividing by multiple fission. *J. Exp. Bot.* **2014**, *65*, 2585–2602. [CrossRef] [PubMed]

46. Kacew, S. Confounding factors in toxicity testing. *Toxicology* **2001**, *160*, 87–96. [CrossRef]

47. OECD. *Test Guidelines for the Chemicals*; OECD Publishing: Paris, France, 1984–2017.

48. Chisholm, S.W.; Brand, L.E. Persistence of cell division phasing in marine phytoplankton in continuous light after entrainment to light: Dark cycles. *J. Exp. Mar. Biol. Ecol.* **1981**, *51*, 107–118. [CrossRef]

49. Singh, S.P.; Singh, P. Effect of temperature and light on the growth of algae species: A review. *Renew. Sustain. Energy Rev.* **2015**, *50*, 431–444. [CrossRef]

50. Banfalvi, G. Overview of Cell Synchronization. In *Cell Cycle Synchronization*; Banfalvi, G., Ed.; Humana Press: Totowa, NJ, USA, 2011.

51. Zones, J.M.; Blaby, I.K.; Merchant, S.S.; Umen, J.G. High-Resolution Profiling of a Synchronized Diurnal Transcriptome from *Chlamydomonas reinhardtii* Reveals Continuous Cell and Metabolic Differentiation. *Plant Cell* **2015**, *27*, 2743–2769. [PubMed]

52. Mitchell, M.C.; Meyer, M.T.; Griffiths, H. Dynamics of carbon-concentrating mechanism induction and protein relocalization during the dark-to-light transition in synchronized *Chlamydomonas reinhardtii*. *Plant Physiol.* **2014**, *166*, 1073–1082. [CrossRef]

53. Garz, A.; Sandmann, M.; Rading, M.; Ramm, S.; Menzel, R.; Steup, M. Cell-to-cell diversity in a synchronized chlamydomonas culture as revealed by single-cell analyses. *Biophys. J.* **2012**, *103*, 1078–1086. [CrossRef]

54. Ehara, T.; Osafune, T.; Hase, E. Behavior of mitochondria in synchronized cells of *Chlamydomonas reinhardtii* (Chlorophyta). *J. Cell Sci.* **1995**, *108*, 499–507.

55. Sorokin, C. Time course of oxygen evolution during photosynthesis in synchronized cultures of algae. *Plant Physiol.* **1961**, *36*, 232–239. [CrossRef]

56. Hlavová, M.; Vítová, M.; Bišová, K. Synchronization of Green Algae by Light and Dark Regimes for Cell Cycle and Cell Division Studies. In *Plant Cell Division*; Humana Press: New York, NY, USA, 2016; pp. 3–16.

57. Jüppner, J.; Mubeen, U.; Leisse, A.; Caldana, C.; Brust, H.; Steup, M.; Herrmann, M.; Steinhauser, D.; Giavalisco, P. Dynamics of lipids and metabolites during the cell cycle of *Chlamydomonas reinhardtii*. *Plant J.* **2017**, *92*, 331–343. [CrossRef] [PubMed]

58. Kalucka, J.; Missiaen, R.; Georgiadou, M.; Schoors, S.; Lange, C.; De Bock, K.; Dewerchin, M.; Carmeliet, P. Metabolic control of the cell cycle. *Cell Cycle* **2015**, *14*, 3379–3388. [CrossRef]

59. Sperber, A.M.; Herman, J.K. Metabolism shapes the cell. *J. Bacteriol.* **2017**, *199*, e00039. [CrossRef] [PubMed]

60. Grigorov, M.G. Analysis of Time Course Omics Datasets. In *Bioinformatic for Omics Data*; Mayer, B., Ed.; Springer: Berlin/Heidelberg, Germany, 2011.

61. Spies, D.; Ciaudo, C. Dynamics in Transcriptomics: Advancements in RNA-seq Time Course and Downstream Analysis. *Comput. Struct. Biotechnol. J.* **2015**, *13*, 469–477. [CrossRef]

62. Borgert, C.J.; Wise, K.; Becker, R. A Modernizing problem formulation for risk assessment necessitates articulation of mode of action. *Regul. Toxicol. Pharmacol.* **2015**, *72*, 538–551. [CrossRef]

63. OECD. Users' Handbook supplement to the Guidance Document for developing and accessing. In *Adverse Outcome Pathways*; OECD: Paris, France, 2018.

64. Bisova, K.; Krylov, D.M.; Umen, J.G. Genome-Wide Annotation and Expression Profiling of Cell Cycle Regulatory Genes in. *Society* **2005**, *137*, 475–491.

65. Shene, C.; Asenjo, J.A.; Chisti, Y. Metabolic modelling and simulation of the light and dark metabolism of *Chlamydomonas reinhardtii*. *Plant J.* **2018**, *96*, 1076–1088. [CrossRef] [PubMed]

66. Johnson, X.; Alric, J. Central Carbon Metabolism and Electron Transport in *Chlamydomonas reinhardtii*: Metabolic Constraints for Carbon Partitioning between Oil and Starch. Eukaryot. *Cell* **2013**, *12*, 776–793. [CrossRef] [PubMed]

67. Peltier, G.; Thibault, P. O2-Uptake in the Light in Chlamydomonas: Evidence for Persistent Mitochondrial Respiration. *Plant Physiol.* **1985**, *79*, 225–230. [CrossRef]

68. Brack, W.; Rottler, H. Toxicity testing of highly volatile chemicals with green algae. *Environ. Sci. Pollut. Res.* **1994**, *1*, 223–228. [CrossRef]

69. Galassi, S.; Vighi, M. Testing toxicity of volatile substances with algae. *Chemosphere* **1981**, *10*, 1123–1126. [CrossRef]

70. Hailing-Sørensen, B.; Nyhohn, N.; Baun, A. Algal toxicity tests with volatile and hazardous compounds in air-tight test flasks with CO_2 enriched headspace. *Chemosphere* **1996**, *32*, 1513–1526. [CrossRef]

71. Lin, J.H.; Kao, W.C.; Tsai, K.P.; Chen, C.Y. A novel algal toxicity testing technique for assessing the toxicity of both metallic and organic toxicants. *Water Res.* **2005**, *39*, 1869–1877. [CrossRef] [PubMed]

72. Yeh, H.J.; Chen, C.Y. Toxicity assessment of pesticides to *Pseudokirchneriella subcapitata* under air-tight test environment. *J. Hazard. Mater.* **2006**, *131*, 6–12. [CrossRef]

73. Chen, C.Y.; Yan, Y.K.; Yang, C.F. Toxicity assessment of polycyclic aromatic hydrocarbons using an air-tight algal toxicity test. *Water Sci. Technol.* **2006**, *54*, 309. [CrossRef] [PubMed]

74. Tsai, K.P.; Chen, C.Y. An algal toxicity database of organic toxicants derived by a closed-system technique. *Environ. Toxicol. Chem.* **2007**, *26*, 1931–1939. [CrossRef] [PubMed]

75. Mayer, P.; Nyholm, N.; Verbruggen, E.M.J.; Hermens, J.L.M.; Tolls, J. Algal growth inhibition test in filled, closed bottles for volatile and sorptive materials. *Environ. Toxicol. Chem.* **2000**, *19*, 2551. [CrossRef]

76. Herman, D.C.; Inniss, W.E.; Mayfield, C.I. Impact of volatile aromatic hydrocarbons, alone and in combination, on growth of the freshwater alga *Selenastrum capricornutum*. *Aquat. Toxicol.* **1990**, *18*, 87–100. [CrossRef]

77. Fairchild, J.F.; Ruessler, D.S.; Carlson, A.R. Comparative Sensitivity of Five Species of Macrophytes and Six Species of Algae to Atrazine, Metribuzin, Alachlor, and Metolachlor. *Environ. Toxicol. Chem.* **1998**, *17*, 1830–1834. [CrossRef]

78. Shurong, D.; Fengfen, Z.; Min, Z. *Toxicity of Fenvalerate to Six Species of Fish and Two Species of Fishfood Organisms*; Ryans, R.C., Ed.; United States Environmental Protection Agency: Guangzhou, China, 1988; pp. 284–298.

79. Hsieh, S.H.; Tsai, K.P.; Chen, C.Y. The combined toxic effects of nonpolar narcotic chemicals to *Pseudokirchneriella subcapitata*. *Water Res.* **2006**, *40*, 1957–1964. [CrossRef] [PubMed]

80. Lee, D.; Fiehn, O. High quality metabolomic data for *Chlamydomonas reinhardtii*. *Plant Methods* **2008**, *4*, 7. [CrossRef] [PubMed]

81. Yang, D.; Song, D.; Kind, T.; Ma, Y.; Hoefkens, J.; Fiehn, O. Lipidomic analysis of *Chlamydomonas reinhardtii* under nitrogen and sulfur deprivation. *PLoS ONE* **2015**, *10*, e0137948. [CrossRef] [PubMed]

82. Schmollinger, S.; Muhlhaus, T.; Boyle, N.R.; Blaby, I.K.; Casero, D.; Mettler, T.; Moseley, J.L.; Kropat, J.; Sommer, F.; Strenkert, D.; et al. Nitrogen-Sparing Mechanisms in Chlamydomonas Affect the Transcriptome, the Proteome, and Photosynthetic Metabolism. *Plant Cell* **2014**, *26*, 1410–1435. [CrossRef]

83. Lv, H.; Qu, G.; Qi, X.; Lu, L.; Tian, C.; Ma, Y. Transcriptome analysis of *Chlamydomonas reinhardtii* during the process of lipid accumulation. *Genomics* **2013**, *101*, 229–237. [CrossRef]

84. Fang, W.; Si, Y.; Douglass, S.; Casero, D.; Merchant, S.S.; Pellegrini, M.; Ladunga, I.; Liu, P.; Spalding, M.H. Transcriptome-Wide Changes in *Chlamydomonas reinhardtii* Gene Expression Regulated by Carbon Dioxide and the CO_2-Concentrating Mechanism Regulator CIA5/CCM1. *Plant Cell* **2012**, *24*, 1876–1893. [CrossRef]

85. Wienkoop, S.; Weiss, J.; May, P.; Kempa, S.; Irgang, S.; Recuenco-Munoz, L.; Pietzke, M.; Schwemmer, T.; Rupprecht, J.; Egelhofer, V.; et al. Targeted proteomics for *Chlamydomonas reinhardtii* combined with rapid subcellular protein fractionation, metabolomics and metabolic flux analyses. *Mol. Biosyst.* **2010**, *6*, 1018–1031. [CrossRef] [PubMed]

86. Xing, J.; Dinney, C.P.; Shete, S.; Huang, M.; Hildebrandt, M.A.; Chen, Z.; Gu, J. Comprehensive pathway-based interrogation of genetic variations in the nucleotide excision DNA repair pathway and risk of bladder cancer. *Cancer* **2012**, *118*, 205–215. [CrossRef]

87. Esperanza, M.; Seoane, M.; Rioboo, C.; Herrero, C.; Cid, Á. *Chlamydomonas reinhardtii* cells adjust the metabolism to maintain viability in response to atrazine stress. *Aquat. Toxicol.* **2015**, *165*, 64–72. [CrossRef]

88. USEPA. *ECOTOXicology Knowledgebase (ECOTOX)*; USEPA: Washington, DC, USA, 2019.

89. USEPA. *Ambient Water Quality Criteria for Chlorinated Benzenes*; Report EPA 440/5-80-028; USEPA: Washington, DC, USA, 1980.

90. Bedient, P.B.; Rifai, H.S.; Newell, C.J. *Ground Water Contamination: Transport and Remediation*, 2nd ed.; Prentice Hall PTR: Upper Saddle River, NJ, USA, 1999.

91. USEPA. CERCLIS3. Available online: https://iaspub.epa.gov/sor_internet/registry/substreg/substance/details.do?displayPopup=&id=83723 (accessed on 9 May 2019).

92. Puzanskiy, R.; Tarakhovskaya, E.; Shavarda, A.; Shishova, M. Metabolomic and physiological changes of *Chlamydomonas reinhardtii* (Chlorophyceae, Chlorophyta) during batch culture development. *J. Appl. Phycol.* **2017**, *30*, 808–818. [CrossRef]

93. Altenburger, R.; Schmitt-Jansen, M.; Riedl, J. Bioassays with Unicellular Algae: Deviations from Exponential Growth and Its Implications for Toxicity Test Results. *J. Environ. Qual.* **2008**, *37*, 16. [CrossRef] [PubMed]

94. Janssen, C.R.; Heijerick, D.G. Algal Toxicity Tests for Environmental Risk Assessments of Metals. In *Reviews of Environmental Contamination and Toxicology*; Springer Nature: Basingstoke, UK, 2003; pp. 23–52.

95. Hörnström, E. Toxicity test with algae—A discussion on the batch method. *Ecotoxicol. Environ. Saf.* **1990**, *20*, 343–353. [CrossRef]

96. Xia, J.; Wishart, D.S. MSEA: A web-based tool to identify biologically meaningful patterns in quantitative metabolomic data. *Nucleic Acids Res.* **2010**, *38*, 71–77. [CrossRef] [PubMed]

97. Booth, S.C.; Weljie, A.M.; Turner, R.J. Computational tools for the secondary analysis of metabolomics experiments. *Comput. Struct. Biotechnol. J.* **2013**, *4*, e201301003. [CrossRef]

98. Rosato, A.; Tenori, L.; Cascante, M.; De Atauri Carulla, P.R.; Martins dos Santos, V.A.P.; Saccenti, E. From correlation to causation: Analysis of metabolomics data using systems biology approaches. *Metabolomics* **2018**, *14*, 37. [CrossRef]

99. Chagoyen, M.; Pazos, F. MBRole: Enrichment analysis of metabolomic data. *Bioinformatics* **2011**, *27*, 730–731. [CrossRef]

100. Sueoka, N. Mitotic Replication of Deoxyribonucleic Acid in *Chlamydomonas reinhardi*. *Proc. Natl. Acad. Sci. USA* **1960**, *46*, 83–91. [CrossRef] [PubMed]

101. Kropat, J.; Hong-Hermesdorf, A.; Casero, D.; Ent, P.; Castruita, M.; Pellegrini, M.; Merchant, S.S.; Malasarn, D. A revised mineral nutrient supplement increases biomass and growth rate in *Chlamydomonas reinhardtii*. *Plant J.* **2011**, *66*, 770–780. [CrossRef] [PubMed]

102. Renberg, L.; Johansson, A.I.; Shutova, T.; Stenlund, H.; Aksmann, A.; Raven, J.A.; Gardeström, P.; Moritz, T.; Samuelsson, G. A metabolomic approach to study major metabolite changes during acclimation to limiting CO_2 in *Chlamydomonas reinhardtii*. *Plant Physiol.* **2010**, *154*, 187–196. [CrossRef] [PubMed]

103. Kiefer, P.; Portais, J.C.; Vorholt, J.A. Quantitative metabolome analysis using liquid chromatography-high-resolution mass spectrometry. *Anal. Biochem.* **2008**, *382*, 94–100. [CrossRef]

104. Iwasa, K.; Murakami, S. Palmelloid formation of *Chlamydomonas* II. Mechanism of palmelloid formation by organic acids. *Physiol. Plant* **1969**, *22*, 43–50. [CrossRef]

105. Messerli, M.A.; Amaral-Zettler, L.A.; Zettler, E.; Jung, S.K.; Smith, P.J.S.; Sogin, M.L. Life at acidic pH imposes an increased energetic cost for a eukaryotic acidophile. *J. Exp. Biol.* **2005**, *208*, 2569–2579. [CrossRef] [PubMed]

106. Küsel, A.C.; Sianoudis, J.; Leibfritz, D.; Grimme, L.H.; Mayer, A. The dependence of the cytoplasmic pH in aerobic and anaerobic cells of the green algae *Chlorella fusca* and *Chlorella vulgaris* on the pH of the medium as determined by 31P in vivo NMR spectroscopy. *Arch. Microbiol.* **1990**, *153*, 254–258. [CrossRef]

107. Arensberg, P.; Hemmingsen, V.H.; Nyholm, N. A miniscale algal toxicity test. *Chemosphere* **1995**, *30*, 2103–2115. [CrossRef]

108. Jutson, M.G.S.; Pipe, R.K.; Thomas, C.R. *The Cultivation of Marine Phytoplankton*; Tsaloglou, M.N., Ed.; Caister Acadmic Press: Poole, UK, 2016; pp. 11–26.

109. Rowan, B.A.; Oldenburg, D.J.; Bendich, A.J. A high-throughput method for detection of DNA in chloroplasts using flow cytometry. *Plant Methods* **2007**, *3*, 5. [CrossRef]

110. R Development Core Team. *R: A Language and Environment for Statistical Computing*; R Foundation for Statistical Computing: Vienna, Austria, 2012.

111. Bligh, E.G.; Dyer, W.J. A Rapid Method of Total Lipid Extration And Purification. *J. Biochem. Physiol.* **1959**, *37*, 911–917.

112. Dunn, W.B.; Broadhurst, D.; Begley, P.; Zelena, E.; Francis-Mcintyre, S.; Anderson, N.; Brown, M.; Knowles, J.D.; Halsall, A.; Haselden, J.N.; et al. Procedures for large-scale metabolic profiling of serum and plasma using gas chromatography and liquid chromatography coupled to mass spectrometry. *Nat. Protoc.* **2011**, *6*, 1060–1083. [CrossRef] [PubMed]

113. Dieterle, F.; Ross, A.; Schlotterbeck, G.; Senn, H. Probabilistic Quotient Normalization as Robust Method to Account for Dilution of Complex Biological Mixtures. Application in 1 H NMR Metabonomics. *Anal. Chem.* **2006**, *78*, 4281–4290. [CrossRef] [PubMed]

114. Chong, J.; Soufan, O.; Li, C.; Caraus, I.; Li, S.; Bourque, G.; Wishart, D.S.; Xia, J. MetaboAnalyst 4.0: Towards more transparent and integrative metabolomics analysis. *Nucleic Acids Res.* **2018**, *46*, W486–W494. [CrossRef] [PubMed]

metabolites

Article

Euglena Central Metabolic Pathways and Their Subcellular Locations

Sahutchai Inwongwan, Nicholas J. Kruger, R. George Ratcliffe and Ellis C. O'Neill *

Department of Plant Sciences, University of Oxford, South Parks Road, Oxford OX1 3RB, UK;
sahutchai.inwongwan@new.ox.ac.uk (S.I.); nick.kruger@plants.ox.ac.uk (N.J.K.);
george.ratcliffe@plants.ox.ac.uk (R.G.R.)
* Correspondence: ellis.oneill@plants.ox.ac.uk; Tel.: +44-(0)1865-275-024

Received: 30 April 2019; Accepted: 11 June 2019; Published: 14 June 2019

Abstract: Euglenids are a group of algae of great interest for biotechnology, with a large and complex metabolic capability. To study the metabolic network, it is necessary to know where the component enzymes are in the cell, but despite a long history of research into *Euglena*, the subcellular locations of many major pathways are only poorly defined. *Euglena* is phylogenetically distant from other commonly studied algae, they have secondary plastids bounded by three membranes, and they can survive after destruction of their plastids. These unusual features make it difficult to assume that the subcellular organization of the metabolic network will be equivalent to that of other photosynthetic organisms. We analysed bioinformatic, biochemical, and proteomic information from a variety of sources to assess the subcellular location of the enzymes of the central metabolic pathways, and we use these assignments to propose a model of the metabolic network of *Euglena*. Other than photosynthesis, all major pathways present in the chloroplast are also present elsewhere in the cell. Our model demonstrates how *Euglena* can synthesise all the metabolites required for growth from simple carbon inputs, and can survive in the absence of chloroplasts.

Keywords: *Euglena*; central metabolic pathway; subcellular location

1. Introduction

Euglenids, a group of unicellular flagellate algae, have long been studied for their biochemistry, physiology, anatomy, and industrial potential, due to the remarkable metabolic plasticity that allows them to grow in a wide range of conditions [1]. *Euglena* can harness energy heterotrophically, mixotrophically, and photo-autotrophically, and its cultivation is relatively easy, fast, and well established. Euglenids can be found in a broad range of ecological niches including fresh water, brackish water, snow, high and low pH conditions, and both aerobic and anaerobic environments [2]. *Euglena gracilis* is the most studied species of *Euglena* and is regarded as a useful model organism for studying cell biology and biochemistry. Euglenids were once considered one of the most ambiguous groups in terms of evolution and metabolic operation, due to the combination of both "plant-" and "animal-" like features [3]. They are now classified into the kingdom Excavata, superphylum Discoba, subphylum Euglenozoa. *Euglena* is one of the very few plastid-containing organisms for which complete loss of the chloroplast is not lethal. Even the human parasitic apicomplexans retain their plastids for the synthesis of isoprenoids, fatty acids, and heme, while in non-photosynthetic, parasitic plants plastids are necessary for aromatic amino acid biosynthesis and are involved in starch synthesis [4]. Whilst these plastid-localised pathways can be targeted to kill such organisms, *Euglena* can survive complete loss of the plastid and the biochemical explanation for this remains to be established.

The genome of *E. gracilis* is estimated to be around 500 Mb in size, with large amounts of highly repetitive sequences [5], which leads to difficulty in genome sequencing and analysis. The structural complexity of the genome has arisen from a series of horizontal gene transfers and endosymbiosis events throughout its evolutionary history, causing difficulty in classifying euglenids using modern molecular techniques [6]. A study of the distribution of the homologues of 2770 expressed sequence tags (ESTs) from *E. gracilis* has shown that euglenids are closely related to the kinetoplastids [7]. Euglenids first split from the ancestral Euglenozoa, a eukaryotic protozoa, around a billion years ago [8]. After the endosymbiotic transfer of genes from a hypothesized, since-lost, red algal endosymbiont to the nuclear genome [9], a eukaryotic green alga endosymbiont was incorporated [10], bringing many genes involved in the function and maintenance of the chloroplast. The transcriptome of *Euglena* suggests that many other genes were acquired from diverse distantly related species and the genetic control mechanisms in *Euglena* involve genes which are as sophisticated as those in animal and plant eukaryotes [11].

Euglena is considered to be a promising organism for industrial application due to its ability to produce various nutrients and bioactive compounds, such as proteins, polyunsaturated fatty acids, vitamin A, vitamin C, and β-1,3-glucan [12]. The application of *Euglena* in environmental engineering has been studied for wastewater treatment systems, energy sources and bioindicators for environmental pollutants. *Euglena* sp. isolated from sewage treatment plants had higher nutrient removal capability and growth rate than other algae [13]. These results indicate that *Euglena* could be considered as a viable source for biofuel production from wastewaters.

There is no doubt that *E. gracilis* is an interesting organism in terms of its evolution, metabolic capacity, and application and has thus been the subject of intense study. Due to its extraordinary metabolic capacity, investigating and understanding the *Euglena* metabolic network could help expand the applications of this organism and shed light on several mysteries of evolution and secondary endosymbiosis. Investigation of the metabolism of *Euglena* requires the definition of the metabolic network, whether at genome scale for flux balance analysis, or at the level of core metabolism for metabolic flux analysis. This would allow the metabolic phenotype of the organism to be investigated in much the same way as in highly compartmented plant cells [14]. In organisms with complex evolution like *Euglena*, even though the central metabolic pathways are conserved, the characteristics and subcellular localisation of the enzymes involved in the pathway can differ. This is particularly true for *Euglena*, where the secondary chloroplast has a relatively recent evolutionary origin (~600 MYA [15]) and a unique third plastid membrane, giving rise to a novel subcellular compartment in this intermembrane space.

Here, we provide an overview of the central metabolic pathways in *Euglena gracilis*, highlighting unique features. We assess the reported subcellular location of enzyme activities and proteins in *Euglena* and propose a model of the organisation of the central metabolic network.

2. Results

2.1. Pathway Localisation from Sequence Information

Even though *Euglena* has long been studied for its biotechnological potential, its genetic and metabolic capacities are poorly established due to the size and complexity of its genome. In the absence of an annotated genome sequence for any species of *Euglena*, transcriptome sequencing has been used as the preliminary alternative to genome structure analysis, with the aim of providing data on gene expression and regulation under different conditions [16,17].

2.1.1. Metabolic Pathways in *Euglena*

The earliest reported extensive transcriptomic analysis of *E. gracilis* studied cells grown in dark and light conditions and illustrated the versatile metabolic capacity of *Euglena* [16]. All the core pathways of carbohydrate metabolism and photosynthesis were identified, including glycolysis, gluconeogenesis, the tricarboxylic acid cycle (TCA), the pentose phosphate pathway (PPP), and the Calvin cycle. In addition, the pathways for production of other major classes of compounds including carotenoids, thylakoid glycolipids, fatty acids, and isoprenoids were also identified in the transcriptome. Besides the evidence for lipid, amino acid, carbohydrate, and vitamin metabolism, the transcriptome also revealed the capacity of *E. gracilis* to produce multifunctional polydomain proteins that relate to those from both fungi and bacteria and may have been obtained by horizontal gene transfer during its evolution [11]. Furthermore, the transcriptome showed the capacity for polyketide and non-ribosomal peptide biosynthesis [18], along with capacities for using the pathways for vitamin C, vitamin E, and glutathione metabolism to respond to stresses. A subsequent comparative study of the transcriptome of *E. gracilis* under aerobic and anaerobic conditions investigated the regulatory system of wax ester metabolism [17]. The metabolic network of *Euglena mutabilis* has been reconstructed using assembled transcript sequences and topology gap filling [19]. The initial draft network was incomplete with many missing reactions and could not simulate the heterotrophic growth of *E. mutabilis* in the dark [19], despite the long documented capacity of this species to do so. In combination, these studies demonstrate that the genomes of *Euglena* have features in common with genomes from both phototrophic and heterotrophic organisms, and these features provide *Euglena* with the metabolic capacity to adapt to a wide variety of conditions. These studies also demonstrate that transcript abundance does not vary greatly under different growth conditions and does not correlate with protein abundance. Thus, exploration of the metabolic capacity of *Euglena* using an exclusively transcriptomic approach is unlikely to be sufficient to understand pathway control.

2.1.2. Metabolic Pathways in the *Euglena* Plastid

The chloroplast genome of *E. gracilis* has been sequenced [20] and is very similar to that of higher plants in its gene content, although the structure and evolution is different [21]. As with other organisms, the acquisition of the plastid came with many gene loses and gene transfers from the endosymbiont to the host genome [22]. The expression level of plastid genes was found to respond to environmental stimuli [23] and the rate of protein synthesis by the *E. gracilis* plastid in the dark is extremely low compared to that in the light [5,24].

As in the primary plastids of other organisms, most of the *Euglena* secondary plastid proteome is encoded in the nuclear genome. However, since the plastid of *Euglena* was acquired through secondary endosymbiosis of a photosynthetic eukaryote, its chloroplasts are surrounded by three membranes [25,26]. Thus, hundreds of plastidic proteins synthesized in the cytosol have to be transported through either three or four membranes to reach their destination in the plastid stroma or the thylakoid lumen [27] and we have no knowledge of the metabolic capabilities of the unique intermembrane space, found in no other group of organisms.

2.1.3. Predicting the Subcellular Location of *Euglena* Proteins

Most of the previously published studies of the subcellular compartmentation of *Euglena* enzymes have relied on subcellular fractionation of organelles and measurement of enzyme activity distributions. Very few studies have exploited complementary molecular techniques to investigate localisation in *Euglena*. In principle, eukaryotic protein subcellular location prediction tools could be useful. To test this, the protein sequences of selected marker enzymes with defined compartmentation were analysed using a subcellular location prediction work flow. These included proteins known to be located in the chloroplast, mitochondria, cytosol, or directed through the secretory pathway. The predicted amino acid sequences of these marker proteins were deduced from the *E. gracilis* transcriptome [16]. In total 28% of these sequences had spliced leader sequences, indicated in bold in Tables 1–3. Two programs were used to predict the subcellular localisation of all the matching *E. gracilis* protein sequences, WoLF PSORT [28], and TargetP 1.1 [29]. Due to the potential presence of plant and non-plant targeting signals on *Euglena* proteins (arising from the complex evolutionary origin of *Euglena* genes), these analyses were conducted using plant, animal, and fungal reference databases in WoLF PSORT and both plant-based and nonplant-based prediction modes in TargetP 1.1. Moreover, since transport of proteins into *Euglena* chloroplasts requires transit via the secretory pathway [27,30,31], any sequence that was predicted to contain a secretion signal based on the plant-based algorithm in TargetP 1.1 was subjected to extended analysis in which the signal sequence was removed and the prediction process repeated to establish the ultimate predicted location of the mature protein.

Mitochondrial Targeting

The mitochondrial marker enzymes are all well-established biochemical markers and are only detected in mitochondrial fractions in *Euglena*, with the exception of isocitrate dehydrogenase which is also detected in the cytosol. At least one isoform of each of these enzymes is predicted to be targeted to mitochondria using TargetP and WoLF PSORT in all modes (see Table 1). However, using the plant-based algorithm in WoLF PSORT there was more support for some of these enzymes being in the chloroplast. One isoform of succinic semialdehyde dehydrogenase, containing a spliced leader sequence, appears to have no targeting signal and so would be predicted to be in the cytosol. One isoform of isocitrate dehydrogenase has no predicted targeting, in line with biochemical evidence for some cytosolic activity of this enzyme.

Proteins without Targeting Signals

Cytosolic marker proteins were selected that are routinely used as marker enzymes in subcellular fractionation studies. Overall, these had less confident predictions and some weak predictions for mitochondrial targeting (Table 2). The exception is thiosulfate sulfurtransferases, for which three isoforms had plastid targeting sequences in WoLF PSORT using the plant mode. Two of these had strong secretion signal predictions in both animal and fungi modes and in TargetP, whilst another isoform has a strong secretion signal prediction in these WoLF PSORT modes. This may indicate that some of these isoforms are targeted to the chloroplast via the endoplasmic reticulum (see below).

Targeting for Secretion

Proteins known to be in the Golgi, and which thus utilise the secretory pathway, were used as benchmarks to test the reliability of secretion signal prediction for *Euglena* proteins (Table 2). They were predominantly identified as being targeted for secretion by TargetP with a high level of confidence, especially using the nonplant algorithm, although mitochondrial targeting was predicted in some instances. WoLF PSORT predicted that these proteins were targeted to the plasma membrane, as they are integral membrane proteins. Some were predicted to also contain secretory signals with high confidence, but not all. One of the mannosyltransferases was predicted to target to the chloroplast using WoLF PSORT in plant mode.

Chloroplast Targeting

A selection of biochemical marker enzymes and components of the photosynthetic apparatus was used to test the ability of these programs to predict targeting to the plastids. TargetP predicted most of these proteins to be either mitochondrial or secreted (Table 3). The only exceptions were for one of the isoforms of fructose-bisphosphate aldolase, and one ribulose-bisphosphate carboxylase/oxygenase (small subunit) that were predicted to be targeted to the chloroplast after removal of the secretory signal peptide. WoLF PSORT on the other hand correctly predicted many soluble enzymes to be targeted to the chloroplast but predicted many of the integral membrane proteins, such as photosystem components, as being targeted to the plasma membrane.

The limitations of the chloroplast targeting prediction of TargetP have been reported before [29]. The predictive power of TargetP 1.1 is based on the presence of N-terminal presequences, including chloroplast transit peptide (cTP), mitochondrial targeting peptide (mTP), or secretory pathway signal peptide (SP) [29]. However, the structure of cTP is not well characterized, especially in *Euglena*, and the prediction performance of chloroplast targeted proteins was reported to be less accurate than that for mitochondria, with occasional poor discrimination between mTP and cTP [32]. This lack of discrimination is partly due to some proteins using the same targeting sequence for both chloroplasts and mitochondria [29]. Thus, using TargetP and WoLF PSORT to predict the location of proteins in *Euglena* might not cover all the possible protein transport systems.

Table 1. Subcellular location prediction of *E. gracilis* mitochondria marker proteins.

Marker Enzyme Name	EC Number	EG Transcript	WoLF PSORT Plant			WoLF PSORT Animal			WoLF PSORT Fungi			TargetP Plant	TargetP Nonplant	TargetP, Signal Cut
Fumarase	**4.2.1.2**	**4764**	**Chl: 8**	-		**Mt: 27.5**	Cyt_Mt: 17	-	Mt: 26	-	-	Mt: 1	Mt: 2	-
		7552	Cyt: 11	-		Cyt: 24	-	-	Cyt: 22	-	-	Cyt: 2	Cyt: 1	-
Succinic semialdehyde dehydrogenase	1.2.1.16	5575	Mt: 10.5	Chl_Mt: 7.5		Mt: 28.5	Cyt_Mt: 16	-	Mt: 22.5	Cyt_Mt: 14	-	Mt: 1	Mt	-
		3780	Chl: 9	-		Mt: 10.5	Cyt_Mt: 10.3	Cyt: 9	Mt: 16	Cyt: 7.5	-	Mt: 5	Mt: 4	-
		5750	Mt: 12.5	Chl_Mt: 7.5		Mt: 27.5	Cyt_Mt: 16.5	-	Mt: 25.5	Cyt_Mt: 14	-	Mt: 3	Mt: 2	-
		6875	Cyt: 7	-		Cyt: 17	Mt: 15	-	Cyt: 22.5	Cyt_Nu: 12	-	Cyt: 3	Cyt: 2	-
Isocitrate dehydrogenase	1.1.1.41, 1.1.1.42	**10115**	Chl_Mt: 7.5	Mt: 7	Chl: 6	Mt: 29	Cyt_Mt: 16.5	-	Mt: 22	-	-	Mt: 1	Mt: 1	-
		10872	Mt: 10	-		Mt: 28	-	-	Mt: 26.5	Cyt_Mt: 14	-	Mt: 3	Mt: 5	-
		10979	-	-		Mt: 24	Cyt_Mt: 15.3	Mt_Per: 12.8	Per: 15	Cyt: 7.5	-	Mt: 2	Mt: 2	-
NADPH:glyoxylate reductase	1.1.1.79	**5154**	Mt: 14	-		Mt: 24	Cyt: 6	-	Mt: 14.5	CytMt:14	Cyt: 12.5	Mt: 3	Mt: 5	-
		13699	Cyt: 7	-		Mt: 27	-	-	Cyt: 12.5	Mt: 10	Cyt_Nu: 7	Cyt: 5	Sec: 2	-

Transcript numbers in bold indicate the presence of the splice leader sequence. PSORT score is the discriminant score, with larger scores having a higher probability. Scores below 5 are not reported. TargetP score is the reliability class is rated from 1 to 5 (1 is the strongest prediction and 5 is the weakest). Chl—chloroplast (green); Cyt—cytosol (grey); E.R.—endoplasmic reticulum (Blue); Mt—mitochondria (orange); Nu—nuclear; Per—peroxisome; PM—plasma membrane (yellow); Sec—secreted or extra cellular (blue). Strength of colour indicates score.

Table 2. Subcellular location prediction of *E. gracilis* cytosol (grey) and Golgi (blue) marker proteins.

Marker Enzyme Name	EC Number	EG Transcript	WoLF PSORT Plant		WoLF PSORT Animal			WoLF PSORT Fungi			TargetP Plant	TargetP Nonplant	TargetP, Signal Cut
Glutamate dehydrogenase	1.4.1.2	11196	Cyt: 8	Chl: 6	Cyt_Nu: 15	Nu: 8.5	-	Cyt: 16	Cyt_Nu: 9.5	Cysk: 7	Cyt: 2	Cyt: 2	-
		10682	Cyt: 8	-	Cyt: 21	Cyt_Nu: 16	Nu: 9	Cyt: 12.5	Mt: 8	Cyt_N u: 8	Cyt: 4	Cyt: 3	-
		7536	Cyt: 11	-	Cyt: 16.5	Cyt_Mt: 13.6	Cyt_Nu: 10.3	Mt: 15	Cyt: 7.5	-	Mt: 5	Cyt: 4	-
		791	Cyt: 6	-	Cyt: 12	Nu: 10	-	Cyt: 11.5	Cyt_Nu: 7.5	Mt: 7	Cyt: 4	Cyt: 5	-
NAD-lactate dehydrogenase	1.1.1.27	11571	-	-	Mt: 10	Cyt: 7	Cyt_Nu: 7	Mt: 19	Cyt: 8	-	Cyt: 3	Cyt: 5	-
thiosulfate sulfurtransferase	2.8.1.1	**32721**	Cyt: 8.5	-	Cyt: 17	Cyt_Nu: 14	Sec: 7	Cyt: 9	Cysk: 9	Mt: 7	Cyt	Cyt: 3	-
		19233	Chl: 13	-	Sec: 14	E.R.: 6	-	Sec: 17	-	-	Mt: 5	Sec: 4	-
		10931	-	-	PM: 15	Mt: 7	-	Cyt: 8.5	Mt: 7	Sec: 7	Cyt: 5	Cyt: 3	-
		27480	-	-	Cyt_Nu: 15.5	Nu: 13.5	Cyt: 12.5	Cyt: 10.5	Cysk: 9	Cyt_N u: 8	Mt: 5	Mt: 5	-
		18090	Cyt: 6	-	Sec: 28	-	-	Sec: 18	-	-	Cyt: 3	Cyt: 2	-
		46995	Chl: 9	-	Cyt: 18	Cyt_Nu: 14.5	Nu: 7	Mt: 15.5	Cyt_Mt: 10.3	-	Cyt: 4	Cyt: 5	-
		1438	Chl: 6	-	Sec: 13	E.R.: 11	Mt: 6	Sec: 9	Mt: 7	-	Mt: 5	Sec: 4	-
α-1,3-Man transferase	2.4.1.258	11655	PM: 6	-	PM: 19	-	-	Mt: 13	PM: 9	-	Sec: 4	Sec: 1	Cyt: 1
α-1,2-Man transferase	2.4.1.259/ 261	**44449**	Chl: 12	-	Sec: 27	-	-	Mt: 10	PM: 10	-	Sec: 5	Sec: 1	Cyt: 1
		15548	PM: 12	-	PM: 15.5	Sec_PM: 10.5	-	PM: 17	-	-	Mt: 2	Sec: 5	-
α-1,6-Man transferase	2.4.1.260	**9095**	PM: 9	-	PM: 13	Sec: 11	-	PM: 19	-	-	Sec: 4	Sec: 1	Cyt: 1
Dolichol phosphate mannose synthase	2.4.1.83	21418	-	-	PM: 11	Cyt: 6	Mt: 6	Cyt: 8	Mt: 7	PM: 6	Cyt: 1	Cyt: 1	-
Dolichyl-phosphate beta-glucosyltransferase	2.4.1.117	15275	PM: 7	-	Sec: 20	PM: 9	-	PM: 12	Sec: 12	-	Sec: 5	Sec: 2	Cyt: 4
α-1,3-Glc transferase	2.4.1.267/ 265	22866	-	-	Sec: 11	PM: 9	Mt: 7	PM: 21	-	-	Cyt: 5	Sec: 5	-
		24290	PM: 7	-	Sec: 16	PM: 11	-	PM: 18	-	-	Mt: 3	Sec: 1	-
		22282	PM: 6	-	Sec: 9	PM: 7	Mt: 7	PM: 17	E.R.: 7	-	Sec: 2	Sec: 2	Cyt: 2
α-1,2-Glc transferase	2.4.1.256	20467	-	-	Sec: 15	PM: 10	-	PM: 17	-	-	Mt: 1	Mt: 3	-
Dolichyldiphosphooligosaccharide-protein glycosyltransferase	2.4.99.18	2741	PM: 13	-	PM: 22	-	-	PM: 17	E.R.: 6	-	Mt: 5	Sec: 2	-
		10761	PM: 11	-	PM: 11	Cyt: 9	-	PM: 26	-	-	Cyt: 2	Cyt: 2	-
		2733	PM: 11	-	PM: 24	-	-	PM: 23	-	-	Cyt: 5	Cyt: 4	-

Transcript numbers in bold indicate the presence of the splice leader sequence. PSORT score is the discriminant score, with larger scores having a higher probability. Scores below 5 are not reported. TargetP score is the reliability class is rated from 1 to 5 (1 is the strongest prediction and 5 is the weakest). Chl—chloroplast (green); Cyt—cytosol (grey); Cysk—Cytoskeleton; E.R.—endoplasmic reticulum (blue); Mt—mitochondria (orange); Nu—nuclear; Per—peroxisome; PM—plasma membrane (yellow); Sec—secreted or extra cellular (blue). Strength of colour indicates score.

Table 3. Subcellular location prediction of *E. gracilis* chloroplast marker proteins.

Marker Enzyme Name	EC Number	EG Transcript	WoLF PSORT Plant		WoLF PSORT Animal			WoLF PSORT Fungi			TargetP Plant	TargetP Nonplant	TargetP, Signal Cut
Ribulose-biphosphate carboxylase (small subunit)	4.1.1.39	7123	Chl: 13	-	Cyt: 15.5	Cyt_Nu: 14.5	Nu: 8.5	Nu: 7.5	Cyt_Nu: 7.5	Mt: 7	Cyt: 2	Cyt: 1	-
		43244	Chl: 8	-	Sec: 30	-	-	PM: 21	-	-	Sec: 3	Sec: 2	Chl: 5
		13866	-	-	Mt: 12.5	Cyt_Nu: 11	Cyt: 10.5	Cyt_Nu: 9.3	Cyt_Mt: 9.3	Cyt: 9	Cyt: 4	Cyt: 3	-
		59484	Chl: 7	-	Sec: 29	-	-	Nu: 10.5	Cyt_Nu: 10	Cyt: 8.5	Sec: 3	Sec: 5	Cyt: 4
NADP- glyceraldehyde-3-phosphate dehydrogenase	1.2.1.9	4454	Chl: 11	-	Mt: 19.5	Mt_Per: 11	Cyt: 6.5	Mt: 18	Cyt: 6	-	Mt: 4	Mt: 4	-
		5575	Mt: 10.5	Chl_Mt: 7.5	Mt: 28.5	Cyt_Mt: 16	-	Mt: 22.5	Cyt_Mt: 14	-	Mt: 1	Mt: 2	-
		9413	Chl: 12	-	Mt: 22.5	Mt_Per: 12	-	Mt: 23	-	-	Mt: 5	Mt: 2	-
		7552	Cyt: 11	-	Cyt: 24	-	-	Cyt: 22	-	-	Cyt: 2	Cyt: 1	-
		15045	Mt: 8.5	Chl_Mt: 7	Mt: 24	-	-	Mt: 26.5	Cyt_Mt: 14	-	Mt: 2	Mt: 2	-
Fructose-biphosphate aldolase	4.1.2.13	**15855**	Cyt: 8	-	Nu: 12.5	Cyt_Nu: 10.5	Cyt: 7.5	Cyt: 16	Mt: 6	-	Cyt: 4	Cyt: 2	-
		25832	Mt: 8.5	Chl_Mt: 7	Mt: 16	Cyt: 14	-	Mt: 24.5	Cyt_Mt: 14	-	Mt: 2	Mt: 4	-
		7827	Chl: 12	-	E.R.: 16	-	-	Sec: 15	-	-	Chl: 5	Sec: 4	-
		5279	Chl: 14	-	Mt: 17	E.R._Mt: 10.5	Per: 6	Mt: 15	Cyt: 6	-	Mt: 4	Mt: 3	-
Pyruvate Pi dikinase	2.7.9.1	1635	Chl: 6	-	PM: 9	Sec: 8	-	Cyt: 11.5	Mt: 9	Cyt_Nu: 7.5	Cyt: 1	Cyt: 3	-
Photosystem II D1	-	13288	PM: 6	-	PM: 12	Lyso: 10	Sec: 6	PM: 21	-	-	Sec: 1	Sec: 1	Cyt: 4
	-	10715	PM: 13	-	PM: 26.5	Sec_PM: 14	-	PM: 27	-	-	Cyt: 2	Cyt: 4	-
Photosystem II D2	-	13288	PM: 6	-	PM: 12	Lyso: 10	Sec: 6	PM: 21	-	-	Sec: 1	Sec: 1	Cyt: 4
	-	10715	PM: 13	-	PM: 26.5	Sec_PM: 14	-	PM: 27	-	-	Cyt: 2	Cyt: 4	-
Photosystem II CP47	-	899	PM: 9.5	-	PM: 31	-	-	PM: 24	-	-	Mt: 5	Sec: 5	Cyt: 2
Photosystem II CP43	-	13640	PM: 10	-	PM: 20	-	-	PM: 22	-	-	Sec: 4	Sec: 1	Cyt: 4
Cytochrome b6f	-	5228	PM: 6.5	-	PM: 12	Mt: 6	-	PM: 13	E.R.: 7	-	Cyt: 4	Cyt: 4	-
	-	23844	Chl: 13	-	Mt: 16.5	E.R._Mt: 13	Mt_Per: 12.5	Mt: 14	-	-	Cyt: 5	Mt: 5	-
Photosystem I A1	-	**158**	PM: 10	-	PM: 23	-	-	PM: 25	-	-	Mt: 5	Mt: 4	-

Transcript numbers in bold indicate the presence of the splice leader sequence. PSORT score is the discriminant score, with larger scores having a higher probability. Scores below 5 are not reported. TargetP score is the reliability class is rated from 1 to 5 (1 is the strongest prediction and 5 is the weakest). Chl—chloroplast (green); Cyt—cytosol (grey); E.R.—endoplasmic reticulum (blue); Lyso—lysosome; Mt—mitochondria (orange); Nu—nuclear; Per—peroxisome; PM—plasma membrane (yellow); Sec—secreted or extra cellular (blue). Strength of colour indicates score.

Apart from the evident limitations of these algorithms as protein localisation prediction tools in *Euglena*, protein targeting into chloroplasts of *Euglena* is likely to be inherently complex. In contrast to plants, the chloroplast of *Euglena* evolved from the secondary endosymbiosis, which led to the chloroplast being surrounded by three membranes [25,26,33]. A recent study of the *E. gracilis* chloroplast proteome identified three classes of chloroplast pre-protein based on targeted signal analysis. Class I and II proteins possess a bipartite topogenic signal (BTS), with Class I proteins composed of a signal peptide (SP) followed by a stop-transfer signal (STS) and a transit peptide (TP), whilst Class II proteins contain only an SP and TP [31,34]. The third class of chloroplast proteins was referred to as unclassified, with no signal sequence detected in the proteins. The transport mechanism used to import proteins from this unclassified category into the plastid remains unknown [30]. The transport of *Euglena* Class I and II pre-proteins into the chloroplast involves the first step of co-translational transport into the endoplasmic reticulum (ER) lumen where the cleavage of the signal peptide occurs (Figure 1). The pre-proteins are subsequently transported to the chloroplasts from the Golgi body via vesicles, which then fuse with the outermost plastid membrane. However, the transport across the inner two membranes of the three-membrane-bound plastids in euglenophytes remains unclear [27,30,34]. The TOC/TIC-like pathway was believed to be involved in the inner membranes transport of the *Euglena* plastid due to the presence of plant-like targeting signal (TP) in the preproteins [35]. However, none of the TOC subunits have been detected in the transcriptome of *E. gracilis*, whereas homologues of several TIC subunits were identified [5]. A recent analysis of the structure of TP sequences in *E. gracilis* has suggested that it is possible for the TP to be recognised by the symbiont-derived ERAD-like machinery (SELMA) transport system, as is the case for diatoms [30,36].

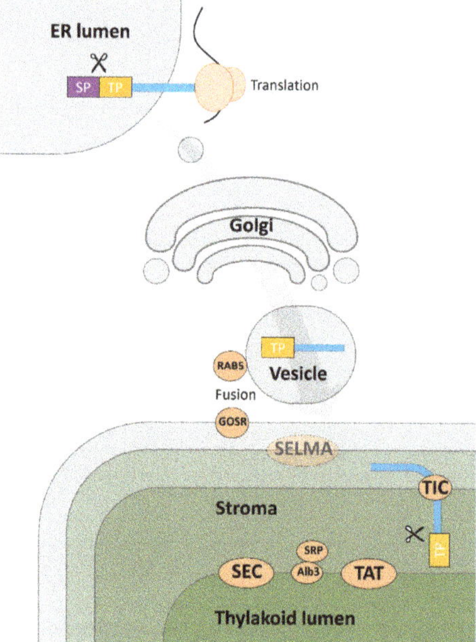

Figure 1. Protein transport into the secondary chloroplast of *Euglena*. Nuclear encoded chloroplast pre-proteins (blue strip) are synthesised into the lumen of the endoplasmic reticulum (ER) where the signal peptide (SP) is cleaved. Pre-proteins with transit peptides (TP) are subsequently transferred to the outermost chloroplast membrane through the Golgi body via vesicles. GOSR and RAB5 GTPase are proposed to mediate the fusion of the vesicle to the outermost membrane. After transport of proteins into the stroma, where the TP is removed, the mature protein can enter the thylakoid lumen via SEC, TAT, or Alb3/SRP pathway. This scheme only considers proteins possessing Class I and II targeting signals, as the transport of those with unclassified signals is not known [34].

It can be concluded that WoLF PSORT and TargetP have limitations with predicting cTPs and do not specifically include protein targeting to the secondary plastid. Predicting chloroplast protein targeting in *Euglena* is likely to require more specific databases or algorithms, since the evolution of the *Euglena* chloroplast is different from that of plants. In contrast, the prediction of mitochondria targeting with high reliability scores, when there is a high degree of agreement amongst the algorithms, can be informative. However, due to the false predictions of chloroplast proteins to other locations, the prediction results cannot be fully relied upon and need to be carefully evaluated in conjunction with evidence from enzymatic and biochemical analyses.

2.2. Pathway Localisation from Biochemical/Proteomic Information

2.2.1. Central Metabolic Pathways of *Euglena*

The central metabolic pathways are essential to all organisms, providing the precursors for other peripheral pathways, especially metabolites with carbon backbones that are derived from carbohydrate metabolism. In addition, under non-photosynthetic conditions, these pathways have a major role in producing the energy and reducing power for the cell. Pathways of carbohydrate metabolism generally consist of glycolysis (Embden–Meyerhof–Parnas pathway), gluconeogenesis, the PPP, the Entner-Doudoroff (ED) pathway, and the TCA cycle. Notably, there is no evidence for the ED pathway in *Euglena*. Results for subcellular location predictions are available in Supplementary Table S1.

Glycolysis and Gluconeogenesis

The intracellular distribution of the glycolytic enzymes in *Euglena* has been studied using fractionation in aqueous and non-aqueous media. This approach showed that most of the glycolytic enzymes are in the cytosol and that several of them are present in both the chloroplast and the cytosol [37,38]. By using sucrose density gradient centrifugation, it was found that phosphofructokinase, pyruvate kinase, triosephosphate isomerase, and aldolase were present in the plastid fraction [39]. In addition, a recent proteomic study reported that several enzymes involved in glycolysis and gluconeogenesis were present in *Euglena* chloroplasts [30].

Hexose-Phosphorylating Enzymes. The activity of hexokinase (EC 2.7.1.1) was three times higher in *E. gracilis* grown on glucose than that on ethanol and acetate [40]. The activity of this enzyme in glucose media was also four times higher in heterotrophic cells than that in autotrophic cells [41]. *E. gracilis* was found to have glucokinase (EC 2.7.1.2) and fructokinase (EC 2.7.1.4) in different locations in both autotrophic and heterotrophic conditions. At 105,000 g separation, the glucokinase was present in the cell pellet while the fructokinase activity was only found in the supernatant [2,42]. Glucokinase is therefore concluded to be in organelles, whilst fructokinase is in the cytosol.

Phosphoglucoisomerase (EC 5.3.1.9). The activity of this enzyme was detected in *E. longa* [2,43], although, the subcellular location has not been reported. Strong targeting signals were not detected in the protein sequences.

6-Phosphofructokinase (ATP-PFK, EC 2.7.1.11) and Diphosphate–Fructose-6-Phosphate 1-Phosphotransferase (PPi-PFK, EC 2.7.1.90). In *E. gracilis*, ATP-PFK was reported to be located in both chloroplasts and the cytosol [39], while PPi-PFK was reported exclusively in the cytosol. During cell growth, the activity of PPi-PFK was 10–30 times higher than the activity of ATP-PFK [44]. No strong targeting signals were detected in these protein sequences.

Fructose Bisphosphate Aldolase (EC 4.1.2.13). There are two classes of aldolase found in *Euglena*: Class I is located in the chloroplast and proplastid, and Class II is located in the cytosol [45]. Class I enzyme peptides were detected in the chloroplast proteome [30] and the Class II cytosolic enzyme was shown to be more active when the *E. gracilis* culture was grown in the dark and is presumed to play the main role in heterotrophic glycolysis and gluconeogenesis [46]. One isoform has no strong targeting signal, whilst two have plastid targeting and one has a strong mitochondrial targeting sequence.

Glyceraldehyde 3-phosphate Dehydrogenase (G3P) Dehydrogenase (EC 1.2.1.12). *E. gracilis* contains both NAD-linked and NADP-linked G3P dehydrogenase, which are found in different subcellular locations [45,47]. The NAD-linked enzyme showed higher activity in heterotrophic conditions and was located in the cytosol. On the other hand, the NADP-linked enzyme was shown to be located in chloroplasts and had higher activity in autotrophic cells [48]. Only the NADP-linked enzyme was detected in the proteome of *E. gracilis* chloroplasts [30].

Triosephosphate Isomerase (EC 5.3.1.1). As with fructose bisphosphate aldolase, two types of the isomerase were identified in *E. gracilis* using enzymatic activity profiling [47]. Type A triosephosphate isomerase was reported to function in the chloroplasts and proplastids of *E. gracilis*, while type B enzymes were located in the cytosol [49]. Sequences matching triosephosphate isomerase could also be detected in the *E. gracilis* chloroplast proteome [30].

Phosphoglycerate Kinase (EC 2.7.2.3)/Phosphoglycerate Mutase (EC 5.4.2.11). The activity of phosphoglycerate kinase was reported in isolated *E. gracilis* chloroplasts [50] and the enzyme was detected in the *E. gracilis* chloroplast proteome [30], although the presence in other subcellular locations has not been investigated. No specific studies of the activity of phosphoglycerate mutase have been reported in *Euglena*. However, the enzyme was recently reported to be present in the *E. gracilis* chloroplast proteome [30]. WoLF PSORT identifies a strong chloroplast targeting sequence on one isoform, with the other three isoforms is predicted to remain in the cytosol.

Enolase (EC 4.2.1.11). The activity of enolase was previously detected in *E. gracilis* but the subcellular location was not described [38,51]. N-terminal targeting peptide analysis of cDNA clones of *E. gracilis* suggested that enolase could be present in both the cytosol and the chloroplast [52]. However, as shown in Section 2.1.3, it is difficult to predict protein targeting into the chloroplasts of *Euglena* and, furthermore, enolase was not found in the chloroplast proteome of *E. gracilis* [30].

Pyruvate Kinase (EC 2.7.1.40). The activity of pyruvate kinase in *E. gracilis* was shown to be highly active in cultures grown on glucose [53]. This enzyme was reported to be located in both proplastids and the cytosol of *E. gracilis*, however, the activity of this enzyme was not detected in the mature chloroplast [39]. WoLF PSORT predicts plastid targeting sequence in two isoforms with very low confidence, whilst one of these has mitochondrial targeting with slightly more confidence, highlighting the challenging nature of predicting subcellular locations.

Fructose-1,6-Bisphosphatase (EC 3.1.3.11). Fructose-1,6-bisphosphatase is involved in gluconeogenesis and has been reported from *Euglena* [39,44]. The cytosolic fructose-1,6-bisphosphatase in *E. gracilis* was detected and characterized [54]. Recently, the enzyme was reported in the *E. gracilis* chloroplast proteome [30], where it is presumably involved in the Calvin cycle. One isoform is predicted not to contain a targeting signal, but the other four are predicted to be variously targeted to the chloroplast, for secretion, or to the plasma membrane, possibly indicating that they all pass through the secretory system to the chloroplast.

Pentose Phosphate Pathway

Oxidative Phase. In contrast to higher plants and green algae, all the enzymes of the oxidative arm of the pentose phosphate pathway in *E. gracilis* were reported to be present in the cytosol, but not the chloroplast. Using non-aqueous fractionation, it was found that two dehydrogenases of the oxidative pentose phosphate pathway were absent from the *E. gracilis* plastid [37]. In separate studies, the activity of 6-phosphogluconate dehydrogenase (EC 1.1.1.44) was confirmed to be in the cytosol [38], and glucose-6-phosphate dehydrogenase (EC 1.1.1.49) was reported to be located in the cytosol [2,38,55–57] and has been used as a cytosolic marker enzyme [58]. Although a single glucose-6-phosphate dehydrogenase was detected in the chloroplast proteome, this fraction was reported to be moderately contaminated with protein from other organelles [30] and thus, subcellular location of the enzyme will need further investigation to confirm its location. This enzyme is specific for NADP in *Euglena* and induced by glucose, with low activity detected under heterotrophic growth in the absence of glucose [53]. There has been no specific study of *Euglena* 6-phosphogluconolactonase (EC 3.1.1.31).

Non-Oxidative Phase. All the enzymes involved in the non-oxidative section of the pentose phosphate pathway have been detected in *Euglena* and most of the enzymes were reported to localize to the chloroplast [2,30]. The activity of ribose 5-phosphate isomerase (EC 5.3.1.6) was reported in isolated *E. gracilis* chloroplasts [59]. The subcellular location of pentose-5-phosphate-3-epimerase (EC 5.1.3.1) has not been reported, although the activity of this enzyme was detected in heterotrophic, autotrophic, and mixotrophic growth conditions, along with the activity of transketolase (EC 2.2.1.1) [60] and transaldolase (EC 2.2.1.2) [47]. Non-aqueous separation techniques showed the presence of transaldolase in *Euglena* chloroplasts and proplastids [39].

Notably, there are two isoforms of each enzyme of the non-oxidative PPP in the *E. gracilis* transcriptome, except transketolase which has three. For three of these enzymes, only one isoform was identified in the chloroplast proteome [30], whereas neither isozyme of transaldolase could be detected. This suggests that the other isoforms are present in another location within the cell and the lack of any detectable targeting signal indicates this is likely to be the cytosol. However, extensive study of this pathway has not been reported and further investigation would be needed to confirm the operation of the pathway in the cytosol.

Anaplerotic Pathway: Dicarboxylic Acid Bypass

Malate dehydrogenase (NADP-specific oxaloacetate-decarboxylating, EC 1.1.1.40) in *Euglena* is located in the cytosol but not in mitochondria, and is specific for NADP and L-malate [2]. The NAD-specific malate dehydrogenase (decarboxylating, EC 1.1.1.39) can only be detected in *E. gracilis* cultured with D-malate [61]. Recently, a proteomic study detected malate dehydrogenase (NADP-specific) in *E. gracilis* chloroplasts [30]. The activity of this enzyme varied widely with light and carbon sources, and has 55 times greater activity in heterotrophic cells than in autotrophic cells. This result suggests a physiological role in *Euglena* for these enzymes in providing NADPH for cytosolic fatty-acid synthesis in the dark [62,63].

Phosphoenolpyruvate carboxylase (PEP carboxylase, EC 4.1.1.31) was shown to have multiple isozymes which were active in different light conditions. It has been reported that PEP carboxylase functions for CO_2 fixation in *E. gracilis* grown in the dark and under CO_2 limited conditions [64,65]. The activity of phospho*enol*pyruvate carboxykinase (PEP carboxykinase, EC 4.1.1.32) in *E. gracilis* is specific for GTP rather than ATP [66]. PEP carboxylase and PEP carboxykinase are discrete, separate enzymes in *E. gracilis* [67]. PEP carboxykinase was reported to be located exclusively in the cytosol and the enzyme could not be detected in cells grown under autotrophic conditions [68]. One isoform is predicted to be localised in the chloroplast by WoLF PSORT with a high degree of confidence, but the locations of the other two isoforms are not predicted confidently. In addition, the activity of PEP carboxykinase was detected in *E. gracilis* cultured with acetate or ethanol, but not with glucose [62]. Pyruvate carboxylase (EC 6.4.1.1) was also reported to be located in the cytosol [69]. The activity of this enzyme was found in cells grown under heterotrophic culture fed with glucose, but not with acetate or in autotrophic cells [2].

TCA Cycle

The reactions of the TCA cycle occur in the mitochondria of *Euglena* in common with all other eukaryotic organisms [2]. Most of the enzymes involved in the TCA cycle are predicted to target to the mitochondria with high reliability (Table S2), in line with previous studies on the localisation of the TCA cycle.

Pyruvate Dehydrogenase (NAD complex 1.2.4.1, NADP+ EC 1.2.1.51). In *E. gracilis* the conventional NAD^+ pyruvate dehydrogenase complex only contributes around 1% of the activity and instead an $NADP^+$-dependent pyruvate dehydrogenase is used to produce the majority of the acetyl-CoA from pyruvate [70]. This latter enzyme has been detected in the mitochondrial fractions of *E. gracilis* [71–73] and all three component polypeptides are predicted to be targeted to the mitochondria. The activity of the NAD complex has not been localised.

Citrate Synthase (EC 4.1.3.7). Citrate synthase activity was detected in both particulate and soluble fractions from bleached *E. gracilis* [38], indicating that the enzyme is located in cytosol and other cell compartments. Testing the activity of this enzyme from different organelle suspensions showed the presence of this enzyme in both mitochondria and microbodies (glyoxysome-like particles) [74,75]. Only one of the four isoforms is predicted to be targeted to mitochondria.

Aconitase (EC 4.2.1.3). The activity of aconitase was detected in *E. gracilis* [76,77]. However, the subcellular location of this enzyme has apparently never been investigated and only one of the two isoforms is predicted to be targeted to mitochondria.

Isocitrate Dehydrogenase (NAD-specific EC 1.1.1.41, NADP-specific EC 1.1.1.42). NAD- and NADP-specific isozymes of isocitrate dehydrogenase have been characterised from *Euglena*. The activity of NAD-specific isocitrate dehydrogenase was detected in mitochondria and cytosol of *E. longa* [38,43]. The NAD-specific isozyme was detected solely in the mitochondria of the streptomycin-bleached *E. gracilis* [75,78,79]. The NADP-specific isozyme was reported in both mitochondria and the cytosol, with the activity of the mitochondrial enzyme being about 25% of that in the cytosol [75,79].

2-Oxoglutarate Decarboxylase (EC 4.1.1.71). *E. gracilis* contains a 2-oxoglutarate decarboxylase that is dependent on thiamine pyrophosphate, in contrast to the more common CoA-dependent 2-oxoglutarate dehydrogenase complex, which was not detected [80]. The thiamine pyrophosphate dependent activity which coverts 2-oxoglutarate to succinic semialdehyde is located solely in mitochondria [81].

Succinic Semialdehyde Dehydrogenase (EC 1.2.1.16). NAD- and NADP-specific succinate semialdehyde dehydrogenase were detected in *E. gracilis* and reported to be in the mitochondria [73,82]. Three isoforms are predicted to be located in the mitochondria, whilst the remaining isoform is not predicted to have a targeting sequence.

Succinate Dehydrogenase (EC 1.3.5.1). As with other eukaryotes, the succinate dehydrogenase in *E. gracilis* is tightly bound to the inner membrane of mitochondria and has been used as a marker enzyme for mitochondria in *Euglena* [83]. [58,74,75,78]. It is predicted to be associated with the plasma membrane by WoLF PSORT, in line with the integral membrane nature of the protein.

Fumarase (EC 4.2.1.2). Using cell fractionation and enzyme activity assays, fumarase is routinely detected solely in *E. gracilis* mitochondria [39,74,75,78] and is commonly used as a soluble mitochondrial marker enzyme [83].

Malate Dehydrogenase (EC 1.1.1.37). In *E. gracilis*, malate dehydrogenase is found in both mitochondria and the cytosol. The cytosolic enzyme had three times higher activity in heterotrophically grown cells than in photoautotrophic cells, whereas the activity of the mitochondrial isoform was largely uninfluenced by variation in growth conditions [62]. *E. gracilis* contains two forms of malate dehydrogenase, NAD-linked and NADP-linked isozymes. Unlike in higher plants, where the NADP-linked malate dehydrogenase is present exclusively in chloroplasts, in *E. gracilis* the majority (81–91%) of both NAD-linked and NADP-linked activity were located in the cytosol with a smaller proportion (13–16%) found in mitochondria. The activity of the NAD-linked isozyme was reported to be about three times higher than that of the NADP-dependent isozyme [84,85].

Glyoxylate Cycle

The glyoxylate cycle is a modified form of the TCA cycle that is found in plants, bacteria, protists and fungi. The cycle has an important role in provision of precursors for gluconeogenesis and allows the cell to use other respiratory substrates when sugars are not available [86]. The subcellular location of the glyoxylate cycle in *Euglena* under different conditions is poorly defined, with studies suggesting that the cycle operates in either mitochondria or discrete microbodies (glyoxysome-like particles). Notably, the presence of microbodies in *E. gracilis* was reported to vary under different conditions [87]. Following cell fractionation on sucrose density gradients, the activities of isocitrate lyase (EC 4.1.3.1) and malate synthase (EC 2.3.3.9), enzymes unique to the glyoxylate cycle, were found in the microbody fraction of *E. gracilis* grown on acetate [75,78]. In contrast, using similar cell fractionation techniques and immunocytochemical analysis, both isocitrate lyase and malate synthase were localised to mitochondria in *E. gracilis* grown on ethanol in which microbodies could not be detected [88].

C2 Metabolism

Ethanol, which can readily diffuse into the cell, is first oxidized to acetaldehyde by alcohol dehydrogenase (EC 1.1.1.1), and the product is then oxidised by acetaldehyde dehydrogenase (EC 1.2.1.10) to produce acetate. Both enzymes are found in *E. gracilis* mitochondria [89–91]. Acetate is taken up either by simple diffusion or active transport through monocarboxylate transporters and is then converted to acetyl-CoA by acetyl-CoA synthetase (EC 6.2.1.1), also located in *E. gracilis* mitochondria [92], and then metabolized through the TCA cycle or channelled into the glyoxylate cycle.

2.2.2. Subcellular Locations of Biomass Production

The composition of *Euglena* biomass is similar to that of many organisms, with storage carbohydrates, proteins and lipids predominating. The amounts of the different components varies substantially depending on the growth conditions, from almost 10% dry weight wax esters [93] under anaerobic growth to over 80% paramylon under aerobic conditions [94].

Carbohydrate Biosynthesis

Unlike most other photosynthetic organisms, such as plants and green algae, *Euglena* stores carbohydrate in the form of a crystalline β-1,3-glucan, called paramylon, instead of starch, and the soluble disaccharide trehalose, instead of sucrose. *Euglena* has a wide range of enzymes involved in carbohydrate metabolism but it is difficult to predict their substrates and products from sequence alone [95].

Paramylon. Paramylon is synthesized from UDP-glucose [96] using the membrane bound paramylon synthetase (beta-1,3-glucan beta-glucosyltransferase, EC 2.4.1.34) that was identified in the *E. gracilis* mitochondrial fraction by measuring activity following differential centrifugation [97] and the genes identified in the transcriptome [98]. Based on transmission electron microscopy, paramylon was synthesised in vesiculated mitochondrial related membrane complexes (chondriomes). The matrix of these vesicles was dense with paramylon granules and extended into the cytosol. The vesicles developed, resulting in the membrane-bound paramylon grains found in the cytosol [41,99,100]. The endo-1,3-β-glucanases (EC 3.2.1.6 and EC 3.2.1.39), exo-1,3-β-glucanases (EC 3.2.1.58), and 1,3-β-glucan phosphorylases (EC 2.4.1.97) involved in glucan metabolism have been characterized [101–103], though the subcellular locations of these enzymes have not been defined. Some of these are predicted to be membrane associated or chloroplast localised.

Trehalose. In *Euglena gracilis*, trehalose synthesis was reported to have a role in the acclimation to osmotic stress [104,105]. Trehalose biosynthesis involves a two-step process through the sequential action of trehalose-phosphate synthase (TPS, EC 2.4.1.15) and trehalose-phosphate phosphatase (TPP, EC 3.1.3.12). It was found that the activities of TPS and TPP could not be separated and so a TPS/TPP enzyme complex of about 250 kDa was suggested to be responsible for trehalose synthesis in *E. gracilis* [106]. In *Arabidopsis*, the bulk of the TPP was reported to be cytosolic [107,108]. However, the subcellular localisation of the TPS/TPP complex in *Euglena* has not been investigated. Analysis of the chloroplast proteome of *E. gracilis* [30] shows no evidence of the TPS and TPP suggesting it is more likely that the TPS/TPP complex is located in the cytosol (or conceivably mitochondria) rather than in chloroplasts. There is no strong targeting signal predicted for this enzyme, supporting the putative cytosolic location.

Amino Acid Biosynthesis

The pathways of amino acid biosynthesis in *Euglena* have been poorly investigated, especially with regard to their subcellular localisation. The recent evidence from the proteomic analysis of *Euglena* chloroplasts suggested that their capacity for synthesis of amino acids is extremely limited, in contrast to plant and algal chloroplasts, which are the major subcellular sites for synthesis of various amino acids [30]. Here we present a summary of the likely subcellular localisation of amino acid biosynthesis in *Euglena*.

Glycine and Serine (Glycolate Pathway Associated). Glycine and serine are synthesised from glyoxylate, an intermediate of photorespiration and gluconeogenesis. Glycolate dehydrogenase (EC 1.1.99.14), the starting enzyme of the glycolate pathway, was reported to be located in both mitochondria and microbodies in *E. gracilis* [78]. Glutamate:glyoxylate aminotransferase (EC 2.6.1.4), which adds the amino group to form glycine [109], is found in mitochondria, the cytosol and microbodies [78,110]. A small proportion of the glyoxylate is converted to glycine by glutamate:glyxoylate aminotransferase in mitochondria, and the majority is split into CO_2 and formate. As in higher plants, the formate is then used to produce serine through condensation with glycine [111,112]. Folate coenzymes, which are involved in this C1 transfer, were reported to be located largely in the cytosol [79]. Glycine can also be produced through the cleavage of threonine by threonine aldolase (EC 4.1.2.5/48) [113], though the subcellular location of this activity has not been reported. The enzymes involved in serine biosynthesis from 3-phosphoglycerate have not been studied in detail in *Euglena*. However, recently, phosphoserine phosphatase was identified in the *E. gracilis* chloroplast proteome, indicating the possibility of a plastidic serine biosynthesis pathway [30].

Methionine, Cysteine, and Threonine. The activity of cobalamin-dependent methionine synthase (EC 2.1.1.13), producing methionine from N^5-methyltetrahydrofolate and homocysteine, was reported to be distributed between the cytosol (68.9%), chloroplast (18.4%) and mitochondria (9.5%) of phototrophic cells. The more stable, Mg-dependent, variant was reported to be found only in the cytosol [114]. Cysteine synthesis in *Euglena* has not been investigated in detail and the subcellular localisations of the enzymes associated with this pathway have not been elucidated. Two enzymes involved in the synthesis of cysteine (serine O-acetyltransferase and cysteine synthase) were reported in the *E. gracilis* transcriptome [113] and isoform A of cysteine synthase was detected in the *E. gracilis* chloroplast proteome [30]. Threonine is synthesized from aspartate via homoserine. Five enzymes involved in threonine biosynthesis in *E. gracilis* were reported to be expressed in different growth conditions [113]. However, the localisations of the enzymes involved in the synthesis pathway have not been elucidated.

Aromatic Amino Acids (Phenylalanine, Tyrosine, and Tryptophan). Chorismate, the precursor to aromatic amino acids, is synthesised from D-erythrose 4-phosphate and phosphoenolpyruvate by the shikimate pathway in seven steps. Five reactions can be catalysed either by separate enzymes, as in plants [115], or by a pentafunctional enzyme, as in fungi [116]. There is evidence for both of these in the *E. gracilis* transcriptome [27].

In green algal and plant cells, the aromatic amino acids are produced exclusively in the plastid but the protein analysis of isolated organelles of *E. gracilis* suggests that the shikimate pathway occurs in both the chloroplast and cytosol [117]. The preferred pathway depends on the growth conditions, with the cytosolic pathway used in the dark and the plastidic pathway in the light [117,118].

Chorismate is then converted into tyrosine and phenylalanine, via prephenate by dehydration, dehydrogenation, and transamination. The enzymes catalysing these reactions are present in *E. gracilis* as unusual domain fusions, also found in thermophilic bacteria [16]. Tryptophan is synthesised from chorismate by a series of reactions via anthranilate. In *E. gracilis* all four of these reactions are carried out by a unique fusion protein rather than a series of separate enzymes, as in other organisms [11,113].

Together the data suggest that aromatic amino acid biosynthesis in *Euglena* is carried out by a combination of plant-, bacterial-, and fungal-like enzymes, as well as unique proteins. The evidence suggests that these pathways are not exclusively located in the plastid, unlike in plants, supporting the dispensability of the plastid for their biosynthesis.

Branched-Chain Amino Acids (Valine, Isoleucine, and Leucine). Pyruvate and α-ketobutyrate are the precursors for valine, leucine and isoleucine biosynthesis in *Euglena*, as in other organisms [119]. In *E. gracilis*, α-ketobutyrate is synthesized by the action of two threonine dehydratases (EC 4.3.1.19 and EC 4.3.1.17) that are located in the cytosol [120]. The subsequent steps are catalysed by acetolactate synthase, dihydroxy-acid reductoisomerase, and branched-amino-acid aminotransferase, all of which are located in the mitochondria [119], suggesting the biosynthesis of branched-chain amino acids is located in mitochondria.

Arginine and Proline. Arginine is synthesised by the sequential transfer of nitrogen onto glutamate semialdehyde. Arginine biosynthesis is likely to occur mostly in the cytosol in *Euglena*, as the majority of ornithine carbomyltransferase is located in the cytosol and smaller portion in mitochondria [2]. Arginine metabolism follows the arginine dihydrolase pathway in which arginine is converted into citrulline and then ornithine, which occurs in the mitochondria [121]. Proline synthesis in *Euglena* has not been investigated. However, proline metabolism is tightly associated with arginine metabolism as ornithine is the precursor for proline synthesis [122], suggesting that synthesis is likely to be located in the cytosol or mitochondria.

Lysine. Bacteria, plants and algae synthesize lysine via the diaminopimelate (DAP) pathway, using aspartate and pyruvate as the precursors. On the other hand, fungi synthesize lysine through the α-aminoadipate (AAA) pathway, which uses 2-oxoglutarate and acetyl-CoA [123,124]. Several enzymes involved in AAA pathway were detected in *Euglena*, including homocitrate synthase (EC 2.3.3.14), homoaconitate hydratase (EC 4.2.1.36) and homoisocitrate dehydrogenase (EC 1.1.1.87) [113]. However, the subcellular location of the AAA pathway has not been reported.

Histidine. Histidinol dehydrogenase, the enzyme catalysing the final step of histidine biosynthesis, has been detected in *E. gracilis* [113,125]. No other enzyme involved in this process was detected and the subcellular localisation of the enzymes involved in histidine biosynthesis have not been investigated.

Glutamate, Glutamine, Alanine, Aspartate, and Asparagine. Aminotransferases and dehydrogenases play the main role in the synthesis of glutamate, alanine, and aspartate from organic acids. For glutamate, the aspartate aminotransferase (glutamate: oxaloacetate aminotransferase) is present in mitochondria, chloroplasts, microbodies, and cytosol, and was shown to be more active in dark growth conditions [74,78]. NADP-specific glutamate dehydrogenase was reported to be located solely in the cytosol of *E. gracilis*, instead of the mitochondria as in other organisms [126]. Similarly, glutamate synthase was reported to be localised to the cytosol in both wild-type and streptomycin-bleached *E. gracilis* strains [127]. Glutamine is synthesized from glutamate using glutamine synthetase, but the properties of this enzyme have not been studied in *Euglena* [128]. Asparagine synthetase, the enzyme that converts aspartate to asparagine, has not been reported from *Euglena*. The activities of alanine aminotransferase and alanine dehydrogenase were detected in *E. gracilis*, but the localisation of these enzymes has not been described [2,115,116].

Tetrapyrrole Biosynthesis. Tetrapyrrole, the core of heme and chlorophyll, is synthesised from δ-aminolevulinic acid (ALA). Heterotrophs tend to synthesize ALA from glycine and succinyl-CoA via the Shemin pathway in the mitochondia [129], whilst photoautotrophs make ALA from glutamate in the C5 pathway, located in the chloroplast [130]. *E. gracilis* is known to utilise both routes [131], and the transcriptome shows a bacterial-derived Shemin pathway and a green algae-related C5 pathway, presumably obtained with the chloroplast [16]. These have been identified in the mitochondria and chloroplasts of *E. gracilis* respectively [132]. This again supports the multiple locations of core metabolic pathways that are plastid localised in other photosynthetic organisms.

Lipid Biosynthesis

The subcellular locations of the enzymes involved in lipid metabolism in *Euglena* are poorly investigated. As in other organisms *Euglena* produces the lipid building block malonyl-CoA from CO_2 and acetyl-CoA using acetyl-CoA carboxylase, which forms a multienzyme complex with phosphoenolpyruvate carboxylase and malate dehydrogenase in the cytosol [133]. Malonyl-CoA is then used to synthesise fatty acid using fatty acid synthases (FAS), of which three types have been reported in *E. gracilis*. FAS I and FAS III were reported to function in heterotrophic growth conditions. The properties of FAS III has not been investigated in detail. The structure of FAS I is similar to yeast and mammalian enzymes, and was located in cytosol [2]. On the other hand, FAS II resembles the plant and bacterial enzymes, and is located in the chloroplasts of *E. gracilis* [134]. In addition to these three types of FAS, a fatty acid biosynthesis system was found in the mitochondria of *Euglena* and is involved in wax-ester synthesis [134].

3. Discussion

By combining multiple strands of evidence, including biochemical, proteomic, and bioinformatic data, we propose a model for the subcellular localisation of the reactions of the network of central carbon metabolism in *E. gracilis* (Figure 2). Many of these pathways are found in similar subcellular locations to those in other, well-characterised organisms. Glycolysis, which catalyses the initial breakdown of sugars produced by photosynthesis or absorbed from the medium, is present in the cytosol and plastids, as commonly found in green plants. The products of this pathway feed into the TCA cycle, which is mitochondrial, as in other eukaryotes. The enzymes commonly associated with microbodies in higher plants are additionally present in the mitochondria, and it is often difficult to separate these two groups of organelles in *Euglena*. The site of synthesis of many amino acids is unclear, though several appear to be synthesised in the mitochondria from TCA cycle intermediates. Lipids can be made in several cellular compartments, though for different purposes, such as the mitochondrial lipids which are directed towards wax ester biosynthesis and plastid lipids that are used to make photosynthetic glycolipids.

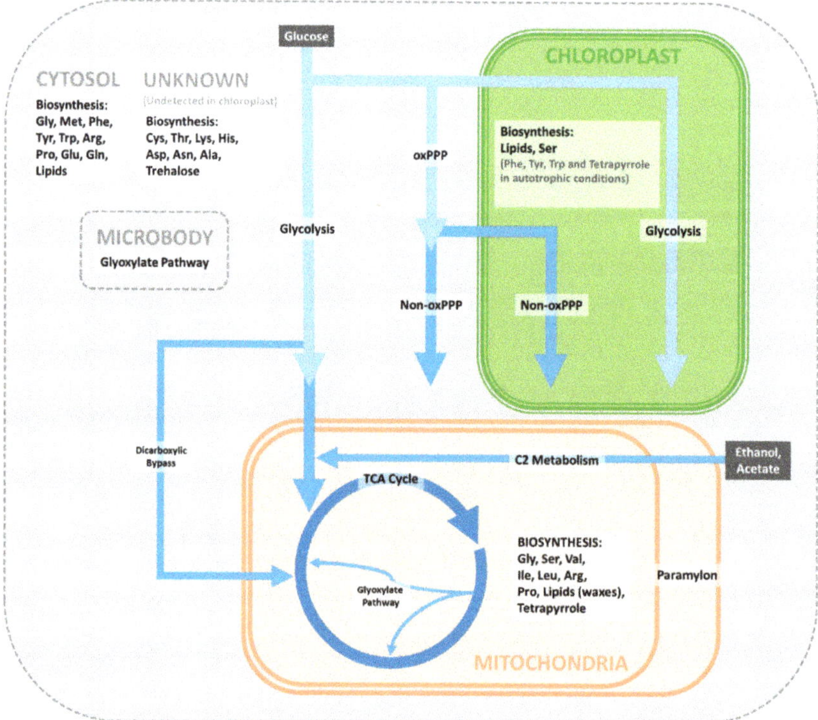

Figure 2. Proposed distribution of central metabolic pathways in *Euglena*. Abbreviations: oxPPP—oxidative pentose phosphate pathway; Non-oxPPP—non-oxidative pentose phosphate pathway.

However, the locations of many metabolic processes in *Euglena* differ substantially from those found in other photosynthetic organisms. For instance, in *Euglena* the complete PPP is present in the cytosol, with a duplicated non-oxidative phase present in the plastid. A plant-like pathway for aromatic amino acid biosynthesis is present in the plastids [117]. However, unlike plants, in *Euglena* an additional pathway, similar to that found in fungi, is located in the cytosol. Tetrapyrroles, essential prosthetic groups of both the respiratory and photosynthetic electron transport chain proteins, are synthesised in both the chloroplast and mitochondria in *Euglena*.

Overall, these results indicate that, aside from the reactions of photosynthesis, all the metabolic pathways found in the *Euglena* plastid are also found elsewhere in the cell. This includes the biosynthesis of isoprenoids, for which two pathways are found in other plastid-containing organisms, the methylerythritol phosphate pathway found in the plastids and the mevalonate pathway in the cytosol. Although we have not found evidence for the location of these pathways in *Euglena*, the methylerythritol phosphate pathway only contributes to carotenoid biosynthesis in *E. gracilis*, and phytol is instead made by the mevalonate pathway [135], unlike in other studied organisms. The unusual and well-established ability of *E. gracilis* to survive on a simple carbon source when their chloroplasts have been destroyed can be rationalised from the subcellular localisation and duplication of these various critical pathways.

The complicated evolutionary history of *Euglena* means it is not trivial to predict the likely subcellular locations of the various metabolic pathways, or to decide whether the pathways will be similar to those in free-living heterotrophs, or plants, or be entirely different. Precise information is missing for some biosynthetic pathways and the lack of understanding of *Euglena* chloroplast protein targeting restricts the prediction of the subcellular location of some *Euglena* proteins. Despite these limitations, overall, the model is similar to plants and green algae, but has some important differences. The development of this model will lead to the ability to predict the metabolic phenotypes of *Euglena* under various growth conditions.

4. Conclusions

The subcellular compartmentation of metabolism has been intensively studied in yeast and in plants. For many, more distantly related organisms, most information is typically inferred by extrapolation from these thoroughly examined species. Drawing on a range of *Euglena* biochemical and proteomic data, we propose a model for the organisation of central metabolism in *E. gracilis*. These analyses reveal unique features of this alga that diverge significantly from expectations derived from well-studied organisms. The most striking difference in *Euglena* is the presence of extra activities of the enzymes of various biosynthetic pathways solely present in the plastids of plants, contributing to the ability of *Euglena* to lose its plastid entirely and survive on simple carbon sources. We propose that this is due to the requirement of the heterotrophic ancestor to synthesise all necessary cellular components before the acquisition of the secondary plastid. In this context, it seems likely that the plastid pathways are replicating pathways that were originally present in the euglenid progenitor.

5. Materials and Methods

5.1. Identification of Euglena Enzymes

The transcriptome of *E. gracilis* was searched for the target proteins using BLASTP with templates that were selected from the corresponding enzymes from other organisms represented in the NCBI databases. Identified *E. gracilis* transcripts were then used as templates to interrogate the NCBI databases, to confirm the correct identification of the proteins. The presence of a spliced leader was confirmed in 39% of all of these sequences, in the range previously reported in Euglena transcriptomes [16,17], by searching for a 10 bp sequence (TTTTTTTCG or ATTTTTTTC) at the 5′ end of the transcript.

5.2. Protein Targeting Prediction for Euglena

A selection of proteins known to be localized to the chloroplast, mitochondria, Golgi or cytosol [2,136,137] were used to validate the use of WoLF PSORT [28] and TargetP 1.1 [29] (Tables 1–3). For proteins predicted to be secreted by TargetP using the plant search parameters, the signal sequence was removed, using the algorithms predicted cleavage site (Figure 3). The remaining sequence was then reanalysed to identify any alterations in targeting and potentially unveil a chloroplast targeting sequence. WoLF PSORT did not predict any secreted proteins using the plant search parameter. Results for metabolic pathway components are available in Supplementary Table S1.

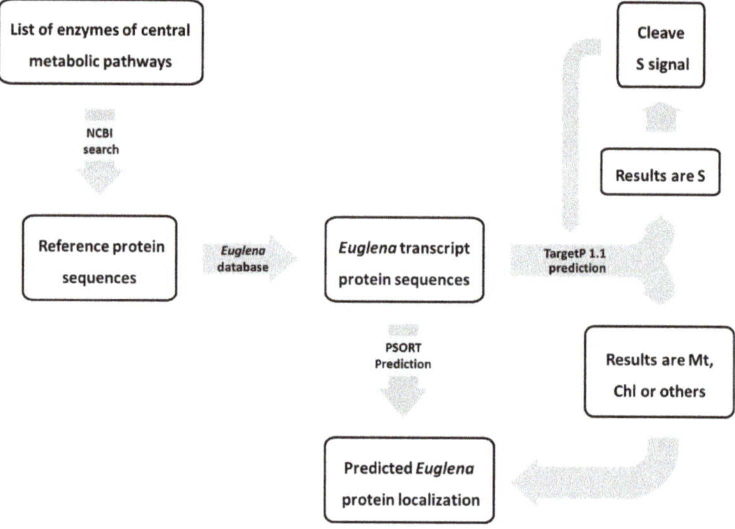

Figure 3. Subcellular location prediction workflow for *Euglena* proteins. Abbreviations: Mt—mitochondria; Chl—chloroplast; others—cytosol; S—secretory pathway.

Supplementary Materials: The following are available online at http://www.mdpi.com/2218-1989/9/6/115/s1, Table S1: Subcellular location prediction of *E. gracilis* metabolic pathway components using WoLF PSORT and TargetP1.1.

Author Contributions: Conceptualization, S.I., N.J.K., R.G.R., and E.C.O.; Data gathering and analysis, S.I.; Visualization, S.I.; Writing—review and editing, S.I., N.J.K., R.G.R., and E.C.O.

Funding: The Development and Promotion of Science and Technology Talents Project (Royal Government of Thailand scholarship) funded this research. E.C.O. is supported by a Violette and Samuel Glasstone Fellowship.

Conflicts of Interest: The authors declare no conflict of interest.

References

1. Einicker-Lamas, M.; Mezian, G.A.; Fernandes, T.B.; Silva, F.L.S.; Guerra, F.; Miranda, K.; Attias, M.; Oliveira, M.M. *Euglena gracilis* as a model for the study of Cu^{2+} and Zn^{2+} toxicity and accumulation in eukaryotic cells. *Environ. Pollut.* **2002**, *120*, 779–786. [CrossRef]

2. Kitaoka, S.; Nakano, Y.; Miyatake, K.; Yokota, A. Enzymes and their functional location. In *The Biology of Euglena*; Buetow, D.E., Ed.; Academic Press Limited: London, UK, 1989; Volume 4, pp. 2–119.

3. Wolken, J.J. *Euglena: An Experimental Organism for Biochemical and Biophysical Studies*; Institute of Microbiology, Rutgers: New Brunswick, NJ, USA, 1961; p. 173.

4. Sato, S. The apicomplexan plastid and its evolution. *Cell. Mol. Life Sci.* **2011**, *68*, 1285–1296. [CrossRef] [PubMed]

5. Ebenezer, T.E.; Zoltner, M.; Burrell, A.; Nenarokova, A.; Novák Vanclová, A.M.G.; Prasad, B.; Soukal, P.; Santana-Molina, C.; O'Neill, E.; Nankissoor, N.N.; et al. Transcriptome, proteome and draft genome of *Euglena gracilis*. *BMC Biol.* **2019**, *17*, 11. [CrossRef] [PubMed]

6. Linton, E.W.; Karnkowska-Ishikawa, A.; Kim, J.I.; Shin, W.; Bennett, M.S.; Kwiatowski, J.; Zakryś, B.; Triemer, R.E. Reconstructing euglenoid evolutionary relationships using three genes: Nuclear SSU and LSU, and chloroplast SSU rDNA sequences and the description of *Euglenaria* gen. nov. (Euglenophyta). *Protist* **2010**, *161*, 603–619. [CrossRef] [PubMed]

7. Ahmadinejad, N.; Dagan, T.; Martin, W. Genome history in the symbiotic hybrid *Euglena gracilis*. *Gene* **2007**, *402*, 35–39. [CrossRef] [PubMed]

8. Parfrey, L.W.; Lahr, D.J.G.; Knoll, A.H.; Katz, L.A. Estimating the timing of early eukaryotic diversification with multigene molecular clocks. *Proc. Natl. Acad. Sci. USA* **2011**, *108*, 13624–13629. [CrossRef] [PubMed]

9. Maruyama, S.; Suzaki, T.; Weber, A.P.; Archibald, J.M.; Nozaki, H. Eukaryote-to-eukaryote gene transfer gives rise to genome mosaicism in euglenids. *BMC Evol. Biol.* **2011**, *11*, 105. [CrossRef] [PubMed]

10. Martin, W.; Somerville, C.C.; Loiseaux-de Goer, S. Molecular phylogenies of plastid origins and algal evolution. *J. Mol. Evol.* **1992**, *35*, 385–404. [CrossRef]

11. O'Neill, E.C.; Trick, M.; Henrissat, B.; Field, R.A. *Euglena* in time: Evolution, control of central metabolic processes and multi-domain proteins in carbohydrate and natural product biochemistry. *Perspect Sci.* **2015**, *6*, 84–93. [CrossRef]

12. Gissibl, A.; Sun, A.; Care, A.; Nevalainen, H.; Sunna, A. Bioproducts from *Euglena gracilis*: Synthesis and applications. *Front. Bioeng. Biotechnol.* **2019**, *7*, 108. [CrossRef]

13. Mahapatra, D.M.; Chanakya, H.N.; Ramachandra, T.V. *Euglena* sp. as a suitable source of lipids for potential use as biofuel and sustainable wastewater treatment. *J. Appl. Phycol.* **2013**, *25*, 855–865. [CrossRef]

14. Kruger, N.J.; Ratcliffe, R.G. Fluxes through plant metabolic networks: Measurements, predictions, insights and challenges. *Biochem. J.* **2015**, *465*, 27–38. [CrossRef] [PubMed]

15. Jackson, C.; Knoll, A.H.; Chan, C.X.; Verbruggen, H. Plastid phylogenomics with broad taxon sampling further elucidates the distinct evolutionary origins and timing of secondary green plastids. *Sci. Rep.* **2018**, *8*, 1523. [CrossRef] [PubMed]

16. O'Neill, E.C.; Trick, M.; Hill, L.; Rejzek, M.; Dusi, R.G.; Hamilton, C.J.; Zimba, P.V.; Henrissat, B.; Field, R.A. The transcriptome of *Euglena gracilis* reveals unexpected metabolic capabilities for carbohydrate and natural product biochemistry. *Mol. Biosys.* **2015**, *11*, 2808–2820. [CrossRef] [PubMed]

17. Yoshida, Y.; Tomiyama, T.; Maruta, T.; Tomita, M.; Ishikawa, T.; Arakawa, K. De novo assembly and comparative transcriptome analysis of *Euglena gracilis* in response to anaerobic conditions. *BMC Genom.* **2016**, *17*, 182. [CrossRef] [PubMed]

18. O'Neill, E.C.; Saalbach, G.; Field, R.A. Chapter five - gene discovery for synthetic biology: Exploring the novel natural product biosynthetic capacity of eukaryotic microalgae. *Methods Enzymol.* **2016**, *576*, 99–120.

19. Prigent, S.; Frioux, C.; Dittami, S.M.; Thiele, S.; Larhlimi, A.; Collet, G.; Gutknecht, F.; Got, J.; Eveillard, D.; Bourdon, J.; et al. Meneco, a topology-based gap-filling tool applicable to degraded genome-wide metabolic networks. *PLoS Comput. Biol.* **2017**, *13*, e1005276. [CrossRef] [PubMed]

20. Hallick, R.B.; Hong, L.; Drager, R.G.; Favreau, M.R.; Monfort, A.; Orsat, B.; Spielmann, A.; Stutz, E. Complete sequence of *Euglena gracilis* chloroplast DNA. *Nucleic Acids Res.* **1993**, *21*, 3537–3544. [CrossRef] [PubMed]

21. Turmel, M.; Gagnon, M.-C.; O'Kelly, C.J.; Otis, C.; Lemieux, C. The chloroplast genomes of the green algae *Pyramimonas*, *Monomastix*, and *Pycnococcus* shed new light on the evolutionary history of prasinophytes and the origin of the secondary chloroplasts of euglenids. *Mol. Biol. Evol.* **2009**, *26*, 631–648. [CrossRef]

22. Martin, W.; Herrmann, R.G. Gene transfer from organelles to the nucleus: How much, what happens, and why? *Plant Physiol.* **1998**, *118*, 9–17. [CrossRef] [PubMed]

23. Geimer, S.; Belicová, A.; Legen, J.; Sláviková, S.; Herrmann, R.G.; Krajčovič, J. Transcriptome analysis of the *Euglena gracilis* plastid chromosome. *Curr. Genet.* **2009**, *55*, 425–438. [CrossRef] [PubMed]

24. Miller, M.E.; Jurgenson, J.E.; Reardon, E.M.; Price, C.A. Plastid translation in organello and in vitro during light-induced development in *Euglena*. *J. Biol. Chem.* **1983**, *258*, 14478–14484. [PubMed]

25. Bachvaroff, T.R.; Sanchez Puerta, M.V.; Delwiche, C.F. Chlorophyll c–containing plastid relationships based on analyses of a multigene data set with all four chromalveolate lineages. *Mol. Biol. Evol.* **2005**, *22*, 1772–1782. [CrossRef] [PubMed]

26. Yoon, H.S.; Hackett, J.D.; Bhattacharya, D. A single origin of the peridinin- and fucoxanthin-containing plastids in dinoflagellates through tertiary endosymbiosis. *Proc. Natl. Acad. Sci. USA* **2002**, *99*, 11724. [CrossRef] [PubMed]

27. Bolte, K.; Bullmann, L.; Hempel, F.; Bozarth, A.; Zauner, S.; Maier, U.-G. Protein targeting into secondary plastids. *J. Eukaryot. Microbiol.* **2009**, *56*, 9–15. [CrossRef] [PubMed]

28. Horton, P.; Park, K.-J.; Obayashi, T.; Fujita, N.; Harada, H.; Adams-Collier, C.J.; Nakai, K. WoLF PSORT: protein localization predictor. *Nucleic Acids Res.* **2007**, *35*, W585–W587. [CrossRef]

29. Emanuelsson, O.; Brunak, S.; von Heijne, G.; Nielsen, H. Locating proteins in the cell using TargetP, SignalP and related tools. *Nat. Protoc.* **2007**, *2*, 953–971. [CrossRef]

30. Novák Vanclová, A.M.G.; Zoltner, M.; Kelly, S.; Soukal, P.; Záhonová, K.; Füssy, Z.; Ebenezer, T.E.; Lacová Dobáková, E.; Eliáš, M.; Lukeš, J.; et al. Proteome of the secondary plastid of *Euglena gracilis* reveals metabolic quirks and colourful history. *bioRxiv* **2019**, 573709. [CrossRef]

31. Durnford, D.G.; Gray, M.W. Analysis of Euglena gracilis plastid-targeted proteins reveals different classes of transit sequences. *Eukaryot. Cell* **2006**, *5*, 2079–2091. [CrossRef]

32. Emanuelsson, O.; Nielsen, H.; Brunak, S.; von Heijne, G. Predicting subcellular localization of proteins based on their n-terminal amino acid sequence. *J. Mol. Biol.* **2000**, *300*, 1005–1016. [CrossRef]

33. Gibbs, S.P. The chloroplasts of *Euglena* may have evolved from symbiotic green algae. *Can. J. Bot.* **1978**, *56*, 2883–2889. [CrossRef]

34. Záhonová, K.; Füssy, Z.; Birčák, E.; Novák Vanclová, A.M.G.; Klimeš, V.; Vesteg, M.; Krajčovič, J.; Oborník, M.; Eliáš, M. Peculiar features of the plastids of the colourless alga *Euglena longa* and photosynthetic euglenophytes unveiled by transcriptome analyses. *Sci. Rep.* **2018**, *8*, 17012. [CrossRef] [PubMed]

35. Sláviková, S.; Vacula, R.; Fang, Z.; Ehara, T.; Osafune, T.; Schwartzbach, S.D. Homologous and heterologous reconstitution of Golgi to chloroplast transport and protein import into the complex chloroplasts of *Euglena*. *J. Cell Sci.* **2005**, *118*, 1651–1661. [CrossRef] [PubMed]

36. Maier, U.G.; Zauner, S.; Hempel, F. Protein import into complex plastids: Cellular organization of higher complexity. *Eur. J. Cell Biol.* **2015**, *94*, 340–348. [CrossRef] [PubMed]

37. Smillie, R.M. Formation and function of soluble proteins in ghloroplasts. *Can. J. Bot.* **1963**, *41*, 123–154. [CrossRef]

38. Bégin-Heick, N. The localization of enzymes of intermediary metabolism in *Astasia* and *Euglena*. *Biochem. J.* **1973**, *134*, 607. [CrossRef] [PubMed]

39. Dockerty, A.; Merrett, M.J. Isolation and enzymic characterization of *Euglena* proplastids. *Plant Physiol.* **1979**, *63*, 468–473. [CrossRef]

40. Graves, J.R.; Lynn, B. Effects of different substrates on glucose uptake and hexokinase activity in *Euglena gracilis*. *J. Protozool.* **1971**, *18*, 543–546. [CrossRef]

41. Bäumer, D.; Preisfeld, A.; Ruppel, H.G. Isolation and characterization of paramylon synthase from *Euglena gracilis* (euglenophyceae). *J. Phycol.* **2001**, *37*, 38–46. [CrossRef]

42. Lucchini, G. Control of glucose phosphorylation in *Euglena gracilis* I. Partial characterization of a glucokinase. *Biochim. Biophys. Acta* **1971**, *242*, 365–370. [CrossRef]

43. Bégin-Heick, N. Oxygen toxicity and carbon deprivation in *Astasia longa*. *Can. J. Biochem.* **1970**, *48*, 251–258. [CrossRef] [PubMed]

44. Miyatake, K.; Enomoto, T.; Kitaoka, S. Detection and subcellular distribution of pyrophosphate: D-fructose 6-phosphate phosphotransferase (PFP) in *Euglena gracilis*. *Agric. Biol. Chem.* **1984**, *48*, 2857–2859. [CrossRef]

45. Willard, J.M.; Gibbs, M. Purification and characterization of the fructose diphosphate aldolases from *Anacystis* is nidulans and *Saprospira thermalis*. *Biochim. Biophys. Acta* **1968**, *151*, 438–448. [CrossRef]

46. Mo, Y.; Harris, B.G.; Gracy, R.W. Triosephosphate isomerases and aldolases from light- and dark-grown *Euglena gracilis*. *Arch. Biochem. Biophys.* **1973**, *157*, 580–587. [CrossRef]

47. Latzko, E.; Gibbs, M. Enzyme activities of the carbon reduction cycle in some photosynthetic organisms. *Plant Physiol.* **1969**, *44*, 295. [CrossRef] [PubMed]

48. Grissom, F.E.; Kahn, J.S. Glyceraldehyde-3-phosphate dehydrogenases from *Euglena gracilis*: Purification and physical and chemical characterization. *Arch. Biochem. Biophys.* **1975**, *171*, 444–458. [CrossRef]

49. Bukowiecki, A.C.; Anderson, L.E. Multiple forms of aldolase and triose phosphate isomerase in diverse plant species. *Plant Sci. Lett.* **1974**, *3*, 381–386. [CrossRef]

50. Forsee, W.T.; Kahn, J.S. Carbon dioxide fixation by isolated chloroplasts of *Euglena gracilis*: I. Isolation of functionally intact chloroplasts and their characterization. *Arch. Biochem. Biophys.* **1972**, *150*, 296–301. [CrossRef]

51. Ammon, R.; Friedrich, G. The behavior of enzymes in *Euglena gracilis*. *Acta Biol. Med. Ger.* **1967**, *19*, 659–672.

52. Hannaert, V.; Brinkmann, H.; Nowitzki, U.; Lee, J.A.; Albert, M.-A.; Sensen, C.W.; Gaasterland, T.; Miklós, M.; Michels, P.; Martin, W. Enolase from *Trypanosoma brucei*, from the amitochondriate protist *Mastigamoeba balamuthi*, and from the chloroplast and cytosol of *Euglena gracilis*: Pieces in the evolutionary puzzle of the eukaryotic glycolytic pathway. *Mol. Biol. Evol.* **2000**, *17*, 989–1000. [CrossRef]

53. Ohmann, E. Die regulation der pyruvat-kinase in *Euglena gracilis*. *Arch. Microbiol.* **1969**, *67*, 273–292. [CrossRef]

54. Ogawa, T.; Kimura, A.; Sakuyama, H.; Tamoi, M.; Ishikawa, T.; Shigeoka, S. Identification and characterization of cytosolic fructose-1,6-bisphosphatase in *Euglena gracilis*. *Biosci. Biotechnol. Biochem.* **2015**, *79*, 1957–1964. [CrossRef] [PubMed]

55. Hovenkamp-Obbema, R. Effect of chloramphenicol on the development of proplastids in *Euglena gracilis*: II. The synthesis of carotenoids. *Z. Pflanzenphysiol.* **1974**, *73*, 439–447. [CrossRef]

56. Kempner, E.S.; Miller, J.H. The molecular biology of *Euglena gracilis*: IV. Cellular stratification by centrifuging. *Exp. Cell Res.* **1968**, *51*, 141–149. [CrossRef]

57. Ohmann, E.; Rindt, K.P.; Borriss, R. Glucose-6-phosphat-dehydrogenase in autotrophen mikroorganismen i. Die regulation der synthese der glucose-6-phosphat-dehydrogenase in *Euglena gracilis* und *Rhodopseudomonas spheroides* in abhängigkeit von den kulturbedingungen. *Z. Allg. Mikrobiol.* **1969**, *9*, 557–564. [CrossRef] [PubMed]

58. Hovenkamp-Obbema, R.; Moorman, A.; Stegwee, D. Aminolaevulinate dehydratase in greening cells of *Euglena gracilis*. *Z. Pflanzenphysiol.* **1974**, *72*, 277–286. [CrossRef]

59. Forsee, W.T.; Kahn, J.S. Carbon dioxide fixation by isolated chloroplasts of *Euglena gracilis*: II. Inhibition of CO_2 fixation by AMP. *Arch. Biochem. Biophys.* **1972**, *150*, 302–309. [CrossRef]

60. Ma, Y.A.N.; Jakowitsch, J.; Deusch, O.; Henze, K.; Martin, W.; Löffelhardt, W. Transketolase from *Cyanophora paradoxa*: In vitro import into cyanelles and pea chloroplasts and a complex history of a gene often, but not always, transferred in the context of secondary endosymbiosis. *J. Eukaryot. Microbiol.* **2009**, *56*, 568–576. [CrossRef]

61. Stern, J.R.; Hegre, C.S. Inducible D-malic enzyme in *Escherichia coli*. *Nature* **1966**, *212*, 1611. [CrossRef]

62. Peak, M.J.; Peak, J.G.; Ting, I.P. Isoenzymes of malate dehydrogenase and their regulation in *Euglena gracilis* Z. *Biochim. Biophys. Acta* **1972**, *284*, 1–15. [CrossRef]

63. Peak, M.J.; Peak, J.G.; Ting, I.P. Light-induced reduction in specific activity of malate enzyme in *Euglena gracilis* Z. *Biochem. Biophys. Res. Commun.* **1972**, *48*, 1074–1078. [CrossRef]

64. Codd, G.A.; Merrett, M.J. The regulation of glycolate metabolism in division synchronized cultures of *Euglena*. *Plant Physiol.* **1971**, *47*, 640–643. [CrossRef] [PubMed]

65. Laval-Martin, D.; Farineau, J.; Pineau, B.; Calvayrac, R. Evolution of enzymes involved in carbon metabolism (phosphoenolpyruvate and ribulose-bisphosphate carboxylases, phosphoenolpyruvate carboxykinase) during the light-induced greening of *Euglena gracilis* strains Z and ZR. *Planta* **1981**, *151*, 157–167. [CrossRef] [PubMed]

66. Briand, J.; Calvayrac, R.; Laval-Martin, D.; Farineau, J. Evolution of carboxylating enzymes involved in paramylon synthesis (phosphoenolpyruvate carboxylase and carboxykinase) in heterotrophically grown *Euglena gracilis*. *Planta* **1981**, *151*, 168–175. [CrossRef] [PubMed]

67. Ohmann, E.; Plhák, F. Reinigung und eigenschaften von phosphoenolpyruvat-carboxylase aus *Euglena gracilis*. *Eur. J. Biochem.* **1969**, *10*, 43–55. [CrossRef] [PubMed]

68. Miyatake, K.; Ito, T.; Kitaoka, S. Subcellular location and some properties of phosphoenolpyruvate carboxykinase (PEPCK) in *Euglena gracilis*. *Agric. Biol. Chem.* **1984**, *48*, 2139–2141. [CrossRef]

69. Yokota, A.; Hosotani, K.; Kitaoka, S. Mechanism of metabolic regulation in photoassimilation of propionate in *Euglena gracilis* z. *Arch. Biochem. Biophys.* **1982**, *213*, 530–537. [CrossRef]

70. Nakazawa, M.; Hayashi, R.; Takenaka, S.; Inui, H.; Ishikawa, T.; Ueda, M.; Sakamoto, T.; Nakano, Y.; Miyatake, K. Physiological functions of pyruvate:NADP+ oxidoreductase and 2-oxoglutarate decarboxylase in *Euglena gracilis* under aerobic and anaerobic conditions. *Biosci. Biotechnol. Biochem.* **2017**, *81*, 1386–1393. [CrossRef]

71. Inui, H.; Ono, K.; Miyatake, K.; Nakano, Y.; Kitaoka, S. Purification and characterization of pyruvate:NADP$^+$ oxidoreductase in *Euglena gracilis*. *J. Biol. Chem.* **1987**, *262*, 9130–9135.

72. Rotte, C.; Stejskal, F.; Zhu, G.; Keithly, J.S.; Martin, W. Pyruvate:NADP oxidoreductase from the mitochondrion of *Euglena gracilis* and from the apicomplexan *Cryptosporidium parvum*: A biochemical relic linking pyruvate metabolism in mitochondriate and amitochondriate protists. *Mol. Biol. Evol.* **2001**, *18*, 710–720. [CrossRef]

73. Hoffmeister, M.; van der Klei, A.; Rotte, C.; van Grinsven, K.W.A.; van Hellemond, J.J.; Henze, K.; Tielens, A.G.M.; Martin, W. *Euglena gracilis* rhodoquinone:Ubiquinone ratio and mitochondrial proteome differ under aerobic and anaerobic conditions. *J. Biol. Chem.* **2004**, *279*, 22422–22429. [CrossRef] [PubMed]

74. Collins, N.; Merrett, M.J. Microbody-marker enzymes during transition from phototrophic to organotrophic growth in *Euglena*. *Plant Physiol.* **1975**, *55*, 1018–1022. [CrossRef] [PubMed]

75. Graves, L.B., Jr.; Trelease, R.N.; Grill, A.; Becker, W.M. Localization of glyoxylate cycle enzymes in glyoxysomes in *Euglena*. *J. Protozool.* **1972**, *19*, 527–532. [CrossRef]

76. Cook, J.R.; Carver, M. Partial photo-repression of the glyoxylate by-pass in *Euglena*. *Plant Cell Physiol.* **1966**, *7*, 377–383.

77. Cook, J.R.; Heinrich, B. Unbalanced respiratory growth of *Euglena*. *Microbiology* **1968**, *53*, 237–251. [CrossRef] [PubMed]

78. Collins, N.; Merrett, M.J. The localization of glycollate-pathway enzymes in *Euglena*. *Biochem. J.* **1975**, *148*, 321–328. [CrossRef]

79. Oda, Y.; Miyatake, K.; Nakano, Y.; Kitaoka, S. Subcellular location and some properties of isocitrate dehydrogenase isozymes in *Euglena gracilis*. *Agric. Biol. Chem.* **1981**, *45*, 2619–2621. [CrossRef]

80. Shigeoka, S.; Onishi, T.; Maeda, K.; Nakano, Y.; Kitaoka, S. Occurrence of thiamin pyrophosphate-dependent 2-oxoglutarate decarboxylase in mitochondria of *Euglena gracilis*. *FEBS Lett.* **1986**, *195*, 43–47. [CrossRef]

81. Shigeoka, S.; Nakano, Y. Characterization and molecular properties of 2-oxoglutarate decarboxylase from *Euglena gracilis*. *Arch. Biochem. Biophys.* **1991**, *288*, 22–28. [CrossRef]

82. Tokunaga, M.; Nakano, Y.; Kitaoka, S. Separation and properties of the NAD-linked and NADP-linked isozymes of succinic semialdehyde dehydrogenase in *Euglena gracilis*. *Biochim. Biophys. Acta* **1976**, *429*, 55–62. [CrossRef]

83. Barry, D.; Merrett, M.J. Malate dehydrogenase isoenzymes in division synchronized cultures of *Euglena*. *Plant Physiol.* **1973**, *51*, 1127–1132.

84. Isegawa, Y.; Nakano, Y.; Kitaoka, S. Submitochondrial location and some properties of NAD$^+$- and NADP$^+$-linked malate dehydrogenase in *Euglena*. *Agric. Biol. Chem.* **1984**, *48*, 549–552. [CrossRef]

85. Miyatake, K.; Washio, K.; Yokota, A.; Nakano, Y.; Kitaoka, S. Occurrence of two NAD$^+$-linked malate dehydrogenase isozymes in *Euglena gracilis*. *Agric. Biol. Chem.* **1985**, *49*, 859–860. [CrossRef]

86. Nakazawa, M. C2 metabolism in *Euglena*. *Adv. Exp. Med. Biol.* **2017**, *979*, 39–45. [PubMed]

87. Graves, L.B.; Hanzely, L.; Trelease, R.N. The occurrence and fine structural characterization of microbodies in *Euglena gracilis*. *Protoplasma* **1971**, *72*, 141–152. [CrossRef] [PubMed]

88. Ono, K.; Kondo, M.; Osafune, T.; Miyatake, K.; Inui, H.; Kitaoka, S.; Nishimura, M.; Nakano, Y. Presence of glyoxylate cycle enzymes in the mitochondria of *Euglena gracilis*. *J. Eukaryot. Microbiol.* **2003**, *50*, 92–96. [CrossRef]

89. RodríGuez-Zavala, J.S.; Ortiz-Cruz, M.A.; Moreno-Sanchez, R. Characterization of an aldehyde dehydrogenase from *Euglena gracilis*. *J. Eukaryot. Microbiol.* **2006**, *53*, 36–42. [CrossRef]

90. Yoval-Sánchez, B.; Jasso-Chávez, R.; Lira-Silva, E.; Moreno-Sánchez, R.; Rodríguez-Zavala, J.S. Novel mitochondrial alcohol metabolizing enzymes of *Euglena gracilis*. *J. Bioenerg. Biomembr.* **2011**, *43*, 519–530. [CrossRef]

91. Ono, K.; Kawanaka, Y.; Izumi, Y.; Inui, H.; Miyatake, K.; Kitaoka, S.; Nakano, Y. Mitochondrial alcohol dehydrogenase from ethanol-grown *Euglena gracilis*. *J. Biochem.* **1995**, *117*, 1178–1182. [CrossRef]

92. Graves, L.B., Jr.; Becker, W.M. Beta-oxidation in glyoxysomes from *Euglena*. *J. Protozool.* **1974**, *21*, 771–774. [CrossRef]

93. Tucci, S.; Vacula, R.; Krajcovic, J.; Proksch, P.; Martin, W. Variability of wax ester fermentation in natural and bleached *Euglena gracilis* strains in response to oxygen and the elongase inhibitor flufenacet. *J. Eukaryot. Microbiol.* **2010**, *57*, 63–69. [CrossRef] [PubMed]

94. Grimm, P.; Risse, J.M.; Cholewa, D.; Müller, J.M.; Beshay, U.; Friehs, K.; Flaschel, E. Applicability of *Euglena gracilis* for biorefineries demonstrated by the production of α-tocopherol and paramylon followed by anaerobic digestion. *J. Biotechnol.* **2015**, *215*, 72–79. [CrossRef] [PubMed]

95. O'Neill, E.C.; Kuhaudomlarp, S.; Rejzek, M.; Fangel, J.U.; Alagesan, K.; Kolarich, D.; Willats, W.G.T.; Field, R.A. Exploring the glycans of *Euglena gracilis*. *Biology* **2017**, *6*, 45. [CrossRef] [PubMed]

96. Marechal, L.R.; Goldemberg, S.H. Laminaribiose phosphorylase from *Euglena gracilis*. *Biochem. Biophys. Res. Commun.* **1963**, *13*, 106–107. [CrossRef]

97. Tomos, A.D.; Northcote, D.H. A protein-glucan intermediate during paramylon synthesis. *Biochem. J.* **1978**, *174*, 283–290. [CrossRef] [PubMed]

98. Tanaka, Y.; Ogawa, T.; Maruta, T.; Yoshida, Y.; Arakawa, K.; Ishikawa, T. Glucan synthase-like 2 is indispensable for paramylon synthesis in *Euglena gracilis*. *FEBS Lett.* **2017**, *591*, 1360–1370. [CrossRef] [PubMed]

99. Calvayrac, R.; Laval-Martin, D.; Briand, J.; Farineau, J. Paramylon synthesis by *Euglena gracilis* photoheterotrophically grown under low O_2 pressure. *Planta* **1981**, *153*, 6–13. [CrossRef]

100. Briand, J.; Calvayrac, R. Paramylon synthesis in heterotrophic and photoheterotrophic *Euglena* (Euglenophyceae). *J. Phycol.* **1980**, *16*, 234–239. [CrossRef]

101. Takeda, T.; Nakano, Y.; Takahashi, M.; Konno, N.; Sakamoto, Y.; Arashida, R.; Marukawa, Y.; Yoshida, E.; Ishikawa, T.; Suzuki, K. Identification and enzymatic characterization of an endo-1,3-β-glucanase from *Euglena gracilis*. *Phytochemistry* **2015**, *116*, 21–27. [CrossRef]

102. Barras, D.R.; Stone, B.A. β-1,3-glucan hydrolases from *Euglena gracilis*: II. Purification and properties of the β-1,3-glucan exo-hydrolase. *Biochim. Biophys. Acta* **1969**, *191*, 342–353. [CrossRef]

103. Kuhaudomlarp, S.; Patron, N.J.; Henrissat, B.; Rejzek, M.; Saalbach, G.; Field, R.A. Identification of *Euglena gracilis* β-1,3-glucan phosphorylase and establishment of a new glycoside hydrolase (GH) family GH149. *J. Biol. Chem.* **2018**, *293*, 2865–2876. [CrossRef] [PubMed]

104. Takenaka, S.; Kondo, T.; Nazeri, S.; Tamura, Y.; Tokunaga, M.; Tsuyama, S.; Miyatake, K.; Nakano, Y. Accumulation of trehalose as a compatible solute under osmotic stress in *Euglena gracilis* Z. *J. Eukaryot. Microbiol.* **1997**, *44*, 609–613. [CrossRef]

105. Porchia, A.C.; Fiol, D.F.; Salerno, G.L. Differential synthesis of sucrose and trehalose in Euglena gracilis cells during growth and salt stress. *Plant Sci.* **1999**, *149*, 43–49. [CrossRef]

106. Fiol, D.F.; Salerno, G.L. Trehalose synthesis in *Euglena gracilis* (euglenophyceae) occurs through an enzyme complex. *J. Phycol.* **2005**, *41*, 812–818. [CrossRef]

107. Schluepmann, H.; Paul, M. Trehalose metabolites in Arabidopsis-elusive, active and central. *Arabidopsis Book* **2009**, *7*, e0122. [CrossRef]

108. Vandesteene, L.; Ramon, M.; Le Roy, K.; Van Dijck, P.; Rolland, F. A single active trehalose-6-p synthase (TPS) and a family of putative regulatory TPS-like proteins in *Arabidopsis*. *Mol. Plant* **2010**, *3*, 406–419. [CrossRef] [PubMed]

109. Murray, D.R.; Giovanelli, J.; Smillie, R.M. Photoassimilation of glycolate, glycine and serine by *Euglena gracilis*. *J. Protozool.* **1970**, *17*, 99–104. [CrossRef]

110. Yokota, A.; Komura, H.; Kitaoka, S. Different metabolic fate of two carbons of glycolate in *Euglena gracilis* Z. In *Advances in Photosynthesis Research: Proceedings of the VIth International Congress on Photosynthesis, Brussels, Belgium, August 1–6, 1983*; Sybesma, C., Ed.; Springer Netherlands: Dordrecht, The Netherlands, 1984; pp. 407–410.

111. Yokota, A.; Kawabata, A.; Kitaoka, S. Mechanism of glyoxylate decarboxylation in the glycolate pathway in *Euglena gracilis* Z. *Plant Physiol.* **1983**, *71*, 772–776. [CrossRef]

112. Yokota, A.; Haga, S.; Kitaoka, S. Purification and some properties of glyoxylate reductase ($NADP^+$) and its functional location in mitochondria in *Euglena gracilis* Z. *Biochem. J.* **1985**, *227*, 211–216. [CrossRef]

113. Hasan, M.T.; Sun, A.; Mirzaei, M.; Te'o, J.; Hobba, G.; Sunna, A.; Nevalainen, H. A comprehensive assessment of the biosynthetic pathways of ascorbate, α-tocopherol and free amino acids in Euglena gracilis var. saccharophila. *Algal Res.* **2017**, *27*, 140–151. [CrossRef]

114. Isegawa, Y.; Watanabe, F.; Kitaoka, S.; Nakano, Y. Subcellular distribution of cobalamin-dependent methionine synthase in *Euglena gracilis* Z. *Phytochemistry* **1993**, *35*, 59–61. [CrossRef]

115. Sulzman, F.M.; Edmunds, L.N. Persisting circadian oscillations in enzyme activity in non-dividing cultures of *Euglena*. *Biochem. Biophys. Res. Commun.* **1972**, *47*, 1338–1344. [CrossRef]

116. Edmunds, L.N.; Walther, W.G.; Sulzman, F.M. At cellular level. *J. Interdiscip. Cycle Res.* **1972**, *3*, 107–108. [CrossRef]

117. Reinbothe, C.; Ortel, B.; Parthier, B.; Reinbothe, S. Cytosolic and plastid forms of 5-enolpyruvylshikimate-3-phosphate synthase in *Euglena gracilis* are differentially expressed during light-induced chloroplast development. *Mol. Gen. Genet.* **1994**, *245*, 616–622. [CrossRef] [PubMed]

118. Richards, T.A.; Dacks, J.B.; Campbell, S.A.; Blanchard, J.L.; Foster, P.G.; McLeod, R.; Roberts, C.W. Evolutionary origins of the Eukaryotic shikimate pathway: gene fusions, horizontal gene transfer, and endosymbiotic replacements. *Eukaryot. Cell* **2006**, *5*, 1517–1531. [CrossRef] [PubMed]

119. Oda, Y.; Nakano, Y.; Kitaoka, S. Utilization and toxicity of exogenous amino acids in *Euglena gracilis*. *Microbiology* **1982**, *128*, 853–858. [CrossRef]

120. Oda, Y.; Nakano, Y.; Kitaoka, S. Occurrence and some properties of two threonine dehydratases in *Euglena gracilis*. *Microbiology* **1983**, *129*, 57–61. [CrossRef]

121. Park, B.-S.; Hirotani, A.; Nakano, Y.; Kitaoka, S. Purification and someproperties of arginine deiminase in *Euglena gracilis* Z. *Agric. Biol. Chem.* **1984**, *48*, 483–489.

122. Koonin, E.V.; Galperin, M.Y. Evolution of central metabolic pathways: The playground of non-orthologous gene displacement. In *Sequence—Evolution—Function: Computational Approaches in Comparative Genomics*; Koonin, E.V., Galperin, M.Y., Eds.; Springer: Boston, MA, USA, 2003; pp. 295–355.

123. Torruella, G.; Suga, H.; Riutort, M.; Pereto, J.; Ruiz-Trillo, I. The evolutionary history of lysine biosynthesis pathways within Eukaryotes. *J. Mol. Evol.* **2009**, *69*, 240–248. [CrossRef]

124. Xu, H.; Andi, B.; Qian, J.; West, A.H.; Cook, P.F. The α-aminoadipate pathway for lysine biosynthesis in fungi. *Cell Biochem. Biophys.* **2006**, *46*, 43–64. [CrossRef]

125. Creaser, E.H.; Varela-Torres, R. Immunological comparisons of histidinol dehydrogenases. *Microbiology* **1971**, *67*, 85–90. [CrossRef] [PubMed]

126. Tokunaga, M.; Nakano, Y.; Kitaoka, S. Subcellular localization of the GABA-shunt enzymes in *Euglena gracilis* strain Z. *J. Protozool.* **1979**, *26*, 471–473. [CrossRef] [PubMed]

127. Miyatake, K.; Kitaoka, S. NADH-dependent glutamat synthase in *Euglena gracilis* Z. *Agric. Biol. Chem.* **1981**, *45*, 1727–1729.

128. Fayyaz-Chaudhary, M.; Javed, Q.; Merrett, M.J. Effect of growth conditions on NADPH-specific glutamate dehydrogenase activity of *Euglena gracilis*. *New Phytol.* **1985**, *101*, 367–376. [CrossRef]

129. Shemin, D.; Rittenberg, D. The utilization of glycine for the synthesis of a porphyrin. *J. Biol. Chem.* **1945**, *159*, 567–568.

130. Beale, S.I.; Castelfranco, P.A. Biosynthesis of delta-aminolevulinic-acid in higher-plants.1. Accumulation of delta-aminolevulinic-acid in greening plant-tissues. *Plant Phys.* **1974**, *53*, 291–296. [CrossRef] [PubMed]

131. Iida, K.; Mimura, I.; Kajiwara, M. Evaluation of two biosynthetic pathways to δ-aminolevulinic acid in *Euglena gracilis*. *Eur. J. Biochem.* **2002**, *269*, 291–297. [CrossRef] [PubMed]

132. Weinstein, J.D.; Beale, S.I. Separate physiological roles and subcellular compartments for two tetrapyrrole biosynthetic pathways in *Euglena gracilis*. *J. Biol. Chem.* **1983**, *258*, 6799–6807.

133. Wolpert, J.S.; Ernst-Fonberg, M.L. Multienzyme complex for carbon dioxide fixation. *Biochemistry* **1975**, *14*, 1095–1102. [CrossRef]

134. Inui, H.; Miyatake, K.; Nakano, Y.; Kitaoka, S. Wax ester fermentation in *Euglena gracilis*. *FEBS Lett.* **1982**, *150*, 89–93. [CrossRef]

135. Kim, D.; Filtz, M.R.; Proteau, P.J. The methylerythritol phosphate pathway contributes to carotenoid but not phytol biosynthesis in *Euglena gracilis*. *J. Nat. Prod.* **2004**, *67*, 1067–1069. [CrossRef] [PubMed]

136. Ferro, M.; Salvi, D.; Riviere-Rolland, H.; Vermat, T.; Seigneurin-Berny, D.; Grunwald, D.; Garin, J.; Joyard, J.; Rolland, N. Integral membrane proteins of the chloroplast envelope: identification and subcellular localization of new transporters. *Proc. Natl. Acad. Sci. USA* **2002**, *99*, 11487–11492. [CrossRef] [PubMed]

137. O'Neill, E. *An Exploration of Phosphorylases for the Synthesis of Carbohydrate Polymers*; University of East Anglia: Norwich, UK, 2013.

 *metabolites*

Article

Far-Red Light Acclimation for Improved Mass Cultivation of Cyanobacteria

Alla Silkina [1], **Bethan Kultschar** [2] **and Carole A. Llewellyn** [2,*]

1   Centre for Sustainable Aquatic Research (CSAR), Bioscience department, College of Science, Swansea University, Singleton Park, Swansea SA2 8PP, UK
2   Department of Biosciences, College of Science, Swansea University, Singleton Park, Swansea SA2 8PP, UK
*   Correspondence: c.a.llewellyn@swansea.ac.uk

Received: 1 August 2019; Accepted: 15 August 2019; Published: 19 August 2019

Abstract: Improving mass cultivation of cyanobacteria is a goal for industrial biotechnology. In this study, the mass cultivation of the thermophilic cyanobacterium *Chlorogloeopsis fritschii* was assessed for biomass production under light-emitting diode white light (LEDWL), far-red light (FRL), and combined white light and far-red light (WLFRL) adaptation. The induction of chl *f* was confirmed at 24 h after the transfer of culture from LEDWL to FRL. Using combined light (WLFRL), chl *f*, *a*, and *d*, maintained the same level of concentration in comparison to FRL conditions. However, phycocyanin and xanthophylls (echinone, caloxanthin, myxoxanthin, nostoxanthin) concentration increased 2.7–4.7 times compared to LEDWL conditions. The productivity of culture was double under WLFRL compared with LEDWL conditions. No significant changes in lipid, protein, and carbohydrate concentrations were found in the two different light conditions. The results are important for informing on optimum biomass cultivation of this species for biomass production and bioactive product development.

Keywords: cyanobacteria; chromatic adaptation; LED; far-red light; growth; photosynthesis; mass cultivation; pigments; *Chlorogloeopsis*

1. Introduction

Cyanobacteria are photosynthetic prokaryotes that are increasingly explored for use in industrial biotechnology. They are extremely diverse and genetically tractable, making them attractive as cell factories, and can adapt to a wide range of extreme habitats, often with the production of unique metabolites [1]. These adaptations can be exploited in industry to increase productivity and for the production of useful compounds such as pigments, mycosporine-like amino acids (MAAs), and fatty acids [2,3].

Having a long evolutionary history, cyanobacteria have evolved with the ability to cope with varying light intensities and wavelengths. They are able to modify their chlorophylls (chls) and carotenoids, as well as rearrange photosystem I (PSI), PSII, and phycobilisomes (PBS) during excess or limited light conditions [4,5]. These rearrangements allow absorption of light to maximise photosynthetic efficiencies. These light-dependent acclimation processes include; complementary chromatic acclimation (CCA), far-red light photoacclimation (FaRLiP), and low light photoacclimation (LoLiP) [6].

Chlorophyll (chl) *a* is the major photosynthetic photo-pigment within almost all organisms that utilise oxygenic photosynthesis [7]. Some cyanobacteria have photoadaptive strategies for absorbing longer wavelengths in the far-red light region (700–750 nm) by the production of chl *d* and *f* [8]. Inducible production of these chls has been seen in a variety of species, such as *Chlorogloeopsis fritschii*, PCC 6912 [9], *Synechococcus sp.* PCC 7335 [10], *Chroococcidiopsis thermalis* PCC 7203, *Leptolyngbya sp.* JSC-1, and *Calothrix sp.* PCC 7507 [4]. This phenomena, FaRLiP, achieves remodeling of PSI and PSII as well as the PBS [11], with production of chl *d*, *f*, and far-red light (FRL) absorbing phycobiliproteins to maximise photosynthesis, productivity, and survival [7,12].

Chlorogloeopsis fritschii (*C. fritschii*) is a subsection V cyanobacterium, first isolated from soils of paddy fields [13]. It has a variety of morphologies and is tolerant to a variety of growth conditions, which are good attributes for an industrial species [14,15]. Previous research on *C. fritschii* has shown the production of chl *d* and *f* under near infrared radiation [9].

Algal biotechnology is a developing area with continued advancements in technologies for cultivation and downstream processing. The main commercial applications of algal biomass are aquaculture feeding, bioremediation, and high value products [16]. The mass production of microalgae species, including cyanobacteria, is investigated around the world. This is because they are rich sources of bio-products such as polysaccharides, lipids, proteins, pigments, and bioactive compounds which can be utilised as feed and food and for pharmaceuticals, cosmetics, and health supplements [17]. The species-specific production of useful metabolites from the algal biomass, including cyanobacteria, has been widely reviewed for industrial biotechnological applications [3,16–18].

Additional research is required to understand the regulation of photosynthesis, photoprotection, and photomorphogenesis in cyanobacteria and the implication of the use of FRL in increasing productivity and/or pigment accumulation as a robust platform in industrial biotechnology [19]. In this study, we characterise the changes in productivity, pigment, and biochemical composition of *C. fritschii* during exposure to light-emitting diode white light (LEDWL) and FRL, followed by a comparison of white light with combined white light and FRL (WLFRL) results. We finish with a discussion on the application within industry.

2. Results

2.1. Growth and Productivity of C. fritschii Under LED White Light and Far Red Light

The two growth conditions (LEDWL and FRL) showed similar growth patterns over the 9 days. Cultures grew in lag phase for the first 4 days in both conditions, followed by an exponential growth phase for 5 consecutive days (Figure 1).

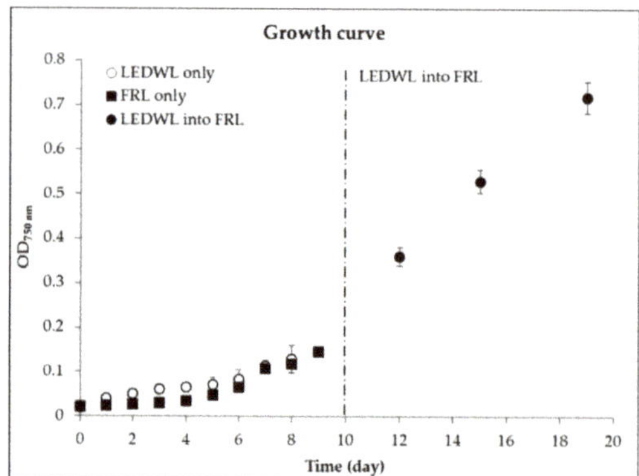

Figure 1. Growth curve, measured using optical density at 750 nm ($OD_{750\ nm}$) of *Chlorogloeopsis fritschii* (*C. fritschii*) under either light-emitting diode white light (LEDWL) or far-red light (FRL) only for 10 days. The dotted line represents the transfer of LEDWL cultures into FRL for a further 10 days.

An overall average growth rate (8 days, 0 to 8) for LEDWL was 0.32 (STDEV = 0.01), in comparison with FRL conditions, which provided an average growth rate of 0.26 (STDEV = 0.02), showing a low light adaptation of *C. fritschii* cultures. No significant difference was found for the accumulation of biomass during the first 8 days under the two light conditions ($p > 0.05$). From day 4, both cultures reached an exponential growth rate. After 10 days of growth under LEDWL, the cultures were transferred from LEDWL to FRL conditions, these cultures were then exposed to FRL for a further 10 days. The average growth rate of this period was 0.27. The growth of *C. fritschii* continued in the exponential phase, with a reduced rate compared to LEDWL.

The final biomass productivity of the culture under LEDWL was 0.014 g $L^{-1}d^{-1}$ (STDEV = 0.001) and 0.03 g $L^{-1}d^{-1}$ (STDEV = 0.001) for WLFRL conditions.

The pigment profile for LEDWL showed the presence of the following pigments: Myxol-quinovoside (myxo), nostoxanthin (nosto), caloxanthin (calo), zeaxanthin (zeax) and echinenone (echin), chl *a*, and β-carotene (Figure 2A). The FRL culture had a similar pigment profile with the exception of the absence of nosto. This could be due to the concentration of this pigment below detection level of the HPLC system. A general trend of accumulation was observed for myxo, nosto, calo, and zeax under LEDWL conditions. Under FRL, the biggest changes were observed for myxo, echin, and β-carotene (Figure 2B).

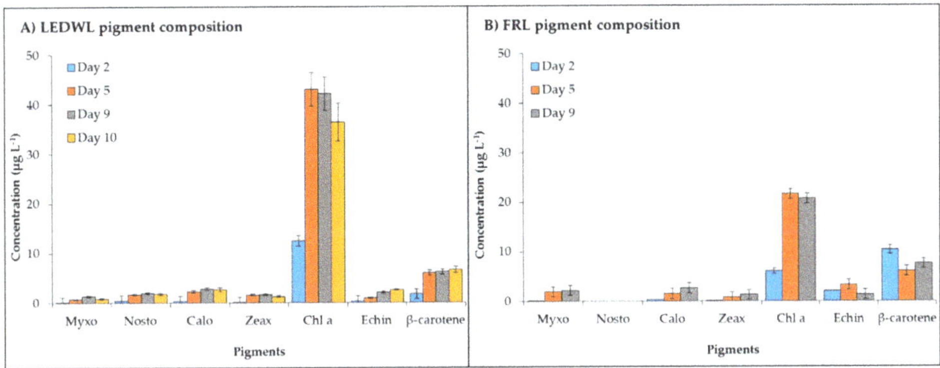

Figure 2. Pigment composition of *C. fritschii* under (**A**) LED white light (LEDWL) and (**B**) far red light (FRL).

Maximum concentration of chl *a* was measured for both light conditions and the cultures grown under LEDWL had double the chl *a* concentration compared to the cultures exposed to FRL. After transferring the cultures from LEDWL to FRL conditions on day 10, the cultures showed a slight decrease in chl *a* concentration. The carotenoids maintained a consistent concentration after the transfer (day 10, Figure 2A).

The final concentration of pigments on day 9 (Table 1) showed a general trend of higher concentration ($\mu g\,g^{-1}$ of dry weight) under LEDWL compared to FRL conditions, except for myxo during FRL conditions, which showed a 1.2 times higher concentration compared to LEDWL. Other pigments, such as calo, zeax, and β-carotene, were 1.3–1.6 times higher in LEDWL cultures with chl *a* and echin concentration was over two-fold higher in LEDWL than FRL cultures (Table 1).

−Chl *d* (detected at 706 nm) was present under both LEDWL and when transferred to FRL conditions (Figure 3) and increased gradually over time. For chl *f*, under FRL there was a ~10-fold increase at day 9 compared to day 2. After transfer of LEDWL exposed cultures to FRL, chl *f* was induced and there was a slight reduction in chl *a* and chl *d* (Figure 3).

Table 1. Concentration (μg g^{-1} of dry weight) of main pigments present at day 9 within *C. fritschii* biomass during LEDWL and FRL conditions, including fold change comparing LEDWL and FRL.

Pigments	LEDWL	FRL	Fold Change
Myxo	3.9 ±.0.3	4.9 ± 0.8 *	1.3
Nosto	5.6 ±0.5	N/A	N/A
Calo	8.3 ± 0.8	6.1±0.8 **	0.7
Zeax	4.8 ± 0.6	2.9±0.6 **	0.6
Chl a	124.7 ± 5.4	47.6±1.6 ***	0.4
Echin	6.1 ± 0.8	3.5±0.6 **	0.6
β-carotene	18.35 ± 1.4	17.2±1.8 *	0.9

Statistical significance was measured using a two-sample t-Test with equal variance, * = $0.05 \leq p > 0.01$, ** = $0.01 \leq p > 0.001$, *** = $p \leq 0.001$.

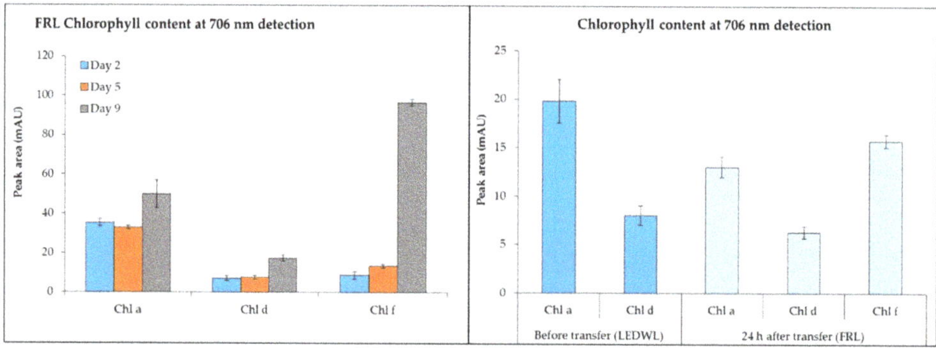

Figure 3. Chlorophyll content (chl *a*, chl *d*, and chl *f*, detected at 706 nm) of *C. fritschii* exposed to FRL at day 2, 5, and 9 and LEDWL (day 9) and 24 h after transfer into FRL (day 10).

2.2. Enhancement of Growth by Combining Two Light Sources (White LED Supplemented with Far Red-Light)

Next, the combination of LEDWL supplemented with FRL compared to LEDWL was investigated. During the first 6 days, the growth under the two light conditions (Figure 4, LEDWL and WLFRL) showed no significant difference ($p > 0.05$). After day 8, the cultures showed a difference in growth performance, with improved results for LEDWL supplemented with FRL (WLFRL). This result was shown in the average growth rate (μ) of 0.39 d^{-1} (STDEV = 0.02) for WLFRL and 0.32 d^{-1} (STDEV = 0.01) for LEDWL growth conditions (STDEV = 0.01). The exponential growth phase for WLFRL was observed over 8 days (day 8 to day 16, $\mu = 0.42$, STDEV = 0.02), whereas the LEDWL condition had a 5 day exponential growth phase ($\mu = 0.33$). The WLFRL light combination resulted in improved growth.

The algal pigments, such as xanthophylls, carotenes, and chlorophylls were detected in both culture conditions (Figure 5). The WLFRL resulted in improved pigment accumulation (Figure 5B) with all pigments considerably increased in their quantity up to the last day of cultivation (day 19). During the exponential growth phase (WLFRL, day 8 to 16), the highest concentration for most of the analysed pigments was observed. The pigments under LEDWL conditions (Figure 5A) showed saturation at day 15, with a slight reduction in concentration by the final day of cultivation (day 19). Final pigment concentrations at day 19 (Table 2) showed an increase in levels under WLFRL conditions compared to LEDWL, with the exception of β-carotene, which showed increased levels in cultures exposed to LEDWL only.

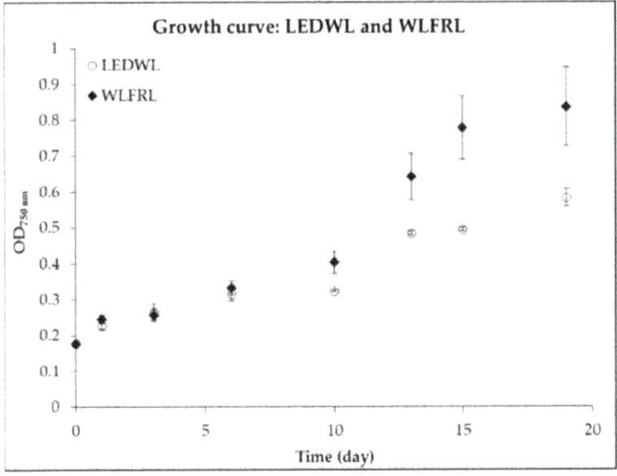

Figure 4. Growth curve of *C. fritschii* under LED white light (LEDWL) and supplemented LED white light with far-red light (WLFRL).

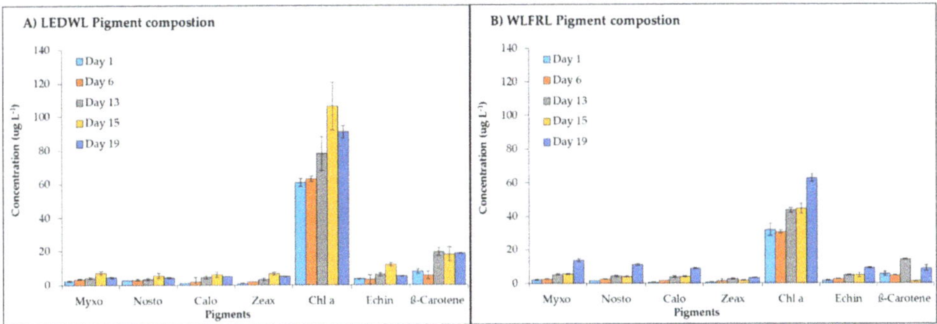

Figure 5. Pigment composition of *C. fritschii* exposed to (**A**) LEDWL only and (**B**) LEDWL supplemented with FRL (WLFRL).

Table 2. Final pigment concentration (μg g^{-1} of dry weight) of main pigments present in *C. fritschii* biomass at day 19 of experiment trial for LEDWL and WLFRL conditions, including fold change comparing LEDWL and WLFRL.

Pigments	LEDWL	WLFRL	Fold Change
Myxo	13.4 ± 1.9	63.6 ± 6.23 ***	4.7
Nosto	12.5 ± 1.1	52.2 ± 4.9 ***	4.2
Calo	15.1 ± 1.9	41.7 ± 2.9 ***	2.8
Zeax	14.8 ± 1.2	15.5 ± 1.2	1.0
Chl a	275.4 ± 12.5	288.5 ± 25.8 *	1.0
Echin	15.7 ± 1.2	41.7 ± 1.6 **	2.7
β-Carotene	56.2 ± 2.7	41.1 ± 1.6 *	0.7

Statistical significance was measured using a two-sample t-Test with equal variance, * = $0.05 \leq p > 0.01$, ** = $0.01 \leq p > 0.001$, *** = $p \leq 0.001$.

The detection of chl *f*, chl *d*, and chl *a* at 706 nm (Figure 6) was investigated under LEDWL supplemented with FRL. An increase in chl *a*, chl *d*, and chl *f* was observed for the cultures grown under supplemented far-red light (WLFRL). Chlorophyll *f* reached its maximum concentration on day 13, after which it gradually reduced to its lowest content at day 19. The same result was observed for chl *d*, with accumulation at day 13; however, the concentration was 5 times less than chl *f*. Chlorophyll *a* consistently increased during the cultivation period and by day 19 reached its maximum concentration (Figure 6).

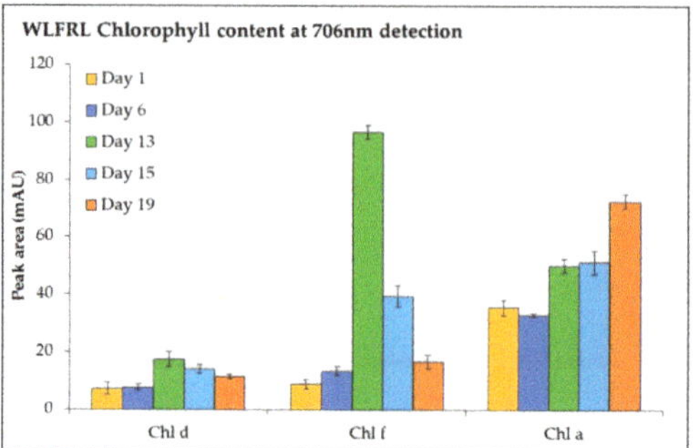

Figure 6. Chlorophyll content (chl *d*, chl *f*, and chl *a*, detected at 706 nm) of *C. fritschii* exposed to LED white light supplemented with far-red light conditions (WLFRL) at day 1, 6, 13, 15, and 19.

2.3. Phycocyanin Concentration During LEDWL and WLFRL Conditions

Cultures grown under LEDWL had high initial concentrations of phycocyanin followed by a reduction of the concentration until day 15. After this, a slight increase in the concentration was observed up to the final day of cultivation (day 20, Figure 7).

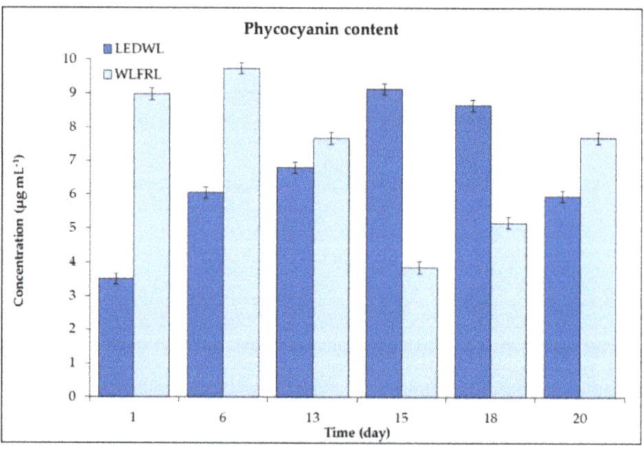

Figure 7. Phycocyanin concentration (μg mL^{-1}) of *C. fritschii* under LEDWL and WLFRL conditions.

The growth conditions under WLFRL had a positive influence on the accumulation of phycocyanin. The concentration increased two-fold in two weeks of cultivation. Maximum concentrations were observed at day 15 and a slight decrease followed until day 20. The maximum concentration observed under the LEDWL and WLFRL was similar at ~9.7–9.7 μg mL^{-1}. At day 13, both cultures, grown on two light conditions (LEDWL and WLFRL), revealed similar concentrations of phycocyanin, thus showing 13 days as an optimum time for the adaptation under both light regimes (Figure 7).

2.4. Biochemical Composition during LEDWL and WLFRL Conditions

Finally, the protein, carbohydrate, and lipid composition of cultures grown under two light conditions (LEDWL and WLFRL) were evaluated (Figure 8). The biomass grown under both light conditions contained 21–25% carbohydrates, 15–22% proteins, and 2–4% lipids. Statistical results (supplementary materials Table S1) showed that the light, time, and combination of both variables (light and time) did not show any significant differences.

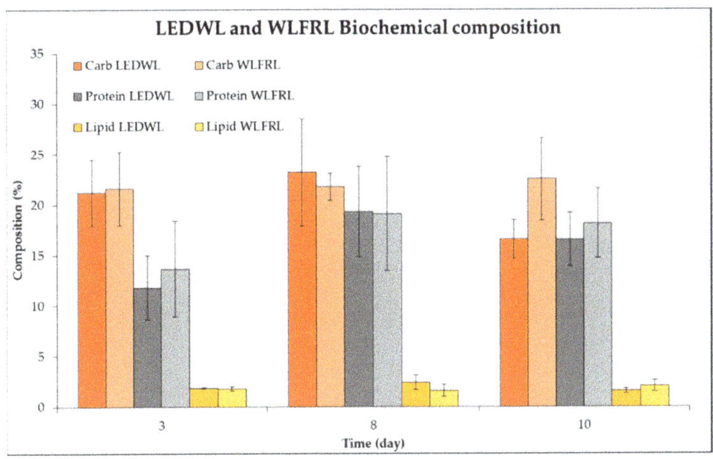

Figure 8. Biochemistry composition (%) of *C. fritschii* grown under LEDWL and WLFRL conditions.

3. Discussion

The discovery of chl *d* and chl *f* in terrestrial cyanobacteria demonstrated that the wavelength range of cyanobacterial photosynthesis could be extend into the far-red region ($\lambda = 700$ to ~800nm) [7,12]. This specific adaptation helps cyanobacteria utilise FRL for growth and photosynthesis [20,21]. It is clear that under FRL, cyanobacteria tend to change their metabolism and perform effective growth and active photosynthesis via metabolomics changes with the development of chl *f* and *d* as accessory pigments in antennae systems. These pigments absorb energy and transfer it to the photosynthetic reactor center (RC). Usually these pigments are not involved in the photosynthetic electron transport chain. Additionally, chl *d* as well as chl *a* can function in the photosynthetic RC [22]. Such a transformation in cyanobacterial metabolism increases the possibilities for absorbing light in longer shifted wavelengths, which is important in cyanobacterial survival. There are many studies on these unique chls, however their full function and role in cyanobacterial metabolism is still not clear, specifically how the growth and productivity will be affected for the mass cultivation of this species for biotechnological purposes.

In our research study, we can confirm that the changes in *C. fritschii* pigment composition (xanthophylls, chlorophylls, and phycocyanin) under FRL combined with LEDWL improved growth and twice increased the biomass productivity in comparison of LEDWL. This FaRLiP process triggered antennal transfer of energy to the photosynthetic RC, confirmed by an increase in chl *f* and *d* concentrations [4,22]. The combination of lights (WLFRL) changed the carotenoids' profile. Under WLFRL, we observed an increased concentration of myxo, nosto, calo, and echin (Table 2). In comparison with mono light adaptation (LEDWL only compared to FRL only), this effect was not seen. An extensive study of carotenoid changes in cyanobacteria by Zakar et al., 2016 [23], confirmed that these pigments are responsible for the light harvesting and photoprotective capacities, showing their essential roles in photosynthetic metabolism [24,25]. The photoprotective mechanism can also occur by cyanobacterial carotenoid-proteins. One of these protein complexes is the orange carotenoid protein (OCP), discovered by David Krogmann [23,26]. The carotenoid composition of OCP is presented by 60% echin, 30% keto-carotenoid 3'hydroechninone, and 10% zeax [27]. The increase in echin and zeax under LEDWL and FRL in our study confirmed the activation of this protein and prevented cellular damage from excessive light. This effect has additionally been confirmed by non-photochemical quenching of the carotenoid-binding protein [28].

The combination of both lights activated different acclimation mechanisms and effective light assimilation for productive photosynthetic efficiency. Two main processes are involved in light adaptation, these are light energy harvest and light energy transfer [6]. In our case, by using both lights, it increased the effectiveness of both light adaptive mechanisms. The light energy harvest was demonstrated by appearance of chl *f*, chl *d*, an increased concentration of carotenoids, and phycocyanin. The light energy transfer was proved by an increased concentration of carotenoids of OCP under WLFRL in comparison with LEDWL conditions. Furthermore, the chl *d* was involved in both processes [29]. This dual mechanism of chl *d* functioning was confirmed by a recent study of transcriptional profiling of *C. fritschii* in FRL for chl *d* synthesis regulation [22]. In summary, we supposed that LEDWL maintained stable growth and that FRL activated the synthesis of chl *d* and chl *f* and restructured the functioning of PSI and PSII [11]. The combination of both lights increased growth, productivity, and oxygenic metabolism within *C. fritschii*.

Mass cultivation of *C. fritschii* is very relevant for applications in biotechnology [30]. The biology of this species has great potential for scale-up and mass cultivation in different latitudes around the world. This thermophilic cyanobacterium has several advantages in terms of large-scale growth. It requires high temperatures, which gives real advantages over other species for mass cultivation. In our study we grew this species under 25 °C, however successful growth in mass scale was shown in Balasundaram et al., 2012, and it can grow at up to 50 °C [31]. The mass cultivation set-up (PBR and raceways) could be placed in desert conditions. It has been confirmed that this species could grow on elevated CO_2 concentrations up to 5% of CO_2 [31,32]. It can therefore be co-located with industries emitting flue gas, e.g., power plant stations [31]. Additionally, this species could be cultivated in African and South East Asian weather conditions, as the biomass contains many valuable compounds for food, feed additivities, and as a whole food and can be used to combat malnutrition [33,34]. The application of this species as a feed for tilapia has been studied, showing that this species has potential in aquaculture [35].

Several advantages of the mass cultivation of this species are related to the aspect of easy downstream processing. This species is auto flocculating and does not require expensive equipment of membrane filtration and/or centrifugation to obtain the algal biomass paste for future processing and preservation [15,16]. This is another aspect of the development of successful mass cultivation of this species in different locations around the world, making this species a model for worldwide application.

The use of mass cultivation of *C. fritschii* in bioeconomy is an important target of algal biotechnology. The understanding of their cell physiology and specific light adaptation will help to improve the biomass and specific compounds production. The main bioactive compounds of *C. fritschii* are presented in Table S2. The principle groups are mycosporine-like amino-acids (MAAs) and pigments. *Chlorogloeopsis* produces chlorophylls, carotenoids, and phycobiliproteins, which contain different colours and can be used as biodegradable dyes [15,36]. Furthermore, bioproducts such as biodegradables and biocompatible plastic could be produced by *Chlorogloeopsis*. Nowadays, these are very important biomolecules, with the potential to be used as a substitute for single use plastic. The reason for this is that petrochemical and non-biodegradable contamination presents a major problem worldwide [37].

Many other applications of *Chlorogloeopsis* and cyanobacteria could be developed. This algal group can produce antimicrobial, antiviral, anticancer, and antiprotozoal compounds for pharmaceutical applications and can be used as a food, feed, and in other value-added products [38]. Further research and product development activities need to be established. In our research, we confirmed that the production of the main group of pigments (chlorophylls, carotenes, xanthophylls, and phycocyanin) could be of potential commercial interest.

4. Conclusions

A combination of LEDWL and FRL showed higher productivity of *C. fritschii*, with an increased concentration of myxo, nosto, calo, and echin. These combined light conditions triggered light harvesting and light energy transfer together with the induction of chls *d* and *f*, giving increased growth, photosynthetic effectiveness, and double the productivity of *C. fritschii* cultures. However, the overall protein, lipid, and carbohydrate composition did not significantly change under WLFRL.

Our results suggested that the overall production of this biotechnologically promising species can be increased by cultivation using additional far-red light.

5. Materials and Methods

5.1. Experimental Design

For the first experiment, three flasks with a total volume of 800 mL each were placed under FRL and LEDWL conditions. The initial cell concentration was 0.5×10^6 cell mL^{-1} (or 750 nm measurements ~0.05).

On day 9, at an OD$_{750}$ of 0.3–0.4, the flasks under FRL were harvested. The cultures grown under LEDWL conditions were sampled in triplicate for growth and pigment analysis and then transferred to FRL (far red light) conditions for a further 6 days. After 24 h of exposure to FRL, the cultures were sampled again in triplicate for pigments and growth analysis.

For the second experiment, three flasks with a total volume of 800 mL of *C. fritschii* culture were placed under LEDWL and FRL combined with LEDWL. The initial cell concentration was 2×10^6 cell mL^{-1} (or 750 nm measurements ~0.2). The duration of the experiment was 20 days.

Every 24 h, the growth parameters, such as cell concentration, biovolume, and OD (optical density) was measured and a collection of the algal biomass (harvested from 15 mL of culture) was made for pigment analysis by HPLC, spectrophotometer, and for biochemical analysis by FTIR.

5.2. Source Organism and Cultivation Conditions

Chlorogloeopsis fritschii was purchased from Pastor Culture Collection (PCC 6912; Paris, France). The master axenic culture was maintained in a temperature and white fluorescent light-controlled room (T = 28 °C) with 16:8 h light: dark cycle.

5.3. Growth Estimation

Every 24 h cell concentration and biovolume by Coulter counter (Multisizer 4, Beckman, USA) measurement was performed to quantify culture growth. Further details are described in Reference [39]. The OD (750 nm) measurements were analysed by UNICAM UV 300 spectrophotometer.

During sampling days, a minimum of 50 mL of culture was taken from each tube and centrifuged (Beckman Coulter Centrifuge, Avanti J-20XP) for 20 min at 8000 rpm. The biomass was washed twice with deionized (DI) water, centrifuged for 20 min at 8000 rpm, then collected and freeze dried (ScanVac Cool Safe, LaboGene, Lynge, Denmark) for 24 h prior to further analysis.

The specific growth rate (μ) was determined for all LED light treatments using Equation (1), where N_0 and N_1 are the cell concentrations (cells mL^{-1}) at times t_0 and t_1, as follows:

$$\mu = \ln N_1 - \ln N_0 / t_1 - t_0. \tag{1}$$

5.4. Dry Weight and Biomass Productivity

Dry weight was measured according to Reference [39]. A known volume of algae was pelleted and washed with deionized (DI) water (three times using 25 mL of DI water each time) prior to being filtered onto pre-weighed and dried filters (Whatman GF/F 47 mm Ø). The filters with algal biomass were then dried and re-weighed until constant weight was reached. Dry weight (g L^{-1}) was then calculated by subtraction of the final filter weight and the pre-filtered weight.

Biomass productivity was calculated as the difference in terms of DW between the sample day and the previous day. The results are expressed in g L^{-1} d^{-1}.

5.5. Pigments Extraction and Measurements

A known mass of frozen cell paste was transferred to an extraction tube containing 1 mL HPLC grade acetone and 0.2 mg zirconium (0.1 mm diameter) beads. The sample was then lysed in Precellys®24, a high-throughput tissue homogenizer, at 6500 rpm 2 × 30 s with a pause of 5 s. The sample was centrifuged (5 min at 20,000× g, Microcentrifuge) and the removed supernatant was used for pigment analysis on HPLC (Agilent HP 1200).

5.6. HPLC

The pigment extract was analysed using a high performance liquid chromatography (HPLC) method described previously (Method C in Airs et al., 2001 [40]). Pigment extracts were injected (100 μL) onto the HPLC column (2 Waters Spherisorb ODS2 cartridges coupled, each 150 × 4.6 mm, particle size 3 μm, protected with a precolumn containing the same phase). Elution was carried out using a mobile phase comprising methanol, acetonitrile, ammonium acetate (0.01 M), and ethyl acetate (Method C in Airs et al. 2001) at a flow rate of 0.7 mL min^{-1}. The photodiode array PDA detector was set to monitor wavelengths at 406, 440, 660, 696, and 706 nm. Carotenoids and chl-*a* were quantified against standards (Sigma) and, for chls *d* and *f*, peak areas were used [7,9].

5.7. Phycocyanin Extraction and Determination

Phycocyanin (PC) was extracted using a modified version [41] of the method developed by Reference [42]. The freeze-dried biomass of each sample was weighed to a known weight on a semi-micro and analytical balance (MSE 124S-100-DU, Sartorius balance, Germany). The sample weight was noted to the nearest 0.1 mg and all PC extractions were conducted in triplicate. The samples were transferred into 15 mL falcon tubes and subjected to a minimum of five freeze-thaw cycles; the samples were immersed in 5 mL of 0.1 mol L^{-1} phosphate buffer (pH = 6) and stored at −20 °C until frozen (~2 h), they were then thawed and subjected to 10 min sonication on ice [43]. The samples were then vortexed for 5 min and then placed back into −20 °C, and the process repeated. After the final freeze-thaw cycle, the cell debris was removed via centrifugation at 8000 rpm for 5 min. The supernatant was recovered and used for PC measurements. Absorbance of the extracts was measured at 592, 618, and 645 nm using a UNICAM UV 300 spectrophotometer. The concentration of the PC was determined using the equations in Reference [44], where OD is the optical density of the pigment at the particular wavelength.

$$PC \ (mg \ mL^{-1}) = [(OD618 \ nm − OD645 \ nm) − (OD592 \ nm − OD645 \ nm) × 0.15] × 0.15 \qquad (2)$$

5.8. Proteins, Lipids, and Carbohydrate Analysis Using Fourier Transformed Infra-Red (FTIR)

FTIR attenuated total reflectance (ATR) spectra were collected using a PerkinElmer Model Spectrum Two instrument equipped with a diamond crystal ATR reflectance cell with a DTGS detector scanning over the wavenumber range of 4000–450 cm^{-1} at a resolution of 4 cm^{-1} as described by References [45,46]. Briefly, ethanol (70%) was used to clean the diamond ATR before the first use and between samples. Approximately 3–5 mg of finely powdered freeze-dried *C. fritschii* biomass was applied to the surface of the crystal and then pressed onto the crystal head. A duplicate (each consisting of an average of 12 scans) of each bioreactor sample was conducted for each light type; therefore, results of 6 ATR spectra were gained and the results were averaged. Background correction scans of ambient air were made prior to each sample scan. Scans were recorded using the spectroscopic software Spectrum (version 10., PerkinElmer, Germany). The contents of lipids, proteins, and carbohydrates in the biomass samples were determined using FTIR, which had previously been calibrated using mix of monosugars (rhamnose, xylose, glucuronic acid, and glucose) for carbohydrates, palmitic acid for lipids, and BSA for proteins at different concentrations. The carrier powder for the FTIR calibration was potassium bromide (KBr) [41].

5.9. Statistical Analysis

Statistical analyses were carried out using the R project software on the OD, pigments concentration, lipids, carbohydrates, and protein content data. Data normality was tested using a Shapiro test.

Non-normal data significance was assessed using GLMs (generalised linear models) furthered by an analysis of variance (ANOVA) on a data set not following a normal distribution. Crossed factor ANOVAS were carried out on normally distributed data. Both statistical methods tested the impact of experimental duration, pigments and intracellular lipids, carbohydrates, and protein content. When statistical significance was found, post hoc Tukey tests were implemented.

All the experiments were performed in triplicate. The standard deviation and means were analysed for significance using the biostatistics software Excel through one-way ANOVA. The Duncan multiple range test was used to compare the significance of difference among tested algae at p values of < 0.05. Results are reported as $\pm$ SD or error bars.

Supplementary Materials: The following are available online at http://www.mdpi.com/2218-1989/9/8/170/s1, Table S1: Statistical analysis of composition of *C. fritschii* culture, Table S2: Bioactive compounds produced by *Chlorogloeopsis* sp.

Author Contributions: A.S. and C.A.L. conceived, designed, and performed the *C. fritschii* growth experiments under different light conditions; the biomass and pigments analysis were performed by A.S. and B.K.; all authors wrote the paper.

Funding: This research was funded by PHYCONET Proof of concept funding "Exploring chlorophyll-f and associated metabolism for improved intensive cultivation of cyanobacteria".

Acknowledgments: The authors would like to thank all CSAR technical staff for the support of the project.

Conflicts of Interest: The authors declare no conflict of interest.

References

1. Al-Haj, L.; Lui, Y.; Abed, R.; Gomaa, M.; Purton, S. Cyanobacteria as Chassis for Industrial Biotechnology: Progress and Prospects. *Life* **2016**, *6*, 42. [CrossRef] [PubMed]

2. Abed, R.M.M.; Dobretsov, S.; Sudesh, K. Applications of Cyanobacteria in Biotechnology. *J. Appl. Microbiol.* **2009**, *106*, 1–12. [CrossRef] [PubMed]

3. Rastogi, R.P.; Sinha, R.P. Biotechnological and Industrial Significance of Cyanobacterial Secondary Metabolites. *Biotechnol. Adv.* **2009**, *27*, 521–539. [CrossRef] [PubMed]

4. Gan, F.; Shen, G.; Bryant, D. Occurrence of Far-Red Light Photoacclimation (FaRLiP) in Diverse Cyanobacteria. *Life* **2014**, *5*, 4–24. [CrossRef] [PubMed]

5. Montgomery, B.L. Shedding New Light on the Regulation of Complementary Chromatic Adaptation. *Cent. Eur. J. Biol.* **2008**, *3*, 351–358. [CrossRef]

6. Ho, M.Y.; Soulier, N.T.; Canniffe, D.P.; Shen, G.; Bryant, D.A. Light Regulation of Pigment and Photosystem Biosynthesis in Cyanobacteria. *Curr. Opin. Plant Biol.* **2017**, *37*, 24–33. [CrossRef] [PubMed]

7. Chen, M.; Li, Y.; Birch, D.; Willows, R.D. A Cyanobacterium That Contains Chlorophyll f - A Red-Absorbing Photopigment. *FEBS Lett.* **2012**, *586*, 3249–3254. [CrossRef] [PubMed]

8. Averina, S.; Velichko, N.; Senatskaya, E.; Pinevich, A. Far-Red Light Photoadaptations in Aquatic Cyanobacteria. *Hydrobiologia* **2018**, *813*, 1–17. [CrossRef]

9. Airs, R.L.; Temperton, B.; Sambles, C.; Farnham, G.; Skill, S.C.; Llewellyn, C.A. Chlorophyll f and Chlorophyll d Are Produced in the Cyanobacterium Chlorogloeopsis Fritschii When Cultured under Natural Light and Near-Infrared Radiation. *FEBS Lett.* **2014**, *588*, 3770–3777. [CrossRef]

10. Ho, M.Y.; Gan, F.; Shen, G.; Zhao, C.; Bryant, D.A. Far-Red Light Photoacclimation (FaRLiP) in Synechococcus Sp. PCC 7335: I. Regulation of FaRLiP Gene Expression. *Photosynth. Res.* **2017**, *131*, 173–186. [CrossRef]

11. Gan, F.; Zhang, S.; Rockwell, N.C.; Martin, S.S.; Lagarias, J.C.; Bryant, D.A. Extensive Remodeling of a Cyanobacterial Photosynthetic Apparatus in Far-Red Light. *Science* **2014**, *345*, 1312–1317. [CrossRef] [PubMed]

12. Chen, M.; Schliep, M.; Willows, R.D.; Cai, Z.-L.; Neilan, B.A.; Scheer, H. A Red-Shifted Chlorophyll. *Science* **2010**, *329*, 1318–1319. [CrossRef] [PubMed]

13. Mitra, A.K. Two New Algae from Indian Soils. *Ann. Bot.* **1950**, *14*, 457–464. [CrossRef]

14. Evans, E.H.; Foulds, I.; Carr, N.G. Environmental Conditions and Morphological Variation in the Blue-Green Alga Chlorogloea Fritschii. *J. Gen. Microbiol.* **1976**, *92*, 147–155. [CrossRef]

15. Balasundaram, B.; Skill, S.C.; Llewellyn, C.A. A Low Energy Process for the Recovery of Bioproducts from Cyanobacteria Using a Ball Mill. *Biochem. Eng. J.* **2012**, *69*, 48–56. [CrossRef]
16. Spolaore, P.; Joannis-Cassan, C.; Duran, E.; Isambert, A. Commercial Applications of Microalgae. *J. Biosci. Bioeng.* **2006**, *101*, 87–96. [CrossRef]
17. Khan, M.I.; Shin, J.H.; Kim, J.D. The Promising Future of Microalgae: Current Status, Challenges, and Optimization of a Sustainable and Renewable Industry for Biofuels, Feed, and Other Products. *Microb. Cell Fact.* **2018**, *17*, 36. [CrossRef]
18. Wijffels, R.H.; Kruse, O.; Hellingwerf, K.J. Potential of Industrial Biotechnology with Cyanobacteria and Eukaryotic Microalgae. *Curr. Opin. Biotechnol.* **2013**, *24*, 405–413. [CrossRef]
19. Montgomery, B.L. The Regulation of Light Sensing and Light-Harvesting Impacts the Use of Cyanobacteria as Biotechnology Platforms. *Front. Bioeng. Biotechnol.* **2014**, *2*, 22. [CrossRef]
20. Miyashita, H. Discovery of Chlorophyll d in Acaryochloris Marina and Chlorophyll f in a Unicellular Cyanobacterium, Strain KC1, Isolated from Lake Biwa. *J. Phys. Chem. Biophys.* **2014**, *4*. [CrossRef]
21. Allakhverdiev, S.I.; Kreslavski, V.D.; Zharmukhamedov, S.K.; Voloshin, R.A.; Korol'kova, D.V.; Tomo, T.; Shen, J.-R. Chlorophylls d and f and Their Role in Primary Photosynthetic Processes of Cyanobacteria. *Biochemistry* **2016**, *81*, 201–212. [CrossRef] [PubMed]
22. Ho, M.-Y.; Bryant, D.A. Global Transcriptional Profiling of the Cyanobacterium Chlorogloeopsis Fritschii PCC 9212 in Far-Red Light: Insights Into the Regulation of Chlorophyll d Synthesis. *Front. Microbiol.* **2019**, *10*, 465. [CrossRef] [PubMed]
23. Zakar, T.; Laczko-Dobos, H.; Toth, T.N.; Gombos, Z. Carotenoids Assist in Cyanobacterial Photosystem II Assembly and Function. *Front. Plant Sci.* **2016**, *7*, 295. [CrossRef] [PubMed]
24. Stamatakis, K.; Tsimilli-Michael, M.; Papageorgiou, G.C. On the Question of the Light-Harvesting Role of β-Carotene in Photosystem II and Photosystem I Core Complexes. *Plant Physiol. Biochem.* **2014**, *81*, 121–127. [CrossRef] [PubMed]
25. Sozer, O.; Komenda, J.; Ughy, B.; Domonkos, I.; Laczk-Dobos, H.; Malec, P.; Gombos, Z.; Kis, M. Involvement of Carotenoids in the Synthesis and Assembly of Protein Subunits of Photosynthetic Reaction Centers of Synechocystis Sp. PCC 6803. *Plant Cell Physiol.* **2010**, *51*, 823–835. [CrossRef] [PubMed]
26. Kay Holt, T.; Krogmann, D.W. A Carotenoid-Protein from Cyanobacteria. *Biochim. Biophys. Acta—Bioenerg.* **1981**, *637*, 408–414. [CrossRef]
27. Sedoud, A.; López-Igual, R.; ur Rehman, A.; Wilson, A.; Perreau, F.; Boulay, C.; Vass, I.; Krieger-Liszkay, A.; Kirilovsky, D. The Cyanobacterial Photoactive Orange Carotenoid Protein Is an Excellent Singlet Oxygen Quencher. *Plant Cell* **2014**, *26*, 1781–1791. [CrossRef]
28. Kirilovsky, D.; Kerfeld, C.A. The Orange Carotenoid Protein: A Blue-Green Light Photoactive Protein. *Photochem. Photobiol. Sci.* **2013**, *12*, 1135–1143. [CrossRef]
29. Mielke, S.P.; Kiang, N.Y.; Blankenship, R.E.; Gunner, M.R.; Mauzerall, D. Efficiency of Photosynthesis in a Chl D-Utilizing Cyanobacterium Is Comparable to or Higher than That in Chl a-Utilizing Oxygenic Species. *Biochim. Biophys. Acta—Bioenerg.* **2011**, *1807*, 1231–1236. [CrossRef]
30. Varshney, P.; Mikulic, P.; Vonshak, A.; Beardall, J.; Wangikar, P.P. Extremophilic Micro-Algae and Their Potential Contribution in Biotechnology. *Bioresour. Technol.* **2015**, *184*, 363–372. [CrossRef]
31. Ono, E.; Cuello, J.L. Carbon Dioxide Mitigation Using Thermophilic Cyanobacteria. *Biosyst. Eng.* **2007**, *96*, 129–134. [CrossRef]
32. Bhola, V.; Swalaha, F.; Ranjith Kumar, R.; Singh, M.; Bux, F. Overview of the Potential of Microalgae for CO_2 Sequestration. *Int. J. Environ. Sci. Technol.* **2014**, *11*, 2103–2118. [CrossRef]
33. Ngo, J.; Serra-Majem, L. Hunger and Malnutrition. In *Encyclopedia of Food Security and Sustainability*; Elsevier: Amsterdam, The Netherlands, 2019; pp. 315–335.
34. Koyande, A.K.; Chew, K.W.; Rambabu, K.; Tao, Y.; Chu, D.T.; Show, P.L. Microalgae: A Potential Alternative to Health Supplementation for Humans. *Food Sci. Hum. Wellness* **2019**, *8*, 16–24. [CrossRef]
35. Merrifield, D.L.; Güroy, D.; Güroy, B.; Emery, M.J.; Llewellyn, C.A.; Skill, S.; Davies, S.J. Assessment of Chlorogloeopsis as a Novel Microbial Dietary Supplement for Red Tilapia (Oreochromis Niloticus). *Aquaculture* **2010**, *299*, 128–133. [CrossRef]
36. Saini, D.K.; Pabbi, S.; Shukla, P. Cyanobacterial Pigments: Perspectives and Biotechnological Approaches. *Food Chem. Toxicol.* **2018**, *120*, 616–624. [CrossRef]

37. Hein, S.; Steinbüchel, A.; Hai, T. Multiple Evidence for Widespread and General Occurrence of Type-III PHA Synthases in Cyanobacteria and Molecular Characterization of the PHA Synthases from Two Thermophilic Cyanobacteria: Chlorogloeopsis Fritschii PCC 6912 and Synechococcus Sp. Strain MA1. *Microbiology* **2015**, *147*, 3047–3060. [CrossRef]

38. Kumar, J.; Singh, D.; Tyagi, M.B.; Kumar, A. *Cyanobacteria: Applications in Biotechnology*; Elsevier Inc.: Amsterdam, The Netherlands, 2018; Volume 7421.

39. Mayers, J.J.; Flynn, K.J.; Shields, R.J. Rapid Determination of Bulk Microalgal Biochemical Composition by Fourier-Transform Infrared Spectroscopy. *Bioresour. Technol.* **2013**, *148*, 215–220. [CrossRef]

40. Airs, R.L.; Atkinson, J.E.; Keely, B.J. Development and Application of a High Resolution Liquid Chromatographic Method for the Analysis of Complex Pigment Distributions. *J. Chromatogr. A* **2001**, *917*, 167–177. [CrossRef]

41. Coward, T.; Fuentes-Grünewald, C.; Silkina, A.; Oatley-Radcliffe, D.L.; Llewellyn, G.; Lovitt, R.W. Utilising Light-Emitting Diodes of Specific Narrow Wavelengths for the Optimization and Co-Production of Multiple High-Value Compounds in Porphyridium Purpureum. *Bioresour. Technol.* **2016**, *221*, 607–615. [CrossRef]

42. Lawrenz, E.; Fedewa, E.J.; Richardson, T.L. Extraction Protocols for the Quantification of Phycobilins in Aqueous Phytoplankton Extracts. *J. Appl. Phycol.* **2011**, *23*, 865–871. [CrossRef]

43. Bermejo Román, R.; Alvárez-Pez, J.M.; Acién Fernández, F.G.; Molina Grima, E. Recovery of Pure B-Phycoerythrin from the Microalga Porphyridium Cruentum. *J. Biotechnol.* **2002**, *93*, 73–85. [CrossRef]

44. Beer, S.; Eshel, A. Determining Phycoerythrin and Phycocyanin Concentrations in Aqueous Crude Extracts of Red Algae. *Mar. Freshw. Res.* **1985**, *36*, 785–792. [CrossRef]

45. Fuentes-Grünewald, C.; Bayliss, C.; Zanain, M.; Pooley, C.; Scolamacchia, M.; Silkina, A. Evaluation of Batch and Semi-Continuous Culture of *Porphyridium purpureum* in a Photobioreactor in High Latitudes Using Fourier Transform Infrared Spectroscopy for Monitoring Biomass Composition and Metabolites Production. *Bioresour. Technol.* **2015**, *189*, 357–363. [CrossRef] [PubMed]

46. Mayers, J.J.; Flynn, K.J.; Shields, R.J. Influence of the N: P Supply Ratio on Biomass Productivity and Time-Resolved Changes in Elemental and Bulk Biochemical Composition of *Nannochloropsis* Sp. *Bioresour. Technol.* **2014**, *169*, 588–595. [CrossRef] [PubMed]

MDPI
St. Alban-Anlage 66
4052 Basel
Switzerland
Tel. +41 61 683 77 34
Fax +41 61 302 89 18
www.mdpi.com

Metabolites Editorial Office
E-mail: metabolites@mdpi.com
www.mdpi.com/journal/metabolites